Parasites in Social Insects

MONOGRAPHS IN
BEHAVIOR AND ECOLOGY

Edited by John R. Krebs and
Tim Clutton-Brock

Five New World Primates: A Study in Comparative Ecology, *by John Terborgh*

Reproductive Decisions: An Economic Analysis of Gelada Baboon Social Strategies, *by R. I. M. Dunbar*

Honeybee Ecology: A Study of Adaptation in Social Life, *by Thomas D. Seeley*

Foraging Theory, *by David W. Stephens and John R. Krebs*

Helping and Communal Breeding in Birds: Ecology and Evolution, *by Jerram L. Brown*

Dynamic Modeling in Behavioral Ecology, *by Marc Mangel and Colin W. Clark*

The Biology of the Naked Mole-Rat, *edited by Paul W. Sherman, Jennifer U. M. Jarvis, and Richard D. Alexander*

The Evolution of Parental Care, *by T. H. Clutton-Brock*

The Ostrich Communal Nesting System, *by Brian C. R. Bertram*

Parasitoids: Behavioral and Evolutionary Ecology, *by H. C. J. Godfray*

Sexual Selection, *by Malte Andersson*

Polygyny and Sexual Selection in Red-winged Blackbirds, *by William A. Searcy and Ken Yasukawa*

Leks, *by Jacob Höglund and Rauno V. Alatalo*

Social Evolution in Ants, *by Andrew F. G. Bourke and Nigel R. Franks*

Female Control: Sexual Selection by Cryptic Female Choice, *by William G. Eberhard*

Sex, Color, and Mate Choice in Guppies, *by Anne E. Houde*

Foundations of Social Evolution, *by Steven A. Frank*

Parasites in Social Insects, *by Paul Schmid-Hempel*

Parasites in Social Insects

PAUL SCHMID-HEMPEL

Princeton University Press
Princeton, New Jersey

Published by Princeton University Press, 41 William Street,
Princeton, New Jersey 08540
In the United Kingdom: Princeton University Press,
Chichester, West Sussex

Library of Congress Cataloging-in-Publication Data
Schmid-Hempel, Paul.
Parasites in social insects / Paul Schmid-Hempel.
p. cm. — (Monographs in behavior and ecology)
Includes bibliographical references (p.) and indexes.
ISBN 0-691-05923-3 (cl : alk. paper).
ISBN 0-691-05924-1 (pbk. : alk. paper)
1. Insect societies. 2. Insects—Parasites. I. Title.
II. Series.
QL496.S36 1998 98-10164
595.717'857—dc21

This book has been composed in Times Roman

Princeton University Press books are printed on acid-free paper and meet the guidelines for permanence and durability of the Committee on Production Guidelines for Book Longevity of the Council on Library Resources

http://pup.princeton.edu

Printed in the United States of America

10 9 8 7 6 5 4 3 2 1

10 9 8 7 6 5 4 3 2 1
(Pbk.)

Contents

Preface and Acknowledgments

Social insects have fascinated people since very early times. For example, the most famous product of bees—honey—was cherished and used in large quantities by Egyptian priests for their religious rites. Sometimes even the dead were preserved in honey (Ransome 1986). On the other hand, social insects can be destructive to the human economy, as the examples of leaf-cutter ants or termites show. The double fascination of useful products and threats persists into modern times and gives social insects a prominent role in the living world.

To biologists, social insects have been fascinating for many different reasons. Darwin stumbled over the problem of sterile individuals and the problem of how altruism may evolve. The powerful tools of kin selection theory (Hamilton 1964) have not only answered some of these questions but social insects have also become a pivotal study subject for the evolution of social behavior in general (Wilson 1975). It is all too easy to forget that several milestones in animal physiology were reached by use of social insects as the test field. We need only mention Karl von Frisch's work on color vision in the honeybee, a milestone in the field of sensory physiology. Oster and Wilson (1978) added a further item to this menu: how the organization of complex biological systems may evolve and be maintained under natural selection. Indeed, the social organization and elaborate caste structure of ants and termites have few counterparts in other organisms. An additional reason social insects are a worthwhile study subject is that many of the relevant phenomena are analogous to those seen in single organisms. For example, there is a close correspondence between selection for insect colonies to be genetically diverse in the face of parasitism and the hypotheses for the evolution of sexual reproduction in single organisms. Throughout the history of biology, social insects have therefore not only fascinated passionate observers of nature but also provided manifold stimulation for developing general concepts. A great number of excellent monographs have been devoted to these subjects (Seeley 1985; Hölldobler and Wilson 1990; Itô 1993a; Bourke and Franks 1995; Crozier and Pamilo 1996), and I refer the reader to them for further study. It is perhaps less well known that social insects, especially the honeybee and its parasites, have also played an important role in the study of insect diseases (see McCoy et al. 1988 for a review).

Unfortunately, with the flurry of studies devoted over the last two decades to the problem of social evolution, kin selection, and within-colony conflicts, the ecology of social insects has become somewhat neglected. The importance of such studies should not be minimized by noting that our view of social insects is therefore somewhat unbalanced. We know, for example, very little about the population biology of colonies in a given area, quite in contrast to the many details of how workers develop and shift through different tasks according to age inside the colony. A brief look at the many lifestyles and diets of social insects will show that the insects' ecology is not only fascinating in its own right, but also important in any given habitat. For example, social insects are prominent and sometimes dominant members in tropical ecosystems, and they are among the few insects that have successfully expanded into arctic or other extreme habitats. So, one aim of this volume is to correct a bias that I feel unhappy about and to highlight an important element in the ecology of social insects within an evolutionary-ecological-behavioral framework.

Research over the past two decades, almost completely outside the social insects, has shown that parasites are much more important factors for the ecology and evolution of their individual hosts, and for entire host populations and communities, than was previously believed (e.g., Andersson 1994; Grenfell and Dobson 1995). The role of parasites in the ecology and evolution of birds, for example, is a very active area of contemporary research. This has led to a more systematic approach to the general question of how parasites affect their hosts, ranging from sexual selection to the problem of trade-offs in host immunocompetence. Also, the broad area of research on how parasites could maintain sexual over asexual reproduction should be viewed in this light. The role of parasites in the biology of social insects has, in contrast, never been systematically addressed and is largely unknown, with the important exceptions of social parasitism, honeybee and fire ant studies, and some additional studies during the last few years. This is somewhat surprising given the attention social insects have received in general, and given the fact that social insects are prime targets for parasites due to their abundance, family structure, and persistent colonies.

This book is an attempt to fill this gap and to summarize the existing knowledge on parasitism in social insects. For reasons just alluded to, I have excluded the social parasites in this volume. Several excellent syntheses on this subject, in addition to the monographs mentioned above, exist (e.g., Buschinger 1986; Bourke and Franks 1991). Rather, the book will concentrate on "real" parasites, that is, infectious diseases, helminths, parasitoids, and the like. It is, however, not intended to duplicate the many excellent accounts on these different groups, such as Godfray's (1994) text on parasitoids. To my disappointment, however, I typically found that social insects hardly appear in these texts, as if they might be so very different from all others and not count in our view of nature. A second goal of this book is of a more conceptual nature—to help organize the field and to identify some basic problems that need further attention.

During my preparations, I have encountered a number of typical problems. First, much of the relevant literature is very widely scattered. Some sources from highly specialized journals or obscure technical reports have remained unavailable. The synopsis provided in appendix 2 is therefore not complete but is almost so, and should contain all of the important reports to date. Second, frustratingly few of the ideas already found in the literature and those formulated here can actually be tested rigorously. This is due to the lack of empirical studies on parasites, but also because the basic knowledge on the population ecology of social insects has barely been collected. It is therefore inevitable that the cited comparisons are very crude and often make very bold assumptions. Nobody could be more painfully aware of the shortcomings than I. However, the ground is barren and therefore such attempts seemed justifiable to me. In several instances, a comparative study provides valuable insight. I have tried to carry this out wherever possible. Due to the scattered nature of the data and the poorly resolved phylogenies of many social insect groups, there is a clear limitation to this approach. I have restricted the use of comparative studies to ants and have used the phylogenies suggested by Hölldobler and Wilson (1990) and Baroni-Urbani et al. (1992). Finally, I have used a number of more explicit theoretical considerations, albeit simple ones. These are intended as starters for the discussion and not as an exhaustive treatment of the subject. As the reader will quickly realize, most of these models can be amended with additional constraints and complications, which would have been outside the scope of this volume and would also far outstrip the available data.

Furthermore, it is not always obvious whether a particular species is a parasite at all. By definition, parasites harm their host in some way. Ideally, this is defined by a reduction of host fitness compared to the unparasitized state. In birds, for example, one can safely assume that if a parasite causes damage to the organ it resides in, then its host will also suffer a fitness loss. In social insects, this is not always so clear-cut. Since workers usually do not reproduce themselves, they can often get diseased and drop out without any noticeable effect on colony performance and reproduction. It is also possible that some parasites, although they may be detrimental to their individual host, can actually benefit the colony in other ways. Mites pose a special problem, since their effects on individual hosts and colonies are sometimes hard to gauge. Most are probably just phoretic or commensalistic. In general, I have accepted the classification of a species as a parasite if an author has reported it as such; I have, however, excluded the mites in the more general, comparative studies. Similarly, the definition of a "social insect" is not always as universally accepted as it might seem. Here, the focus is more on advanced sociality and less on communally nesting species and more primitive associations. Last but not least, the subject of this book is quite complex because it touches a large variety of issues, from kin selection theory to population genetics to immunology or epidemiology. Therefore, these subject areas will have to be briefly exposed where necessary to help the reader appreciate the arguments. I hope that I have done some justice to the pivotal ideas in these fields.

The book is organized as follows. The first chapter is devoted to a summary of some relevant biological characteristics of social insects; it is intended mainly for readers who are not so familiar with social insects. Chapter 2 summarizes the relevant natural history of the parasites that are typically found in social insects. A short overview over special characteristics and some elements of the systematics of the different groups are given to facilitate the connection to the specialized literature on the subject. Chapter 3 is a second look at the natural history. However, this time a particular but crucially important aspect is considered: how parasites enter the colony and become transmitted further. Chapter 3 therefore complements the preceding ones by classifying and discussing the various attack routes of parasites. Chapter 4 is in some sense a reconsideration of Oster and Wilson's (1978) classical treatise on caste and ecology in social insects. Here, the consequences of social organization for the parasites and, vice versa, the effects of parasites on colony organization are considered. Chapter 5 is a major section, because the genetics of host-parasite interactions comes into play. Because social insects typically live together in genetically closely related groups, it seems particularly relevant to consider how parasitism blends with the breeding strategies of the social insect host. This field is probably one of the few where current research is quite active. Chapter 6 addresses the ecological dynamics of host-parasite interactions in social insects. Frustratingly little is known on this subject. The few studies on the spread of disease within honeybee colonies cannot count, because chapter 6 is concerned with the dynamics of populations of colonies. Hence, a large part of this section will be devoted to developing some preliminary ideas and concepts. Chapter 7 takes us down to the problem of how host and parasite coevolve. This is captured in the problem of parasite virulence and host resistance. Finally, chapter 8 touches on some of the areas that consider how parasites could select for sociality and how kin recognition comes into the picture.

When I started this adventure, I expected that it would be a short exercise since little is known on the subject and no meaningful patterns would be found. Fortunately, I was wrong. It is true that the database is small and widely scattered. But there are a number of excellent studies that needed to be discussed. It is also true that no elegant, unifying concept such as kin selection or sex allocation theory is available for studying disease in social insects. Perhaps some steps have now been taken in this direction. But the book will serve its purpose if it will stimulate others to dwell into this topic and to make most of its contents eventually superfluous. There are certainly more than just a few open questions left. Social insects are a study subject that has an enormous potential to understand how parasites coevolve with their hosts. In many respects, social insect societies are also a model of social organization in living organisms in general, including our own species. Hence, it is certainly more than worthwhile to tackle questions that unite two of the most successful strategies of life: sociality and parasitism.

I am grateful to the many people who have made this book possible. Dozens of very helpful comments on single chapters or all of the manuscript were provided by Brenda Ball, Koos Boomsma, Andrew Bourke, Deborah Gordon, Jukka Jokela, Stella Koulianos, Francis Ratnieks, David Roubik, Horst Schwarz, Regula Schmid-Hempel, and an anonymous reviewer. I have not always followed their advice and so the errors and shortcomings do remain my own. My students have kept me in a constant state of creative unrest and luckily helped me to resist the fatal attraction of an everyday university routine. I gratefully acknowledge the help of Renate Brunner, Kristine Campbell, Roland Loosli, Hanne Magro, and Christine Reber in gathering literature and for manuscript preparation. I would like to especially mention the financial support by the Swiss National Science Foundation, which has supported my research over the years and thus has added to the study of parasites in social insects. I am grateful to the editors of Princeton University Press's Monographs in Behavior and Ecology, John Krebs and Tim Clutton-Brock, for suggesting this contribution and for keeping an interest in the project that—as usual—took much longer than anticipated. Emily Wilkinson, Sam Elworthy, and Alice Calaprice from Princeton University Press were supportive in every way. Last but not least, I have to thank my wife, Regula, for her encouragement, support, and patience during the whole process.

Zurich, January 1998

Parasites in Social Insects

1 The Biology of Social Insects

Social insects are socially living insects. There are, however, controversial views about how to define and describe sociality (e.g., Sherman et al. 1995; Crespi and Yanega 1995). For the purpose of this book, these controversies add little but to remind us that "social insects" are very diverse. Moreover, many species of mammals, birds, fish, bryozoa, anthozoa, shrimps (Duffy 1996), and spiders are social, too. The biology of many social insect species is sometimes more reminiscent of group-living birds or mammals than of other social insects. For example, females of some halictid bees (Packer 1993) or wasps in the genus *Ropalidia* (Gadagkar et al. 1993) are not simply locked into an inflexible social behavior, but have quite similar options with respect to cooperation or selfishness as do birds or mammals (Brown 1987). Females can join other females, nest communally in groups, become a "queen," or help as a "worker." They may even breed alone and start a nest independently. On the other hand, leaf-cutter ants or fungus-growing termites possess highly evolved and sophisticated social behaviors that have no equivalent in any other group of organisms. Here, I consider social insects as those that are generally classified as primitively or advanced eusocial (table 1.1); yet I will occasionally look at other systems as well but will generally exclude communally nesting species (such as found in many andrenid bees). A hallmark of social insects as envisaged here is that they live in colonies and behave cooperatively.

Taxonomically, most social insects (some 19,000 known species) are found in the order Hymenoptera, this is, in the ants (all known species are social or parasites of other ants, Hölldobler and Wilson 1990), bees (Michener 1974), and wasps (Ross and Matthews 1990) (table 1.2). But truly social species are also known in the sphecid wasps (*Microstigmus,* Matthews 1991), in beetles (ambrosia beetles, Kent and Simpson 1992), and in the Thysanoptera (the thrips, Crespi 1992). Some authors include sphecid wasps in the superfamily Apoidea (e.g., Gauld and Bolton 1988). All known members of the order Isoptera (the termites) are social (ca. 2,300 species). Some excellent reviews of the biology of ants (Hölldobler and Wilson 1990), bees (Michener 1974, Roubik 1989), wasps (Spradbery 1973, Ross and Matthews 1990), and termites (Krishna and Weesner 1969, 1970) exist. The general treatise of Wilson (1971) on social insects is still unsurpassed. Hence, this first chapter is not intended to duplicate these works, but to provide a sketch of some prominent biological

Table 1.1

A Simple Overview of the Levels of Sociality in Insects

	Trait		
Level of Sociality	*Cooperative Brood Care*	*Reproductive Castes*	*Overlap between Generations*
PARASOCIAL SEQUENCE			
Solitary	–	–	–
Communal	–	–	–
Quasisocial	+	–	–
Semisocial	+	+	–
Eusocial	+	+	+
SUBSOCIAL SEQUENCE			
Solitary	–	–	–
Primitively subsocial	–	–	–
Intermediate subsocial I	–	–	+
Intermediate subsocial II	+	–	+
Eusocial	+	+	+

NOTES: The parasocial and subsocial sequences refer to two major evolutionary routes through which eusociality is thought to have arisen (after Wilson 1975). "+" means the trait is present, or "–" is absent. Reproductive castes refer to the fact that only some individuals are reproductive.

characteristics of social insects to introduce the subjects of the following chapters. I have arranged them by topic rather than by taxonomic group in order to emphasize how parasites will encounter social insects, independent of the taxonomic group of the host. The chapter should be useful primarily for readers who are not so familiar with social insect biology.

1.1 The Individual and the Colony Cycle

Parasites attack individual hosts. Even in social species, therefore, the individual and its life history are important. As a colony grows in number, the individuals inside are born, then develop and die (fig. 1.1). A typical social insect colony usually contains more than one class of individuals, i.e., it has several castes. A caste is a set of individuals of a particular morphological type and/or an age group that performs a distinguishable, separate task in the colony (Oster and Wilson 1978, p. 19). Such tasks can include nursing the brood, foraging, or

Table 1.2

The Taxonomic Distribution of Eusociality

Order	*Family*	*Social Species*
INSECTS		
Isoptera	Hodotermitidae Indotermitidae Kalotermitidae Mastotermitidae Rhinotermitidae Serritermitidae Stylotermitidae Termopsidae Termitidae	Ca. 2300 species, all eusocial
Thysanoptera	Phlaeothripidae	Subfertile soldiers in 1 genus (Crespi 1992)
Homoptera	Pemphigidae	6 genera with soldiers known (Aoki 1977)
Hymenoptera	Anthophoridae	7 genera (Allodapini)
	Apidae (honeybees, stingless bees, bumblebees)	6 eusocial Apini (*Apis*) 4–500 eusocial Meliponini 300 primitively eusocial Bombini
	Halictidae (sweat bees)	6 genera (Halictini, Augochlorini)
	Sphecidae (digger wasps)	1 species (*Microstigmus comes*) (Matthews 1991)
	Vespidae	Ca. 9400 species (Polistinae, Stenogastrinae, Vespinae)
	Formicidae	Ca. 8800 species
Coleoptera	Curculionidae	1 species (*Austroplatypus incompertus*) (Kent and Simpson 1992)
OTHER TAXA		
Arthropoda	Aranea	1 species? (*Anelosimus eximius*) (Vollrath 1986)
	Crustacea	1 species (*Synalpheus regalis*) (Duffy 1996)
Mammalia	Rodentia	Several species of mole rats (Sherman et al. 1991; Burda and Kawalika 1993)

SOURCES: After Wilson 1971, Spradbery 1973, Michener 1974, Snelling 1981, Ross and Matthews 1990, Crespi and Yanega 1995, Sherman et al. 1995, and Crozier and Pamilo 1996.

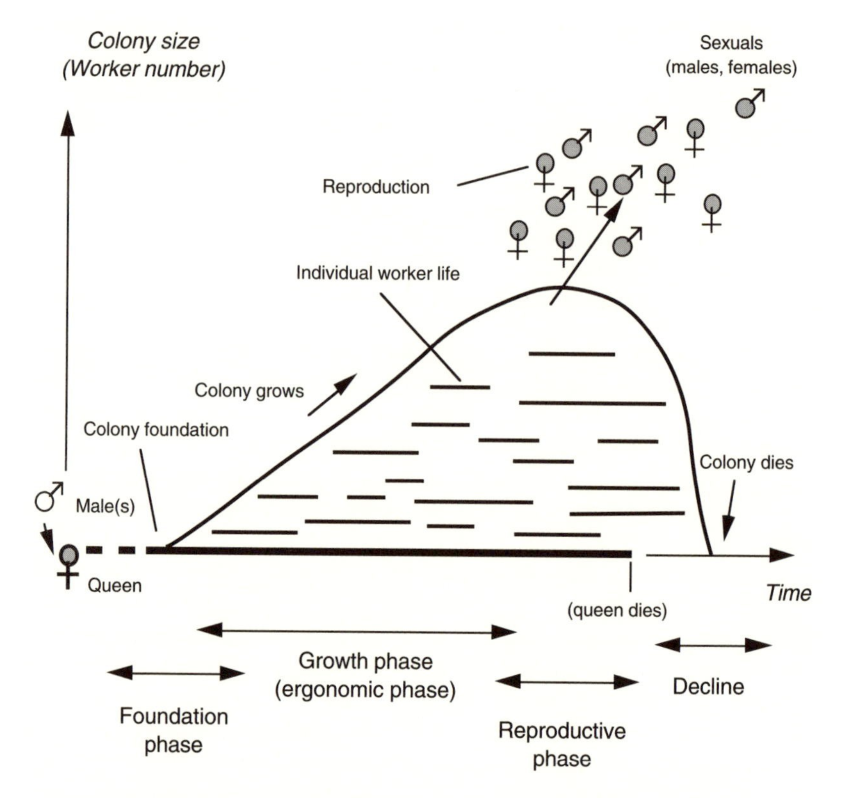

Figure 1.1 A schematic view of how individuals and colonies of social insects grow, mature, and die. The scheme depicted here is modeled after an annual species of social wasps (e.g., paper wasps) or bees (e.g., bumblebees). The queen founds the colony after the diapause (broken line) and produces workers. Toward the end of the cycle, sexuals (males and females) are produced. They (normally) leave the colony, mate, and start the next generation. The shorter horizontal arrows characterize the birth, life, and death of individual workers in the colony. The thick horizontal line indicates the life of the queen. Decline of the colony typically follows after queen death. The colony cycle is characterized by the number of workers present in the colony (colony size). The designation of the different phases follow Oster and Wilson (1978). Social insects show many variations on this basic scheme. Examples include more than one queen present (polygyny), perennial life cycles, and reproduction by fission.

milling leaves for the fungus garden (as in the leaf-cutter ants). Soldiers, usually large and behaviorally aggressive workers, are an example of a morphological caste, which is part of a polymorphic caste system. Young workers usually tend the brood—an instance of caste based on age rather than morphology. Although the significance of age, as compared to other factors such as opportunities to work, as a determinant of task attendance is controversial (see discussion in Bourke and Franks 1995), age-related changes are widespread. Such differences lead to division of labor (polyethism) within the colony, a further hallmark of social insect biology. Most importantly, only some individuals actively reproduce (the reproductive division of labor). Variation in how fast the individuals develop or how many workers are produced are important determinants for the macroscopic differences in the lifestyle among species of social insects, for colony sizes, and for reproductive timing.

The Life History of Individuals

The individual life of workers is characterized simply in figure 1.1 by arrows connecting birth and death. But these arrows hide important differences. In the social Hymenoptera (the bees, ants, and wasps), individuals have a holometabolous development (fig. 1.2). After the egg stage, several larval instars, dependent on being provisioned by adults, occur. In ants, there are usually four stages (3–6); in wasps there are five, and perhaps four or five in bees (Michener 1974). In some species, the larvae have been shown to actively contribute to the colony's economy by providing food or necessary digestive enzymes to the adults. Another larval function is the production of silk in weaver ants (*Oecophylla* spp.) for the construction of nest chambers from leaves (Hölldobler and Wilson 1990). Eventually, the last larval instar of hymenopteran insects pupates and ecloses as the adult (the imago)—the typical ant, bee, or wasp as we know it. However, imagoes are not necessarily "adult" in the sense of being fertile (the usual definition of adulthood in population biology). Rather, the colony produces adults that are either "workers" (typically nonreproducing adults) or reproductively competent sexual females (gynes, queens) and males (drones in hymenoptera). This developmental path also means that the adult ecloses with its final morphology. It no longer grows or changes shape, although many other age-related processes still occur. For instance, the fat body generally becomes smaller as the worker ages.

In contrast, development in the social Isoptera (the termites) is hemimetabolic (fig. 1.2). The juveniles live through a series of larval instars (the nymphs) and finally molt into the imago, the adult form. The nymphs already resemble the adults and engage in the colony's activities. The hemimetabolous development thus allows for high degrees of flexibility. In the termite *Trinervitermes*, for example, a male first instar larva can molt into a soldier larva, then into a small worker, and later again into a large soldier as it passes along its individual developmental sequence. A female larva can develop into a large worker and remain so until the final seventh instar (Watson et al. 1984).

The adult life span of workers (table 1.3) can be quite short and normally does not match queen longevity. Therefore, as the colony grows with the queen as the long-lived, permanent inhabitant, a turnover of workers occurs. Risks related to foraging activity are presumably the most important factors that set adult

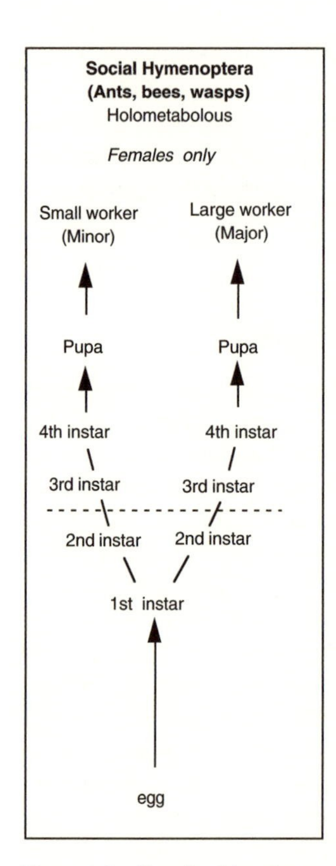

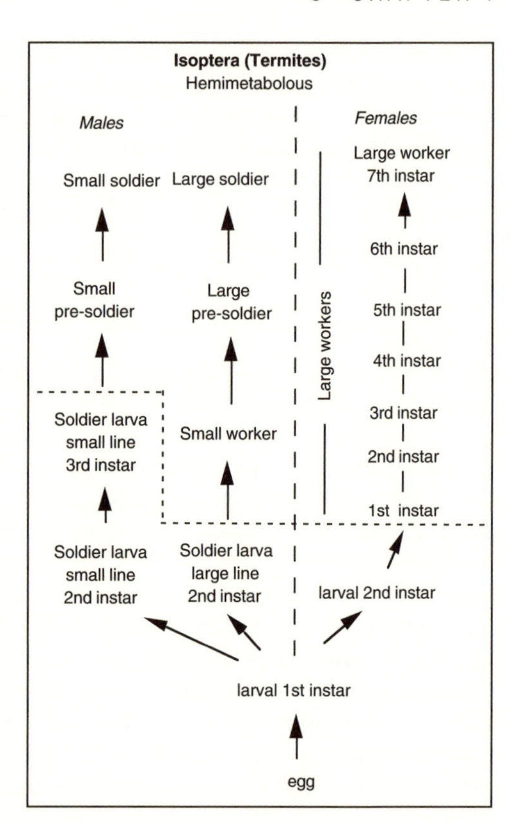

Figure 1.2 Sketch of developmental routes followed by (left panel) colony members of social Hymenoptera (holometabolous development in ants, bees, and wasps) and by (right panel) termites (Isoptera) (hemimetabolous development). In Hymenoptera, only females form the worker caste. They develop through a number of instars during which differentiation into small or large workers takes place, the exact stage varying among taxa (for example, as in ants). The system in termites is more complex: females can develop into large workers, males into small workers or soldiers (as in *Trinervitermes*). The stages below the dotted line are helpless and fully dependent on brood care; above the line, they can actively contribute to work in the colony, except for pupae (modified after Oster and Wilson 1978).

worker life span. In the desert-dwelling ant *Cataglyphis bicolor,* for example, workers live for about four weeks as adults in the nest before they start their foraging activities. The length of the subsequent foraging career is limited by predation of spiders and robber flies and is estimated to be around six days (Schmid-Hempel and Schmid-Hempel 1984). Life span is also often correlated with worker body size so that larger workers are usually longer-lived (e.g., in *Solenopsis:* Porter and Tschinkel 1985; Tschinkel 1993). In addition, as colony

Table 1.3

Longevity of Workers of Social Insects

Species	Life Span	References
ANTS		
Aphaenogaster rudis	3 years	Hölldobler and Wilson 1990
Cataglyphis albicans	32 days	Schmid-Hempel 1983
Cataglyphis bicolor	34 days	Schmid-Hempel 1983
Leptothorax lichtensteini	3 years	Hölldobler and Wilson 1990
Monomorium pharaonis	66 days	Hölldobler and Wilson 1990
Myrmecia nigriceps	2.2 years	Hölldobler and Wilson 1990
Myrmica rubra	2 years	Hölldobler and Wilson 1990
Pogonomyrmex owyheei	42 days	Porter and Jorgensen 1981
Pogonomyrmex barbatus,	Up to 30 days	Gordon and Hölldobler 1987
Pogonomyrmex rugosus		
BEES		
Allodape angulata	1 year	Skaife 1953
Apis mellifera, summer	32 days	Seeley 1985
Bombus fervidus,	20–30 days	Goldblatt and Fell 1987
Bombus pennsylvanicus		
Bombus morio	54 days	Brian 1983, p. 175
Bombus terrestris	25 days	Pers. obs.
Dialictus versatus	21 days	Michener 1969
WASPS		
Mischocyttarus drewseni	21 days	Jeanne 1972
Polistes fadwigae	75 days	Yoshikawa 1963
Vespa orientalis	46 days	Ishay et al. 1968
Vespa simillima	13 days	Ross and Matthews 1990, p. 250
Vespa tropica	35 days	Ross and Matthews 1990, p. 251
Vespa vulgaris	25 days	Ritchie 1915
TERMITES		
Coptotermes acinaciformis	2 years	Gay et al. 1955, cited in Wilson 1971
Coptotermes lacteus	2 years	Gay et al. 1955, cited in Wilson 1971
Cubitermes ugandensis	267 days	Williams 1959
Reticulitermes lucifugus	5 years	Buchli 1958

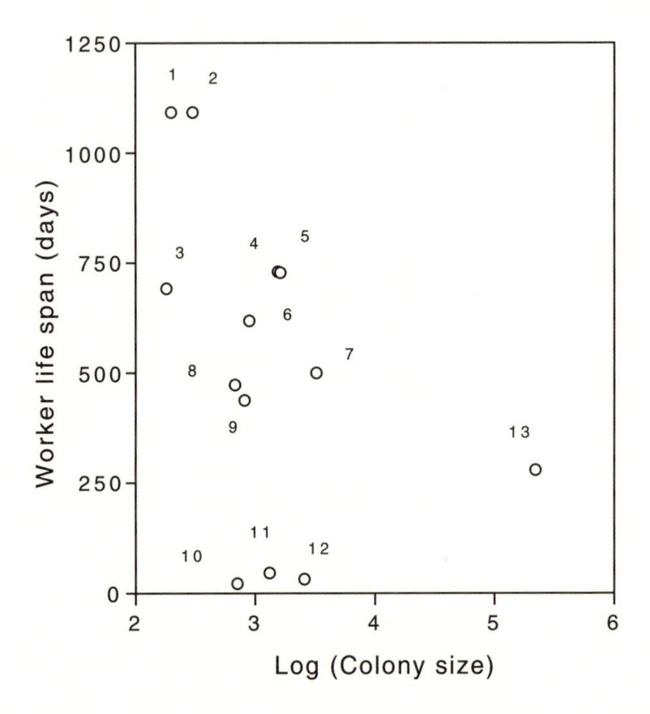

Figure 1.3 Relationship of life span of workers (in days) and typical size (worker number) reached by the mature colony for various ant species. (Data from Hölldobler and Wilson 1990 and table 1.3.) The formal statistics are as follows: Spearman's $r_s = -0.3824$, $N=13$ species, $P=0.19$. Species are: 1 *Leptothorax nylanderi,* 2 *Aphaenogaster rudis,* 3 *Myrmecia vindex,* 4 *Myrmica rubra,* 5 *Myrmica laevinodis,* 6 *Myrmecia gulosa,* 7 *Myrmica sabuleti,* 8 *Myrmecia pilosula,* 9 *Myrmecia nigrocincta,* 10 *Cataglyphis albicans,* 11 *Monomorium pharaonis,* 12 *Cataglyphis bicolor,* 13 *Solenopsis invicta.*

size increases, the egg-laying rate of the queen increases too, but worker longevity appears to decrease (e.g., Hölldobler and Wilson 1990, p. 170; Ross and Matthews 1990, p. 251) (see fig. 1.3 for ants). Matsuura (1991) noted the same trend when he compared five species of social wasps of the genus *Vespa* in Japan. Although there is no relation to the degree of sociality itself, highly advanced species tend to have larger colonies and thus have, on average, shorter-lived workers. These dynamical views of colony development, set by worker life span and turnover, will obviously become important when the spread of parasites within the colony is considered (chap. 4).

The Colony Life Cycle

The colony life cycle is determined by the foundation of the colony, subsequent growth, and production of sexuals. The basic cycle depicted in figure 1.1 shows

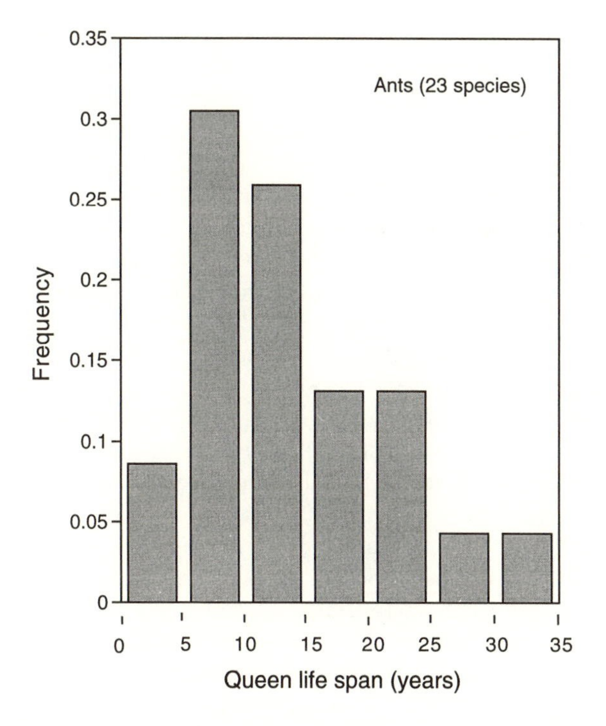

Figure 1.4 The distribution of life spans (in years) for ant queens (N=23 species). (Data from Hölldobler and Wilson 1990 and Brian 1965.)

many variations. For example, all ants and termites are perennial and have more than one reproductive season. Figure 1.1 also illustrates that the growth of the colony and its composition at any one time is affected, among other things, by how long its members live. Often, the longevity of the queen determines the longevity of the colony. Typically, queens are considerably longer-lived than workers (fig. 1.4). They can, in fact, live for several decades, e.g., in *Camponotus, Formica,* and *Lasius* up to 18–29 years (Hölldobler and Wilson 1990). Queens of the sweat bee *Halictus marginatus* have been recorded to live 5–6 years (Plateau-Quénu 1962), those of the stingless bee *Melipona quadrifasciata* around 3 years (Kerr et al. 1962). Queens of the honeybee live many years (Jean-Prost 1956; Seeley 1985; Winston 1987, p. 56). However, they are usually superseded after 1–2 years and leave their colony, i.e., the old queen is replaced by one of her daughters. When supersedure occurs, the colony changes its genetic profile. Interestingly, there seems to be no obvious relationship between queen longevity and degree of sociality achieved (Hölldobler and Wilson 1990, p. 170). Sometimes, real "queens" may not be present at all, as, for example, among the approximately one hundred species of ponerine ants where all workers are fertile and cooperate (Peeters 1993).

Colony founding in social insects occurs in a number of different ways. Most species are capable of independent founding of their own, but some species depend on being able to occupy a host nest to start their own colony ("dependent founding," as in several *Formica* or *Lasius* ants). For example, in the leaf-cutter ants (*Atta*), a single fertilized queen starts a colony on her own in a closed brood chamber (haplometrosis). She lives from her flight muscles that are metabolized during this time. She also carries a priming piece of the fungus from her parental nest to start the new colony's garden. If successful, her colony will eventually contain several millions of workers, her daughters, all of them occupying an impressive nest several meters in height and diameter. In the termites, the reproductive male-female pair (the "king" and "queen") starts the colony. In *Calotermes flavicollis,* the first 10–20 young workers emerge after 8 weeks of being fed with glandular secretions and wood paste, the latter being the normal diet of the species. After one year, 50–60 workers are present, two-thirds of which are soldiers. The proportion of soldiers later drops to 3% as the colony grows (Brian 1965). Across species, the time from nest foundation to first worker emergence varies considerably. Primitive ants, for example, usually take a long time to form a colony. In *Myrmecia forficata* the female lays eight eggs, of which usually only three survive, and which take 4 months to develop. *M. regularis* takes 8 months to produce the first workers (Haskins and Haskins 1950).

Colony founding by single queens (haplometrosis) is often reinforced by aggressiveness against potential cofounders. This period of colony foundation and early growth is particularly critical, since most colonies will not survive (table 1.4; see also fig. 6.1). Associations of foundresses (pleiometrosis) are therefore thought to have evolved to increase success during this period (e.g., Ross and Matthews 1990). This is not uncommon in wasps or ants (e.g., in *Acromyrmex versicolor,* where several unrelated females cooperate: Rissing et al. 1986; Rissing and Pollock 1987). If cofounding by several queens persists into the later life stage of the colony, and if it is combined with reproductive activity, it leads to polygyny, i.e., to the presence of several functional queens. Bourke and Franks (1995) provide an excellent discussion of life histories in ants. Pleometrosis seems overall rarer in termites, perhaps because colony foundation already requires a couple, and hence further joiners may not be as valuable, or the conflicts of interest may be less likely to be settled through cooperation. Cofounding pairs in termites are in fact normally hostile toward each other.

The emergence of the first brood marks the start of the colony as a social group. The following early colony growth phase is also often characterized by the production of small workers (the nanitics). Soon afterwards, the colony enters a growth phase with the full economy of the society established (termed the "ergonomic phase" by Oster and Wilson 1978). During this period, the colony has a fully developed division of labor, often based on age-related polyethism,

Table 1.4

Colony Survival in Social Insects

Species	*Observation on Colony Survival*	*References*
BEES		
Halictus duplex	20% of started nests survive to social stage.	Sakagami and Hayashida 1961
Lasioglossum (Dialictus) zephyrun	A few percent survive first 2–3 months.	Batra 1966
Bombus pascuorum	Of 80 nests: 25 died early, 23 destroyed by predators or fire, 32 (29%) survived.	Cumber 1953 as *B. agrorum*
Apis mellifera	78% of established (old) colonies survive winter in upstate New York, but only 24% of new (young) colonies. Mean colony longevity is 5.6 years.	Seeley 1978
Apis mellifera	45% of old colonies, only 8% of young colonies survive winter in Ontario.	Morales 1986, in Winston 1987
WASPS		
Polistes sp.	3% (2 of 69) colonies survive to produce workers and sexuals.	Yoshikawa 1954
Dolichovespula sp.	1 out of 12 colonies survive to produce workers and sexuals.	Brian and Brian 1952
Vespula analis	36 of 59 colonies lost queens before growth phase, 3 ceased activity, 2 destroyed by predators; 30% (18) were successful.	Matsuura 1984
ANTS		
Formica ulkei	Of 56 mounds, 18 survived for 2 years, 13 for 6 years.	Talbot 1961
Formica exsecta (= F. opaciventris)	Annual colony mortality: 8–9%; annual birthrate: 5–16%.	Scherba 1961, 1963
Lasius flavus	6 of 18 nests survived for 8 years.	Waloff and Blackith 1962
Pogonomyrmex owyheei	Colonies live for 14–30 years (average 17 years).	Porter and Jorgensen 1988

and continues to grow in numbers (fig. 1.5). Where worker polymorphism exists, large workers are often more likely to forage while smaller ones will care for the brood. The large-sized workers (the majors) also serve as soldiers in the defense of the colony against enemies or of a foraging trail against competitors. Morphological castes are weak or absent in bees and wasps.

The number of workers that are typically found in a colony during the ergonomic phase is often taken as a measure of colony size (table 1.5). In an extensive review, Schmid-Hempel et al. (1993) found that colony size correlates not only with a number of fitness measures, but also with individual behavior or social organization. In particular, colony size at the time of reproduction, i.e., when sexuals are raised, is a good predictor of the number of sexuals that can be produced altogether, and thus a shorthand measure of reproductive success (Cole 1984; Schmid-Hempel et al. 1993; see also chap. 4). Obvious advantages to being a large colony involve competitive superiority or higher levels of resiliency against environmental fluctuations. In fact, in wasps (Jeanne 1991) and in ants (Kaspari and Vargo 1995) the typical colony size increases at higher latitudes. This has been interpreted as an adaptation to variable and harsh environments.

Toward the end of the cycle, the colony will reproduce (fig. 1.1). During this period, sexuals, i.e., drones and young queens (the gynes), emerge. This reproductive phase normally follows the ergonomic phase. In some perennial species, where growth is periodically interrupted by some kind of diapause, e.g., hibernation, the reproductive phase may come at other times (e.g., in the honeybee soon after hibernation). The sexuals of otherwise flightless groups, such as ants and termites, are winged and disperse to mate. In some cases, e.g., in slavemaking ants, intranest mating seems to be the rule (Buschinger 1989). As in other organisms, only a few daughter queens will ultimately enjoy reproductive success. For instance, Wildermuth and Davis (1931, cited in Wilson 1971) estimated that only 0.1% of all queens of *Pogonomyrmex badius* will be successful at having offspring. In termites (*Nasutitermes*), where colonies typically release huge numbers of sexuals, the chances of success may be as low as 1 : 10,000 for any queen or king (Wilson 1971, p. 444). In other species, such as the honeybee, many fewer gynes are produced and hence the chances of success are considerably higher.

Some highly social species with elaborate colony structures and division of labor (e.g., army ants, honeybees, stingless bees) start new colonies by budding or fission. The swarming of honeybees is the best-known example of this kind. If the colony is large enough, the mother queen leaves the nest with a part of the worker force (the prime swarm). If the colony is strong enough, other, newly emerged daughter queens may leave the colony with additional parts of the worker force (afterswarms). These will find a new nest site, mate, and establish a colony. Finally, one of the daughter queens that stayed back in the old nest will manage to kill her remaining rivals and thus inherit the parental nest with its worker force.

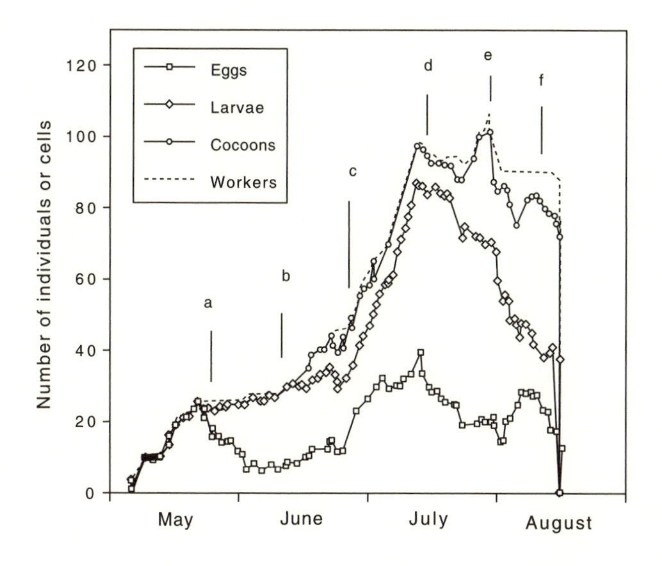

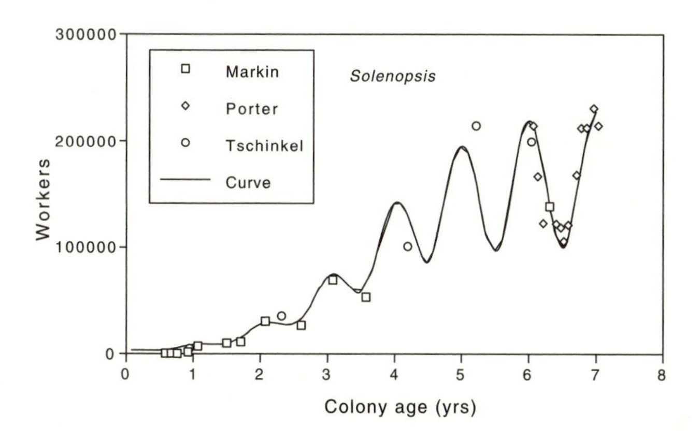

Figure 1.5 Colony development of (a) an annual social insect, the wasp *Parapolybia indica.* The letters indicate first larva hatched (=a), first cell capped (b), first adult emerged (c), a critical number of workers died (d), first male emerged (e), and foundress dies (f). (Reproduced from Gadagkar 1991. Used by permission of Cornell University Press.) (b) Development in a perennial species, the fire ant *Solenopsis invicta.* The single data points refer to counts in three different studies (Markin, Porter, and Tschinkel, as indicated in the graph). The curve is a fit to the data. (Reproduced from Tschinkel 1993 by permission of Ecological Society of America.)

Table 1.5

Colony Sizes of Social Insects

Species	*Colony Size (Number of Workers)*	*References*
ANTS		
Anomma wilverthi	20,000,000	Wilson 1971, tab. 4.6
Aphaenogaster rudis	303	Hölldobler and Wilson 1990
Atta colombica	2,500,000	Wilson 1971
Cataglyphis albicans	700	Schmid-Hempel 1983
Cataglyphis bicolor	2600	Schmid-Hempel 1983
Eciton burchelli	425,000	Wilson 1971, tab. 4.6
Formica fusca	500	Brian 1965
Formica rufa	30,000	Brian 1965
Lasius flavus	10,000	Brian 1965
Monomorium pharaonis	2500	Hölldobler and Wilson 1990
Myrmecia nigrocincta	821	Haskins and Haskins 1980
Solenopsis invicta	50,000	Porter and Tschinkel 1985
BEES		
Apis mellifera	50,000	Seeley 1985
Bombus terrestris	200	Pers. obs.
Bombus morio	2000	Brian 1983
WASPS		
Dolichovespula sylvestris	95	Brian 1965
Paravespula germanica	1613	Brian 1965
Paravespula vulgaris	1000	Brian 1965
Protopolybia pumila	7000	Brian 1965
Vespa crabro	100	Brian 1965
Vespa tropica	313	Ross and Matthews 1990
TERMITES		
Coptotermes formosanus	395,800	Gu-Xiang and Zi-Vong 1990
Incisitermes minor	9200	Nutting 1969
Macrotermes sp.	Ca. 2,000,000	Lüscher 1955
Trinervitermes geminatus	19,000–52,000	Sands 1965
Zootermopsis laticeps	2400	Nutting 1969

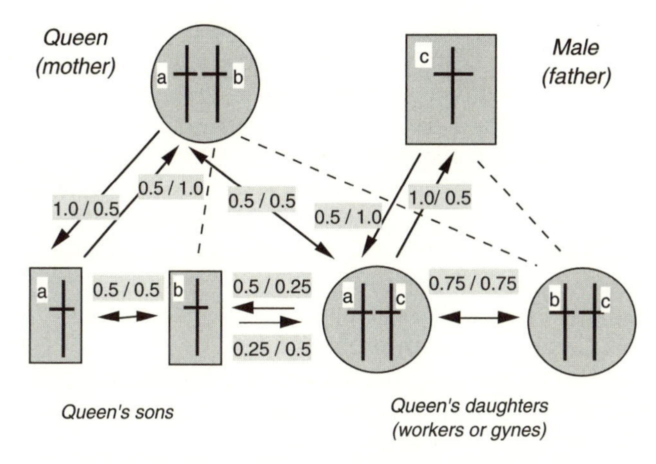

Figure 1.6 A simple pedigree of relatedness for a colony of social Hymenoptera, assuming one queen per colony, singly mated, and with no male production by the workers. Arrows indicate transmission of genes, symbolized by one locus with alleles a, b, and c, and as applicable to the class of individuals represented by the circles (females) or rectangles (males). Small figures give the respective values for regression relatedness and for life-for-life relatedness, respectively, for the corresponding pairs of individuals. For definitions of these terms, see box 5.1. The graph illustrates several asymmetries in the relatedness between pairs of individuals (classes), depending on who is the actor and the recipient of an interaction (direction of arrow). Individuals homozygous at the sex locus will develop into diploid males (not shown).

Sex and Caste Determination

In almost every case studied so far, sex of offspring in social insects is genetically determined. In the social Hymenoptera, sex follows the haplo-diploid mechanism where unfertilized, haploid eggs develop into males and fertilized, diploid eggs into females. In almost all cases, sex is furthermore determined by the alleles at a single locus, or at least this is strongly suspected. Hemi- or homozygous individuals are male, heterozygous individuals are female (Crozier and Pamilo 1996). When there are few sex alleles, or when the effective population size is small, the sex locus may become homozygous; these individuals develop into diploid males. Diploid males are known from a number of species in various hymenopteran taxa (see Crozier and Pamilo 1996, their table 1.4). For some partly social groups, such as the Euglossinae (the orchid bees), diploid males may impose limitations on the evolutionary path to sociality (Roubik et al. 1996). More importantly still, the mode of sex determination in the social Hymenoptera creates high values as well as asymmetries in the genetic relatedness among colony members that has important consequences for cooperation and conflict within the society—visible, for example, as biases in the sex ratio of offspring (Trivers and Hare 1976; Boomsma and Grafen 1991; Sundström 1994) (fig. 1.6).

On the other hand, the development of a female larva into different worker morphs or into a reproductive form (a daughter queen, here called gyne) is environmentally determined. It depends on the amount and quality of food given to the larva. Well-fed larvae or those given special nutrition (such as "royal jelly" in the honeybee) develop into large workers or into reproductives. Only a few exceptions from this pattern are known. In the stingless bees, genus *Melipona,* queens appear to be full heterozygotes of an independently segregating two-locus system (Kerr 1950, 1969). Food shortage will nevertheless cause such heterozygous larvae to develop into workers (Kerr and Nielsen 1966), thus demonstrating the important role of environmental effects on the worker/female determination. Genetic determination has also been suggested for the European slave-making ant *Harpagoxenus sublaevis,* where worker-like reproductive females (the ergatomorphs) differ from the typical winged queens by a single recessive allele (Winter and Buschinger 1986).

1.2 Populations of Colonies

The macroscopic population biology of social insects—the demography and dynamics of populations of colonies—is not as well known as the behavior of individual workers or the internal dynamics of colonies. Nevertheless, a number of studies in different taxa have shown that colony mortality is quite substantial, especially early in their life cycle (table 1.4, fig. 6.1). The study of the life cycles of ants and termites is naturally more difficult to carry out, because the cycles may last many years and often unfold in difficult-to-access underground nests. Long-term studies, such as the one carried out in 1931–1941 on the ant *Formica ulkei* (Scherba 1958), are consequently rare. Scherba estimated the annual birthrate of new colonies to be around 9.1% and the mortality rate to be 16%. Colonies probably lived for 20–25 years. These data match those reported by Talbot (1961) for the same species (table 1.5) and Porter and Jorgensen (1988) for harvester ants. Detailed studies on how important characteristics, such as the colony's foraging range or interactions with neighbors, develop over the life cycle of the colony are unfortunately also very rare (Gordon 1992, 1995). Although the data are far from complete, the general pattern is that colony mortality early in life is quite high, while established colonies have a good chance to persist for a long time. The survival rate for entire colonies is therefore close to what ecologists call Type I survival, i.e., a rapid early decline followed by a low mortality rate afterwards. Enemies, adverse weather conditions, or food shortages are the selective events that are usually thought to lead to colony death.

Only a few studies so far have asked what may limit the population size of social insects. Physical conditions are of course limiting for the distribution of many species. For example, in the ant *Myrmica rubra,* the number of sunny days per year is a good predictor for its occurrence (Elmes and Wardlaw 1982). Summer temperatures may often limit colony growth, whereas winter temperatures limit overwinter colony survival (e.g., as in the meat ant *Iridomyrmex purpureus:* Greenslade 1975a,b). Besides these factors, nest sites are limited in some species (Pickles 1940). This has been demonstrated by Herbers (1986): the population density of the ant *Leptothorax* increased when additional nest sites were experimentally added.

1.3 Nesting

Nesting habits of social insects vary widely (fig. 1.7). Some nests are in the open, as found in army ants or Asian honeybees; other nests are protected by especially constructed envelopes (some wasps), and still others are ground nests in preexisting burrows (e.g., bumblebees) or nest cavities dug out by the colony's workers (e.g., desert ants such as the appropriately named *Cataglyphis*). Many species are cavity dwellers in trees (e.g., the tropical stingless bees, Meliponini). The sophistication of nest organization varies in similar ways. In ants, the brood is typically arranged in loose piles. Eggs, larvae, and pupae are scattered on the floor of the nest chamber. In the more advanced bees and in the wasps, the brood is placed singly in cells that are regularly spaced, as in the brood comb of the honeybee with its hexagonal cells. Sometimes a single colony occupies more than one nest ("polydomy" leading to a "polycalic" society). Contact among nests is maintained by the workers that transport the brood from one nest to another. Limitation in available nest sites is the most frequently cited explanation as to why social insect colonies are mono- or polydomous (Herbers 1986).

Colonies move to a new site much more frequently than previously thought. Herbers (1985) observed in North America that nests of fourteen ant species are regularly moved over the duration of a season. Usually, the colony expands during the summer and uses several nests, then contracts to a few hibernation sites in winter. In some species a typical nest lasts only for 2–3 weeks. It is likely that the causes for nest relocations are often associated with a change in microclimatic conditions, predation, food supplies, and, indeed, parasites (see chap. 4).

The spatial activity of colonies also includes the maintenance of territories. Colonies in most populations of social insects are in fact overdispersed. For example, ants possess a variety of behavioral repertoires to defend territories, including a rich array of pheromones to mark their boundaries or to alarm nestmates to defense (Hölldobler and Wilson 1990). In termites, the avoidance of other colonies' feeding galleries may separate neighbors. Territories are typically absent in the flying social insects, i.e., in bees and wasps. Territorial conflicts sometimes lead to the killing of adversaries. In some groups, e.g., the ant *Formica polyctena* (Mabelis 1979), the killed enemies are taken home as prey.

a

b

c

d

e

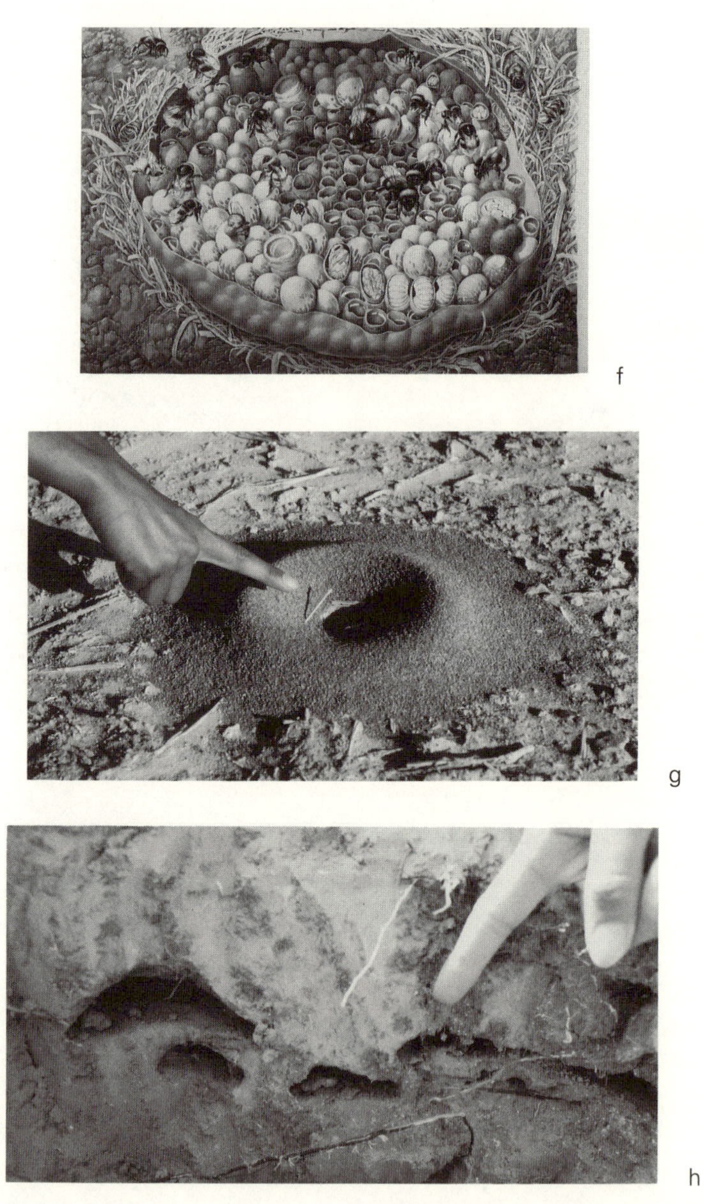

Figure 1.7 Nests of social insects. (a) An open nest of the wasp ***Mischocyttarus,*** elongated cells pending from support structure; (b) an arboreal nest of the ant *Azteca,* with cover; (c) an open nest of the wasp *Polistes,* cells in a cluster; (d) massive mound of the termite *Nasutitermes* with solid outer walls; (e) mound of the wood ant *Formica;* (f) subterranean nest of a bumblebee, *Bombus* (original drawing provided by M. Pirker, reproduced by permission); (g) and (h) subterranean nests of the desert ant *Cataglyphis,* with entrance and brood chambers.

This pattern is more generally found in aggressive disputes between species rather than within species (Hölldobler and Wilson 1990, p. 414). In addition, ants, bees, and wasps regularly engage in robbery, i.e., they attack neighboring nests and steal food or brood.

Finally, spatial activities also include the observation that the reproductives are dispersing either before or after mating. Some of the phenomena associated with this process are quite spectacular. For example, the mating swarms of the seed-harvester ants *Pogonomyrmex* in the southwestern deserts of North America consist of thousands of winged males and females, whirling and diving in the air. After mating, queens will take off and disperse to found a new colony (Hölldobler and Wilson 1990). Similarly, termite sexuals leave the colony in the thousands and are harvested because they are considered to be a delicacy by local people. More subtle, but no less spectacular, are long-range dispersals such as reported for bumblebee queens that migrate for hundreds of kilometers along the Scandinavian coast (Mikkola 1984).

1.4 Summary

Social insects comprise about 20,000 species from several orders. Their mode of life, colony organization, reproductive pattern, and nesting habits differ widely. In many ants, bees, and wasps, typically, a queen founds a colony which then grows in worker numbers and eventually reproduces daughters (the gynes) and sons (the drones). Many bees and wasps are annual, while ants and termites are perennial. In termites, colonies are founded by a male-female pair. Deviations from the simple scheme include, for example, polygyny (more than one functional queen per colony) and dependent colony founding (in a host nest). Individual development in ants, bees, and wasps follows the holometabolous path with several larval instars that metamorphose into the adult form. Termites are hemimetabolous, where individuals develop through several nymphal stages that are part of the workforce of the colony. Queens can potentially be very long lived, while workers are replaced as the colony grows and eventually reproduces. While the behavior and turnover of workers within the colony is reasonably well studied, the dynamics of populations of colonies is often unknown. Sex in hymenopteran social insects is determined by a haplo-diploid mechanism that generates close and asymmetric genetic relationships among colony members.

2 The Parasites and Their Biology

In the ancient Greek world, a *Parasitos* was a person who received free meals from a rich patron in exchange for amusements and conversations (Brooks and McLennan 1993). Unfortunately, the term is not as easy to define in biology. Webster's International Dictionary defines it as "an organism living in or on another living organism, obtaining from it part or all of its organic nutriment, commonly exhibiting some degree of adaptive structural modifications, and causing some real damage to its host." Obviously, this definition leaves out phenomena such as social parasitism and leaves open problems such as defining real damage. However, it seems almost impossible to give a universal definition, and, since social parasitism is not discussed in this book, we may just as well stick to such vague descriptions.

In this chapter, an overview over the different parasite groups and their biology is given. The main emphasis is on those associated with social insects. A summary of the known parasites of social insects by taxonomic group is given in the lists in Appendixes 2.1 to 2.11. Not surprisingly, the number of parasite species (here called "parasite richness") described from the different social insect taxa is related to how well they are studied in general, i.e., the sampling effort (fig. 2.1). In the subsequent analyses, sampling effort will be controlled for by analyzing the residuals from the regression shown in figure 2.1 (the "standardized parasite richness") rather than the raw values themselves (Walther et al. 1995). Using the raw data, however, would not alter the major conclusions. The number of recorded parasite species per social insect host species is approximately Poisson-distributed, with an average of 2.63 ± 0.19 (S. E.; $N=488$) recorded parasites (1.31 ± 0.14 pathogens, 1.32 ± 0.10 parasitoids) per species (fig. 2.2). The record holder is the honeybee, *Apis mellifera,* with over seventy recorded parasite species in the database, but also with some 170 studies that are published and registered in *Biological Abstracts* each year. Another major reason why the European honeybee is so well known is that it is a managed species in most parts of the world. Beekeepers often spread diseases by exchanging frames and hive bodies that contain parasites. These are later noticed and thus become known to science.

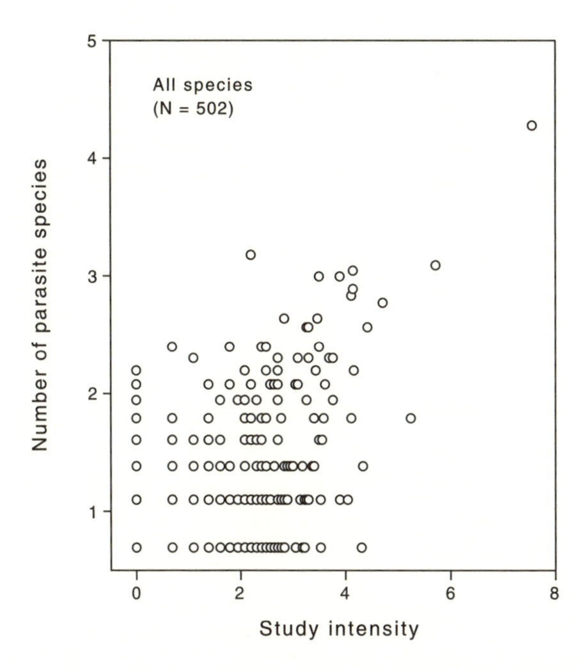

Figure 2.1 The number of parasites recorded per host species (parasite richness) in social insects. The abscissa is the number of published studies on the host species during 1985–1995 from a survey in *Biological Abstracts*. This quantity is used as a measure of sampling effort for the species. Only hosts where at least one parasite has been reported are included. Data ln-transformed to normalize variances. Parasite data are taken from the lists in appendix 2, except for socially parasitic host species, mites, and sphecid hosts (*Microstigmus*). The point on the upper right hand is the honeybee with 71 recorded parasites and 1899 published studies. The regression is: $Y=0.762+0.252\,X$, $r^2=0.331$, $N=502$ species; $F=249.06$, $P<0.0001$; with $Y=\ln$ (parasite richness+1), $X=\ln$ (studies+1). The residuals from this regression are used in subsequent analyses of this book ("standardized parasite richness").

2.1 Viruses

A common problem with investigating viral diseases of insects is the custom to name a virus after its host and thus to ignore possible multihost relationships. The methodological and conceptual progress of molecular epidemiology should reduce this problem in the future. Currently, more than twenty groups of viruses are known to be insect pathogens (Tanada and Kaya 1993). Among them, DNA viruses are prominent and include, among other groups, the nuclear polyhedrosis viruses, granulosis viruses, Polydnaviridae (multipartite, double stranded), Poxviridae (Entomo-poxvirinae), Iridoviridae (nonenveloped,

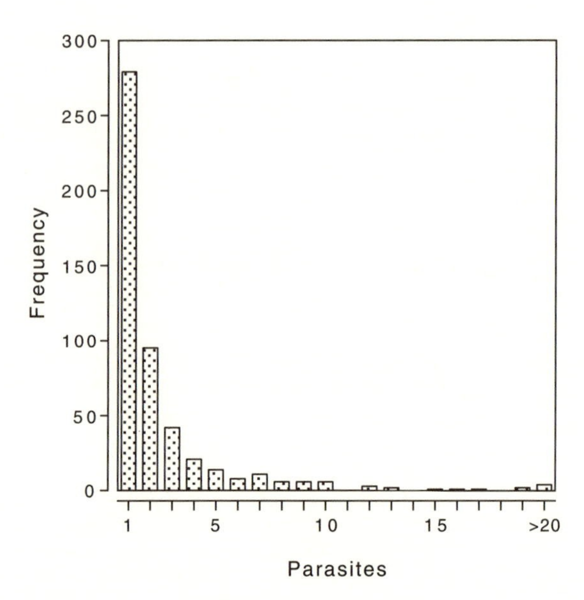

Figure 2.2 Distribution of the reported parasite species per social insect host species, where at least one parasite species has been reported. Data are taken from the records listed in appendix 2, except mites. The mean parasite richness is $x = 2.63 \pm 4.3$ (S.D., $N = 502$ host species; range 1–71).

double-stranded viruses), and Parvoviridae (nonenveloped, single stranded). In the group, the Baculoviridae (BV) have received most of the attention because of their ease of detection and potential as microbial control agents. They account for some 90% of the reported cases in the Hymenoptera outside *Apis mellifera* (Evans and Entwistle 1987, their table 10.11). Baculoviruses are only found in arthropods. Known RNA viruses include the cytoplasmic polyhedrosis viruses, the Picornoviridae, and many others. Unfortunately, among the major social insect groups, next to nothing is known for the Isoptera (termites) and their relation to viruses.

The infective cycle of viruses always involves the attachment of the infective particles to the host cells, uptake into the cell, and uncoating of the virus, followed by expression and replication of the viral genome that leads to the production of progeny. Enzymes both from the virus and the host cell are involved in this process. In insects in general, the primary route of viral infection is the alimentary tract, i.e., when the animal feeds on infected material. Also, larvae are in general more affected by viruses than adults. The parasite usually can penetrate the gut wall only in the midgut section, because the other parts of the alimentary tract are of ectodermal origin and provide an effective barrier against infections. Apparently the pH value of the intestinal tract is important for the establishment of the parasites. Under the appropriate conditions, the virions or infective stages degrade (e.g., virions of Baculoviridae are susceptible to alkaline conditions: Evans and Entwistle 1987, p. 302). Typically, guts of phytophagous insects have alkaline conditions while those of predatory or scavenging species have low pH values (Evans and Entwistle 1987). Another important element is the presence of the peritrophic membrane that is secreted by and lines the cells of the midgut and that can act as a shield to infections. Viruses are also transmitted transovarially, i.e., via the eggs or reproductive tract to offspring.

Iridescent viruses (IV, Iridoviridae) are known from the honeybee but have not been reported in other Hymenoptera. The name alludes to the typical iridescence observed in infected tissue in many of these viral infections (Tanada and Kaya 1993). Not much is known about the routes of infection. The virus typically resides in the alimentary tract, fat body, hypopharyngeal gland, and ovaries. Thus, it is reasonable to conclude that transmission and infection occur via food, feces, and gland secretions, and the virus can be passed on transovarially to offspring. In contrast to many other honeybee viruses, though, mites seem not to be involved as vectors. The honeybee IV is associated with "clustering disease" of the honeybee (and known to occur naturally only in *Apis cerana*), characterized by unusual inactivity and formation of small clusters of workers (Bailey and Ball 1978). High temperatures usually impede replication of iridoviruses (Tanada and Kaya 1993, p. 258).

Cytoplasmic polyhedrosis RNA viruses have been reported from a wide range of insect hosts (but only 3% of these in Hymenoptera; none in Isoptera: Tanada and Kaya 1993, p. 277). They seem to play no role in social insects. Other small RNA viruses, e.g., Birnaviridae or Picornaviridae, have broad host ranges and often cause inapparent infections. In the honeybee, almost twenty different RNA viruses have been identified (Ball and Bailey 1991).

Picornaviridae that attack insects are well known, e.g., the cricket paralysis virus (CrPV). This virus replicates in the cytoplasm of epidermal cells of the alimentary canal and also in nerve cells of ganglia (Tanada and Kaya 1993, p. 301). Picornaviridae are related to the mammalian polio virus. In social insects, acute and slow bee paralysis virus (APV, SPV), bee virus X (BVX), and sacbrood virus (SBV) are known in the honeybee. The normal pathway of infection is by oral ingestion or through the integument. Symptoms of bee paralysis virus infections are trembling movements combined with sprawled legs and wings, symptoms commonly called paralysis. Sometimes, hairs are lost and the animals get a shiny, black appearance. The disease is actually caused by chronic bee-paralysis virus (CBPV, not a Picornavirus), while APV persists only as an inapparent infection. APV may thus be rather common but often remains undetected because of few visible effects. For example, figure 2.3 shows how the occurrence of APV is revealed by infecting test bees with the extract of seemingly

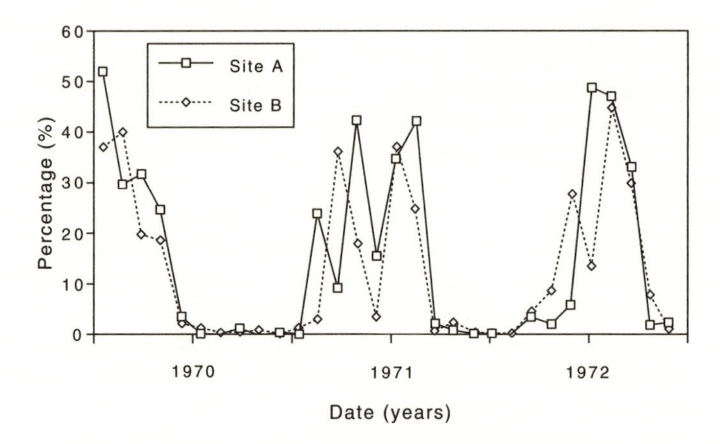

Figure 2.3 Percentage of test honeybee workers killed by acute bee paralysis virus (ABPV) when injected with extracts from seemingly healthy adult bees. Sources of extracts were colonies from two sites (site A, site B) in Rothamsted, England. (Reproduced from Bailey and Ball 1991, by permission of Academic Press.)

healthy bees. The mite *Varroa jacobsoni* acts as a vector and activator of the virus (Bailey and Ball 1991; chap. 3.2). Varroa can induce APV and SPV to multiply to lethal levels. Both viruses, however, do not normally cause mortality under natural conditions (B. Ball, pers. comm.). CBPV generates two different classes of syndromes: trembling-crawling bees, or those that are unable to fly and become hairless-black. It is thus thought that diseases like "Waldtrachtkrankheit" (type 1 syndrome) and "hairless-black syndrome" ("Schwarzsucht") (the type 2 syndrome) are caused by the same agent. Diseased individuals die within a few days, and the affected colonies may collapse within a week (Bailey and Ball 1991, p. 11). The viral particles can be found in the ganglia cells of the thorax and abdomen, in the alimentary tract, and in mandibular and hypopharyngeal glands (Bailey and Ball 1991; Tanada and Kaya 1993). CBPV also has an associated small RNA virus ("chronic bee paralysis virus associate": CBPVA). It is a satellite virus that interferes with the replication of CPBV (Bailey and Ball 1991, p. 16).

Sacbrood virus (SBV) is well known from the European *Apis mellifera* and Asian *A. cerana* and affects larvae. They become unable to pupate, since the endocuticle of the last instar cannot be shed, perhaps because the formation of chitinase is prevented. SBV is widely distributed and, for example, is the most common virus in parts of Australia (up to 90%; Hornitzky 1987). It can be found in up to 30% of the colonies in Britain. A slightly different strain from Thailand seems to have recently spread over much of the region in the Himalayas (Burma to India). It possibly exhibits a four-year cycle (Verma et al. 1990).

Deformed wing virus in the honeybee leads to deformed emerging adults; it is related to Egypt bee virus. The virus multiplies slowly and is vectored by *Varroa jacobsoni* (Bailey and Ball 1991). Several viruses are associated with the microsporidian honeybee parasite *Nosema apis*. These include black queen cell virus (BQCV), causing dark walls in queen cells. BQCV is also common in workers. Bailey et al. (1981), for example, found an average of 30% infection in workers in twenty-five investigated colonies at Rothamsted. It does not readily infect larvae and adults, even after experimental inoculations (Bailey and Woods 1977). Filamentous virus (FV) multiplies in fat bodies and ovarian tissue of adult honeybees. FV and also bee virus Y (BVY) are loosely associated with *N. apis,* although BVY can exist on its own. They produce no serious symptoms. Bailey and Ball (1991) suggest that infection by *Nosema* reduces host resistance to a point where the viruses are able to infect via the alimentary tract. As *Nosema* develops in the epithelial cells of the midgut, this is conceivable. Both BQCV and BVY have been claimed to add to the pathogenic effect of *Nosema apis.*

Bee virus X (BVX), distantly related to BVY, is another established honeybee virus. It has no relationship to infections by *Nosema* but is associated in a loose way with infections by the amoeba *Malpighamoeba mellifica,* since both the amoeba and the virus are transmitted in the same way—via fecal contamination. BVX develops winter epizootics that take a heavy toll. Cloudy wing virus, Kashmir bee virus, slow paralysis virus, Arkansas virus, and Egypt bee virus should also be added to the list of known honeybee viruses. Some 15% of colonies have been found infected by the cloudy wing virus in Britain. Its prevalence is presumably dependent on the occurrence of (unknown) chance events that affect the rate of spread (Bailey et al. 1981). Kashmir bee virus is remarkably virulent since it kills a bee within three days of infection. It does not only occur on the Indian subcontinent, where it can overlap with its other host *Apis cerana,* but also in New Zealand and Australia. The latter strains are also found to be highly pathogenic. Anderson and Gibbs (1982, cited in Bailey and Ball 1991) suggested that the virus was acquired in Australasia from a natural reservoir of stingless bees, but this has not been substantiated and is now considered unlikely (B. Ball, pers. comm.).

Varis et al. (1992) sampled dead honeybees from England and Finland and checked them for the presence of different viruses. Thirty bees from each sample location were screened by immunodiffusion tests for black queen cell virus (BQCV), bee viruses X (BVX) and Y (BVY), chronic paralysis virus (CBPV), acute paralysis virus (APV), Kashmir bee virus, sacbrood virus (SBV), cloudy wing virus (CWV), and deformed wing virus (DWV), while filamentous virus (FV) was identified with electron microscopy. Hence, their procedure could only reveal the fraction of the viruses found in dead but not live bees, and therefore does not give a true estimate of parasite prevalence. Interestingly, the viruses occurred in similar proportions in both areas, except for FV. In Britain, FV is one of the most common and least harmful viruses (Bailey 1982). It usually decreases in summer (similar to *N. apis*). BVY was found in about 5% of the samples. It seems remarkable that the prevalence of FV is higher at northern latitudes (Finland: ca. 25%; Britain: 5%) (see also fig. 2.5 for *Crithidia*)—in particular, as winter mortality of hives in Finland (15%) is generally higher than in Britain. CWV was observed in approximately 25% of the samples. In Britain it is thought to increase colony mortality (e.g., Bailey et al. 1983a). Only CPV and BQCV (ca. 1% of cases) were found as additional viruses.

Although only the honeybee and to some extent the fire ant are reasonably well screened for the occurrence of viruses, the few available reports from other species cover distant taxa (Appendix 2.1). For example, Wigley and Dhana (1988, cited in Glare et al. 1993) identified cricket paralysis virus and Kashmir bee virus in the social wasp *Vespula germanica* (Vespidae). Entomopox-like viruses were found in bumblebees (Clark 1982), and acute bee paralysis virus has also been identified in their pollen loads (Bailey and Gibbs 1964). Clark (1982) collected bumblebees from flowers between April and September in Maryland in eastern North America. Of a total of 592 workers examined (*Bombus pennsylvanicus, B. fervidus*), 29 contained entomopox-like viruses (4.9%). A further 49 workers from *B. bimaculatus* and *B. vagans* were negative. Clark extracted small amounts of infected hemolymph with microcapillaries through the intersegmental membrane of the abdomen to test for infectivity on healthy hosts. Unfortunately this procedure gave no conclusive result because the workers soon died in the laboratory. All infected bees behaved normally when collected in the field and survived as well as healthy ones in the laboratory. In addition to hemolymph, the hypodermis and the salivary glands also contained viruses. This is somewhat unusual since in other insect families the viruses are more likely to be found in the fat body. The host range of entomopoxviruses seems to be relatively narrow, with little cross-infection between insect families.

The fire ants have been found to be infected by a nonoccluded baculovirus. Infection occurs in the cells of the fat tissue (Tanada and Kaya 1993). Avery et al. (1977; see also Jouvenaz 1986) reported viruslike particles in a species of the *Solenopsis saevissima* complex in Brazil (corresponding to *S. invicta* or *S. richteri* in North America) and also in *S. geminata* in Florida. They were not similar to those seen by Steiger et al. (1969) but more close morphologically and developmentally to those reported from the Rhinoceros Beetle *Oryctes* and the Whirligig Beetle *Gyrinus*. The effects on host mortality could not be reliably determined. An association of these viruslike particles with the microsporidian *Thelohania* that also infects the ants was observed (Avery et al. 1977). Steiger et al. (1969) observed virus-like particles in the cytoplasm of nerve and glia cells as well as in fat bodies of wood ants (*Formica lugubris*).

From these reports, it is safe to assume that the apparent lack of viruses in many social insects, other than honeybees and fire ants, is due to the level of investigation (see fig. 2.1) rather than to a genuine absence of viral parasites. In addition, in many cases, the seemingly benign effects of viruses, their need to be activated to express pathological effects, and their association with other parasites cloud causes and effects; the pronounced seasonal cycles (fig. 2.3) also make detection less obvious. Many viruses may also be contracted from reservoirs in coexisting nonsocial insect species and hence appear irregularly in the record. In conclusion, then, it would be timely to carry out systematic surveys for viruses in social insect populations of many different taxa.

2.2 Bacteria and Other Prokaryotes

For epidemiological surveys of bacterial infections, routine techniques to isolate, cultivate, and test the probes, such as serological tests or growth tests on media, must be employed. This is usually quite involved. It is possible that molecular probes will become more important in the near future and will thus help to accelerate data collection. In addition, insects normally have small numbers of bacteria in their gut. Only under adverse conditions, such as low temperature or a deficient diet, will the bacteria multiply and lead to pathogenic conditions. All these factors will again make detection difficult. Hence, it is not surprising that our knowledge of bacterial parasites of social insects is quite limited.

Mollicutes (Spiroplasmataceae)

Mollicutes are bacteria-like organisms and were not really recognized until the 1970s (Whitcomb et al. 1973). They lack real cell walls, certain cell components, and fibrils but are nevertheless mobile. All are placed in the family Spiroplasmataceae; identification is made with a serogroup classification system (Whitcomb et al. 1987). Mollicutes have been isolated from insects in the orders Hymenoptera, Hemiptera, Diptera, Coleoptera, and from plants. Overall, spiroplasmas are parasites, commensals, and perhaps also mutualists. In insects, spiroplasmas can be found in the gut lumen or hemolymph of the animal. Normally, they cause a disruption of the infected cells and degeneration of neural axons. Plants are systemically infected, but the spiroplasmas also appear on flower surfaces. Plant spiroplasma seem indeed able to infect visiting insects. These may serve as vectors, such as leafhoppers on maize (Davis et al. 1981). Social insects that regularly visit and utilize plants—besides bees, there are also the seed-harvesting ants and termites—should be prone to acquiring the plant pathogens.

Spiroplasma melliferum, serologically related to *S. citri*, infects the honeybee (workers and drones) (Clark 1977). Infection occurs by os from where the parasite eventually enters the hemocoel. Infected bees become sluggish and die within a week (range 3–20 days). The prevalence is seasonally variable, with peaks in May–June after the swarming period. The same or at least antigenically close forms have been observed on flower surfaces, from where the bees seem to acquire the infection (Clark 1977; Davis et al. 1979; Raju et al. 1981). *Spiroplasma apis* has also been claimed as the causative agent of a lethal disease ("May disease") of honeybees in France. Several other mycoplasmas in honeybees and flowers have been characterized (Junca et al. 1980). In their detailed study, Clark et al. (1985) identified twenty-eight different strains of *Spiroplasma,* some of them from bumblebees, in the hemolymph of *Bombus impatiens* and the intestines of *B. pennsylvanicus.* It is unknown whether *Bombus* is the major host in the wild. This seems unlikely, however, because it also occurs in many solitary bees, e.g., *Osmia cornifrons,* some Andrenidae, and the bee *Anthophora* (Raju et al. 1981). Rather, there may be a reservoir of different host species in the field that share the parasites (similar serological types), perhaps via shared use of common flower resources (Durrer and Schmid-Hempel 1994).

Bacteria

In contrast to the general ignorance of bacterial parasites of social insects in general, the classical bacterial infections of the honeybee—European and American foulbrood—are probably among the best investigated diseases of social insects. Only after a considerable amount of work was *Melissococcus* (=*Streptococcus* (*Bacillus*)) *pluton* unambiguously identified as the causative agent of European foulbrood (Bailey 1963). The bacteria enter the host by os where the bacterium multiplies and remains confined to the peritrophic membrane of the gut wall. As the infection progresses, the larvae change color but also show abnormal movements and occupy unusual positions within their cells. Host death may occur from the accumulated toxic products of the bacteria which can diffuse through the intestinal wall to other organs and tissues. It is more likely, however, that death results from starvation, as bacteria do not invade the tissues but utilize food within the gut (B. Ball, pers. comm.). European foulbrood (*Melissococcus pluton*) kills its host in the larval stage (4–5 days old). *M. pluton* is sometimes associated with secondary invading bacteria, such as *Streptococcus faecalis.* This bacterium is presumably picked up by the foragers in the field, where it occurs in a number of insect species and on flowers (Mundt 1961). It gets established when European foulbrood breaks out in the colony (Bailey 1981). *S. faecalis* and several Enterobacteriaceae are typical gut inhabitants of many insects. According to Krieg (1987), most of the regular pathogens seem to be effectively eliminated by the immune mechanisms of the insect hemolymph after invasion. Facultative parasites, on the other hand, are often immunoresistant and can therefore induce bacteremias within the hemocoel, followed by septicemia and disease. This also makes it difficult sometimes to distinguish between primary infections and secondary invaders as the major cause of disease.

American foulbrood (*Paenibacillus* [formerly *Bacillus*] *larvae*) usually leads to death after the larva has spun its cocoon, but otherwise the course of the infection is similar to European foulbrood. Bailey and Ball (1991) give a detailed account of the disease: the larvae become infected through spores in the food. The vegetative forms of *P. larvae* seem to emigrate through the gut wall to infect vital tissues, e.g., fat body, trachea, and hypodermal cells. The symptoms are visible 10–15 days after infection. Young larvae of fewer than 2 days of age are very susceptible, but larvae become more resistant later on. The presence of *P. larvae* can go unnoticed for some time as colonies show few symptoms, if any. Typically, only after an infection has built up, already killing a few hundred larvae, an outbreak will follow, eventually killing the entire colony. *P. larvae* has an enormous epidemiological potential. Sturtevant (1932) estimated that some 2.5×10^9 spores are produced by each host. Spores are shed by the infected bee over long periods of time and are very resistant to sunlight, heat, or chemicals (Wilson 1972). They can remain infective for more than 35 years (Haseman 1961). However, the normal incidence of the disease is fairly low, because spores are removed by adult bees and because only young larvae are susceptible. These diseases are seasonally prevalent and disappear by midsummer (Bailey and Ball 1991).

Serratia marcescens (originally thought to be *Pseudomonas apiseptica*) are occasional pathogens of honeybees. Because infection is not possible by feeding but occurs after spraying, infection may occur through the tracheal openings. If a bacteremia develops, the worker bee is killed. The infection is correlated with the presence of mites. Therefore, Wille and Pinter (1961) suggested that mites act as vectors for the disease. On the other hand, the bacterium has been reported to be abundant in soil near infected apiaries (Burnside 1928) and thus could be a soil organism that is occasionally contracted by honeybee workers. It would be interesting to know whether soil-dwelling social bees are more often affected by such parasites. Other rare bacterial infections of the honeybee include *Bacillus pulvifaciens* (Hitchcock et al. 1979), *Pseudomonas fluorescens*, *P. aeriginosa,* and *Yersinia pseudotuberculosis* (isolated from hemolymph)(Landerkin and Katznelson 1959; Horn and Eberspächer 1976). *Pseudomonas aeriginosa* enters by os and, after penetrating the gut wall, can develop a fatal septicemia in the hemocoel (Landerkin and Katznelson 1959). *Hafnia alvei* is an opportunistic parasite of vertebrates that occurs in soil and water. It was observed by Strick and Madel (1988) to infect honeybee colonies when the mite *Varroa jacobsoni* was present. In experimental infections, the mite transferred the bacteria from the pupal cuticle to the hemolymph, causing infection and death. *Aerobacter cloacae* causes B-melanosis, a disease affecting the ovaries of honeybee queens. It has, however, also been found in bumblebee queens (Steinhaus 1949; Skou et al. 1963; Fyg 1964), where it may or may not cause disease.

In addition to those clearly pathogenic bacteria, a number of studies have turned up a range of bacteria with uncertain effects, most of them presumably harmlessly saprophagous, or even mutualistic. For example, Gilliam et al. (1990) analyzed microorganisms associated with pollen, honey, and brood provisions in the stingless bee *Melipona fasciata.* Three different species were identified: *Bacillus megaterium, B. circulans,* and *B. alvei* (the latter is also a secondary associate of European foulbrood). These bacteria (close relatives of the serious disease *P. larvae*) produced a variety of enzymes (e.g., esterases, lipases, proteases) that are able to convert food into more digestible products. More interestingly still, these bacteria also produce antibiotics and fatty acids to inhibit competing microorganisms. This suggests that these bacteria are helpful to their host. The authors do indeed note that other microorganisms are conspicuously absent in food stores of bees. Machado (1972, cited in Roubik 1989) demonstrated that experimental elimination of bacteria from pollen stores ultimately led to the death of the colony of the tropical stingless bee *Melipona quadrifasciata.* There are also several cases where microorganisms provide nutrients to their host (see examples cited in Roubik 1989, p. 229); an almost classical example is, of course, the termites and their gut associates. Bacteria have also been found in natural populations of the fire ant *Solenopsis invicta* (Jouvenaz et al. 1980) although so far they have not been identified (Miller and Brown 1983). *Serratia marcescens* was observed in the wasp *Vespula germanica* (Glare et al. 1993) but is probably rare.

Miller and Brown (1983) experimentally fed different species of bacteria (*Serratia, Enterobacter, Pseudomonas*) to fire ants *Solenopsis richteri* and *S. invicta.* These inoculations were able to establish infections and to kill the ants within a short time through the accumulation of toxic products. Although this evidence does not show that *Solenopsis* is a natural host of these bacteria, it does at least demonstrate that the bacteria are not necessarily very host-specific. This adds to the general observation made above for the honeybee: bacterial parasites of social insects may often be acquired as opportunistic infections by organisms that inhabit soil and plants, or that are prevalent in coexisting insect species.

Again, the available records seem to suggest that bacteria are parasites mainly of the honeybee. This is very unlikely to reflect the real distribution and prevalence of bacterial infections in social insects in general. One can expect certain classes of social insects frequently to come into contact with bacterial reservoirs and thus be subject to regular infection. This is most likely the case for social bees or seed-harvester ants, which may contract bacteria that are normally associated with plants, or for scavengers, predatory ants, and wasps that forage for insect prey harboring bacteria. Also, it is likely for soil-dwelling species that may contract soil-inhabiting bacteria in opportunistic infections. These expectations would deserve attention, since a conspicuous lack of bacterial parasites in prone species may indicate special mechanisms of prevention (e.g., antibiotic substances produced by glands or hygienic behaviors; see chap. 7), and/or selective forces that prevent specialization of bacteria on difficult hosts.

Rickettsias

Rickettsias are small prokaryotes that are closely associated with arthropods (Krieg 1987). In vertebrates, some well-studied rickettsias are transmitted by arthropod vectors, i.e., mites and fleas. Given the manifold associations of mites, for example with social bees, and the fact that normally many social bee species occur together in the same habitat, it may be worthwhile to pay more attention to possible rickettsia-infections in the future. Otherwise, the situation in social insects is completely unknown. Perhaps *Wolbachia,* an apparently widespread rickettsia with effects on offspring sex (Werren et al. 1995) is more common than hitherto appreciated. It seems that these pathogens are not very specific and are therefore infective for a wide range of host species (Krieg 1987). In fact, in a very recent study, T. Wenseleers (pers. comm.) used molecular tools and tested positive at high percentages for rickettsia and *Wolbachia*-types of microorganisms in a large number of ant species from formicinae, ponerinae, dolichoderinae, and myrmicinae from Europe and southeast Asia. Clearly, such observations are highly relevant for our understanding of parasitism in these groups.

2.3 Fungi

Fungal infections can be conspicuous when hyphae grow on the insect's surface. Many entomopathogenic fungi, however, produce no externally visible structures. Fungal growth is favored by high humidity, moisture, and warm temperatures. In addition, soils with high organic content are more favorable than sandy soils for the persistence of fungal diseases of soil-inhabiting insects (Bünzli and Büttiker 1959, cited in Benz 1987). In appropriate soils, therefore, fungal spores survive (Benz 1987) and hence sustain epidemiological cycles that span successive generations of otherwise annual social insects. A spectacular case is *Massospora cicadina* infecting the periodic cicadas in North America. Since the adult host appears only after long intervals (e.g., every 17 years for *Megicicada septendecim*) and since it does not infect the immature stages, the fungus has to survive this period in the soil (e.g., Soper et al. 1976). In terms of population regulation, entomopathogenic fungi seem especially important for aphids, leafhoppers, grasshoppers, and their allies.

The history of discoveries of fungal diseases is quite remarkable and interesting in its own right (McCoy et al. 1988). The recognition that fungi attack insects goes back to the early civilizations in China. Two thousand years ago, the Chinese identified *Cordyceps* species from cadavers of silkworm. The spectacular growth of the fungus on the cadavers was thought to convey immortality to the dead. Stone effigies of the infected insects were therefore used in the funeral ceremonies. *Cordyceps*-infected caterpillars are still used in Chinese medicine as an aphrodisiac or to cure opium addiction. The fungus *Beauveria bassiana,* widely used today as a biological control agent, was first recognized a thousand years ago in China. In European science, entomogenous fungi were first described in the eighteenth century from silkworms and honeybees. Reaumur in 1726 mentioned the characteristic growth of *Cordyceps* in a noctuid moth and described it as "Chinese plant worm." This aptly describes the large (tree-like) fungal stalk that develops from the cadaver. In fact, at the time, such infections were often characterized as "vegetable wasps" and "plant worms," implying that these "plants" had wasps or caterpillars in the place of roots (e.g., Torrubia in 1754). The landmark study for general insect pathology also involved a fungus: Agostino Bassi's experimental demonstration in 1834 that a microorganism can infect an insect (the silkworm infected by *Beauveria bassiana*).

Fungi may consist of a single cell only, such as in yeasts. Alternatively, branched filaments or hyphae form a mycelium. Fungi have interesting reproductive cycles involving sexual and asexual modes. Zoospores, i.e., motile asexual reproductive cells, are known from the subdivision Mastigomycotina. Fungi that are especially important to social insects mostly belong to the subdivisions Zygomycotina (e.g., *Erynia*), Ascomycotina (*Cordyceps*), and Deuteromycotina (*Beauveria, Metarhizium*). The insect host is usually infected by infective propagules that are termed spores or conidia (in Zygomycotina, Deuteromycotina), zoospores (Mastigomycotina), or plantons or ascospores (Ascomycotina). Figure 2.4 illustrates a typical cycle of an insect fungus.

Fungi gain access to the host through the cuticle and sometimes through natural body openings (e.g., tracheae). The infection process must involve adhesion of the fungal spore (or appropriate propagule) to the host's surface, penetration of the body wall, and development of the fungal tissue inside the host, which normally leads to host death. The germination of the spore leads to the formation of a germ tube that penetrates the cuticle. It seems therefore obvious that thin-walled integuments such as are usually found in larvae or newly formed pupae are easier to breach. Optimal humidity conditions (often very high) and temperatures (depending on the type of fungus) are necessary for germination. As with many other microorganisms, extended exposure to ultraviolet light destroys spores (e.g., Krieg et al. 1981; Brooks 1988). The process is also sensitive to the presence of secreted substances on the host's body surface, e.g., fatty acids and other fungicidal substances (Tanada and Kaya 1993, p. 324; Beattie et al. 1985; see also chap. 7). The actual process of penetration involves enzymatic (e.g., proteases, chitinases) and physical means (pressure by the growing tip) as well as the formation of special supporting structures (appressoria, e.g., in *Metarhizium*) that fixate the hyphen on the surface. Fungi also enter through the alimentary tract where moisture is not a problem, e.g., as known from *Beauveria bassiana* in the fire ant *Solenopsis* (Broome et al. 1976) and in termites (Kramm et al. 1982). However, the chemical environment of the alimentary tract may be very aggressive and destroy the growing hyphae. An obvious target is the tracheal openings. This is also utilized by *Beauveria* to infect insects. Eventually, fungal infections spread to the host's hemocoel.

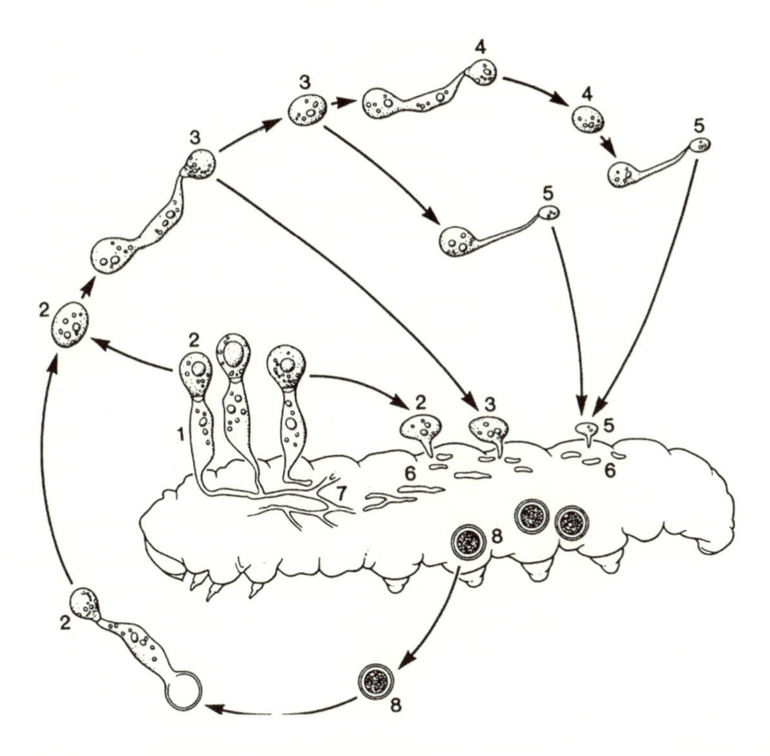

Figure 2.4 Life cycle of an entomophthorous fungus. (1) The endostroma in the insect host produces conidiophores. A primary (2), secondary (3), and tertiary (4) conidium is formed when no further hosts are available. (5) Capilliconidia are the major infective units. After successful infection, hyphal bodies (6) are formed that develop into mycelia and stroma (7), which also produce resting spores (8). (Reprinted from Tanada and Kaya 1993 by permission of Academic Press.)

As with any other antigen, once the fungus is in the hemocoel it is attacked by the immune system, and hemocytes may or may not be able to phagocytose or encapsulate it. The fungus in turn may simply outgrow this defense or produce mycotoxins. The successful fungus will grow hyphal bodies that then invade specific tissues. In the normal course of events, this will sooner or later lead to host death. The fungus then continues to grow saprophytically and eventually produces reproductive spores that are dispersed. Hence, sporulation usually occurs on the dead host but may sometimes also involve live ones. Because fungi are often saprophagous only, the simple observation of mycelial growth on dead hosts does not mean that the fungus did actually kill the insect. Spore dispersal finally completes the fungal life cycle. As briefly mentioned before, some species, such as *Cordyceps*, produce spectacular vertical stalks that carry the spore-producing organs high above ground into the air.

Except from one case in black flies, transovarial transmission seems unknown from fungi (Tanada and Kaya 1993, p. 322). In fact, fungal pathogens attack larvae and adults but rarely the egg stage, even in nonsocial hosts. On the other hand, species such as *Beauveria bassiana* and *Metarhizium anisopliae* have a very broad range of insect hosts, while others are much more host-specific. Obviously, host specificity is affected by the specificity of the adhesion step, i.e., the first step in the infection process, which may be passive or more specific, based on surface recognition molecules. In addition, most entomopathogenic fungi are facultatively parasitic. Entomopathogenic fungi are generally quite virulent, and host death is the usual outcome. For example, high pathogenicity is known from ant hosts (Sanchez-Pena et al. 1993). Similarly, case mortality a few days postinfection exceeds 50% in *Metarhizium anisopliae* attacking the termite *Cryptotermes brevis* (Kaschef and Abou-Zeid 1987). Similar claims have been made for fungi that kill *Zootermopsis angusticollis* (Rosengaus and Traniello 1993), although the authors did not provide direct evidence for actual infections. There are, however, fungal infections that remain chronic with few symptoms and do not appear to be lethal (Balazy et al. 1986; Sanchez-Pena et al. 1993). In some cases, infected insects move to elevated places before dying, a behavior that promotes spore dispersal (e.g., in ants: Loos-Frank and Zimmermann 1976).

There are many reports of fungal infections in tropical ecosystems. But there is no general agreement as to whether fungi are also important parasites of social insects, especially of ants. Whereas Hölldobler and Wilson (1990) state that fungal infections are less common than expected, Evans (1982; see also Evans and Samson 1984) maintains that fungal diseases of ants are much more common, and that they may even be important for the regulation of invertebrate populations of tropical forests. There is indeed some circumstantial evidence suggesting that fungal pathogens are an ever-present and virulent threat, particularly to soil-nesting social insects such as termites, ants, and many bees. For example, most fungi seem not to be very host-specific and also can infect many nonsocial hosts in the same habitat. In addition, many fungi, e.g., the Entomophthorales, produce spores that can remain infective in the soil for many years (Soper et al. 1976). Together with their capacity to utilize a wide range of hosts and to move through the soil by growing hyphae (Gottwald and Tedders 1984; Carruthers and Soper 1987, p. 379), this indicates that the force of infection by fungal diseases is particularly intense for soil-dwelling social insects. As far as can be deduced from the data base in appendix 2, there is no difference in total parasite richness nor in fungal parasite species richness among the four major groups of social insects (fig. 2.5). If a tendency can be inferred from figure 2.5b, the terrestrial groups—ants and termites—indeed seem to have fewer fungal parasites, contrary to what one would expect with the arguments given above. Because of the uncertainty in the data, any interpretation of these differences must be viewed with caution.

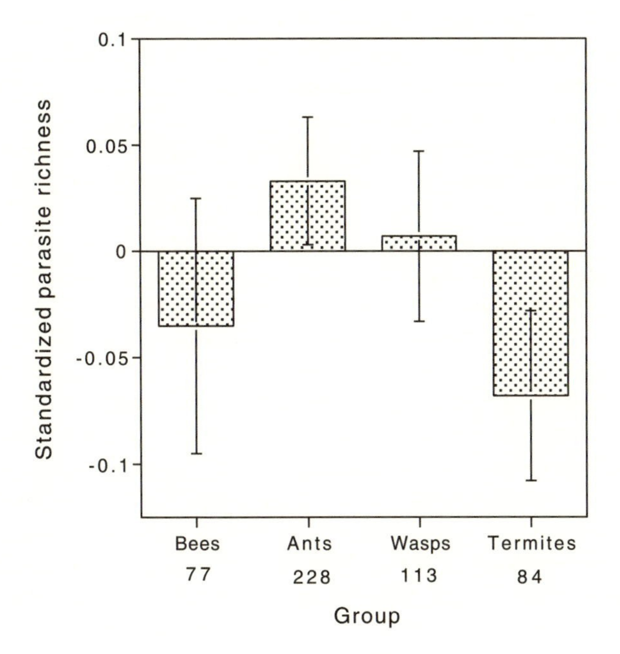

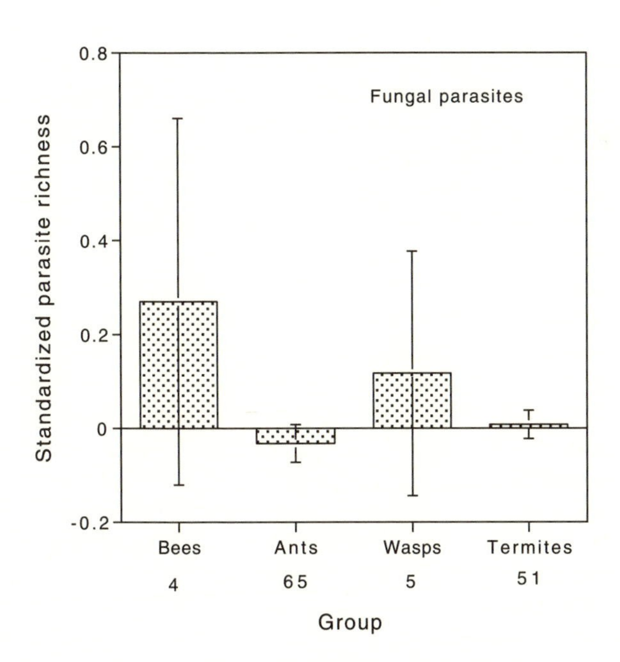

Figure 2.5 Parasite load (standardized parasite richness from fig. 2.1) in the four major groups of social insects. (a) For all recorded parasites (for parasites as in fig. 2.1). There is no difference among groups (Kruskal-Wallis: $H=2.363$, $df=3$, $P=0.50$, $N=502$ species). (b) Fungal parasites only (from appendix 2.3, and groups as in fig. 2.1) ($H=3.462$, $df=3$, $P=0.33$, $N=125$ species). Small numbers are number of host species in each group. Bars denote S. E.

Chalkbrood of the honeybee, caused by the fungus *Ascosphaera apis,* is a particularly well-studied example. This fungus infects drone and worker brood, but drones often suffer the most, because they are usually located at the periphery of the brood nest and are more likely to be chilled (so that the fungus develops better). Adults are not affected (Bailey and Ball 1991). Infection occurs via contaminated food. The hyphae then grow out from the gut into the body cavity, particularly at the hind end of the body. The larvae die within two days after the cells have been capped and later turn into shrunken, hardened cadavers. Bailey and Ball (1991) note the similarity of these infections with those of the American foulbrood, *Paenibacillus larvae,* in terms of etiology and infective processes. Quite similar diseases are also known from solitary bees, e.g., in the leaf-cutter bee *Megachile* and in *Osmia,* although they are caused by different species of fungi (Vandenberg and Stephen 1982; Skou and King 1984). Because of the nesting habits of *Megachile,* the young, emerging leaf-cutter bees have to chew their way through the nest tunnel above until they can reach the entrance of the burrow and leave their natal site. In this process, they come into contact with the fungus-infected, presumably moribund or dead larvae ahead of them that were not hatched in due time. This can lead to mechanical contamination of the emerging bee, and to dispersal of fungal spores to the bee's food plants (Vandenberg et al. 1980). Some of those spores will then probably be transmitted to other leaf-cutter bees or to social bees (e.g., honeybees) when they visit the same plants. Hence, infection by *Ascosphaera* in leaf-cutter bees may represent a natural reservoir for the managed bees. Roubik (1989, p. 230, based on Gilliam 1978 and pers. comm.) observed that distinct *Ascosphaera* are associated with over forty different species of native North American bees such as *Anthophora, Nomia,* and *Megachile.*

Aspergillus flavus and some other species (e.g., *A. fumigatus*) infect brood and adults of the honeybee and cause "stonebrood" ("Steinbrut"). Most larvae die after the cells have been capped. As the name indicates, the infected hosts turn into hard, stone-like mummies. Infection is by os, the internal tissues are quickly consumed, and the mycelium usually breaks out at the anterior end of the body. Infection is also possible in adult bees. In general, the infection seems transient, and only very large inocula may kill an entire colony (Bailey and Ball 1991, p. 61). Fyg (1964) summarized a number of diseases of the honeybee queen, including some caused by fungal infections. These include melanosis ("Ei-Schwarzsucht") characterized by black patches surrounding the egg and their subsidiary cells in the ovaries. The queen may start to produce unfertilized eggs (i.e., drones), or stop egg-laying and become superseded. The parasite also infects the poison sac and gland. (A different but similar infection, B-melanosis, is caused by a bacterium, *Aerobacter cloacae.*) Several other melanosis-like symptoms have an unclear origin and may represent tumor growth rather than parasitoses.

Skou and Holm (1980) later investigated the risk of transmission of melanosis during the instrumental insemination of honeybee queens. On one hand, they found that diseased queens had filled spermathecae but apparently were unable to use the sperm. On the other hand, there was a correlation among apiaries between the prevalence of diseased queens and the prevalence of melanosis symptoms in the colony workers. There was no evidence that experimental insemination would be more likely to lead to infections. However, a fungal parasite was actually transferred during such procedures. During their studies, Skou and Holm (1980) also isolated a wide array of fungal infections mostly from the alimentary tract of their queen samples: *Saccharomycopsis* (= *Candida*) *lipolytica* (the most frequent infection, associated with the reproductive tract and poison apparatus), *Hansenula anomala, Candida parapsilosis* (= an anamorph of *Lodderomyces elongisporus*), *Candida apicola, Candida guilliermondii,* and *Saccharomyces rouxii.* Perhaps some of these infections are opportunistic when queens are stressed, for example when fed antibiotics (Gilliam et al. 1974). Almost all of these fungi remained inapparent after experimental infection. Similarly, Prest et al. (1974) isolated a variety of fungi from discolored honeybee worker, drone, and queen larvae and pupae—*Penicilium cyclopium, P. funiculosum, Rhodotorula* sp., *Cladosporum cladosporoides, Phoma* sp., *Chactophoma* sp., *Beauveria tenella,* and *Aspergillus niger*—each causing a different, characteristic discoloration. It is unclear, however, how these reports can be translated to natural populations of social insects, because only the honeybee has been investigated in this detail. The lesson to be learned from these studies is clearly that a number of fungal infections of social insects may never be detected because they are inapparent or cause few, easily overlooked symptoms. For instance, some fungal infections affect the reproductive tract (e.g., melanosis) and are associated with the production of unfertilized eggs or sterilization of the colony queen. These changes would—if undetected—probably be treated as a case of adaptive sex allocation or life-history strategy.

In an extensive survey, Skou et al. (1963) found 16 out of 55 inspected *Bombus*-queens to be infected by the yeast *Candida* and 8 out of 27 with *Acrostalagmus.* They report that diseased queens go into hibernation late but emerge earlier. Since emergence depends in part on how long the fat storage lasts, it may be that the infection is primarily associated with a reduction in the size of the fat body, which in turn triggers a too early emergence. Given the prevalence, yeasts could be an important component of overwinter mortality. In many cases the pathogenic effects of yeasts appear to be mild or unknown. Gilliam (1973) in honeybees and Jouvenaz (1990; also Jouvenaz and Kimbrough 1991) in fire ants both report that some external stress is important for yeasts to be present or to have an additional effect on the colony.

Cordyceps species are important, sometimes obligatory parasitic fungi for ants and probably are the main group of entomopathogenic fungi in tropical habitats (e.g., De Andrade 1980; Evans 1982; Evans and Samson 1982, 1984; Samson et al. 1982, 1988). This has been studied in some detail for the ant *Cephalotes atratus* (Evans 1982; Evans and Samson 1982). In general, these fungi are potentially quite pathogenic: Van Pelt (1958, in Evans and Samson 1984) exposed *Camponotus* ants to infection by *Cordyceps unilateralis*. They died 3 days later and the fungal stromata appeared after a further two days. Paulian (1949) noted that the African ponerine ant *Paltothyreus tarsatus* is often killed by a *Cordyceps* fungus (preceded by a typical "topping" behavior). The fungus also attacks *Pachycondyla* (*Mesoponera*) *calfraria* and *Polyrhachis militaris*. Apparently, the phenomenon is quite common in shaded forests. On a single tree in the Amazonian rain forest that was sampled at regular 2-week intervals over 2 years, Evans (1982) found a total of 2376 infected *Cephalotes* ants (unfortunately, the total sample is not reported). Some of the fungi also occur in other insects in the same habitat, e.g., in flies. Moreover, outbreaks (rather than endemic infections) of diseases in forest ecosystems in subtropical and warm-temperate habitats have often involved *Cordyceps* species, especially *C. unilateralis* in formicine ants (Samson et al. 1988, p. 143).

These fungi also cause abnormal behaviors of the infected host. In particular, erratic movements, aggregation around trees, extended self-grooming, hiding beneath bark, and a reversal of preferences for arboreal or ground-level activities are often found (Samson et al. 1988). Finally, the ants become sluggish, seek shelter and attach themselves with their mandibles to the tree bark. The fungal mycelium then grows and firmly fixes the ant's body on the substrate. For example, *Cephalotes* ants infected with the fungus *C. kniphofioides* invariably die on tree bark. Thereafter, many fungal synemmata grow over the substrate to form reproductive tissue that is infective for others. Other nestmates may "tend" the infected workers hidden below bark. Also, scavengers such as *Solenopsis* ants remove these corpses. These latter behaviors thus increase the danger for another worker of becoming infected (Evans 1982). Infected ants have also been seen to cling to grasses (Evans and Samson 1984). Infected *Dolichoderus* ants, in contrast, die under litter and the fungus then forms spore-carrying stalks that eventually emerge from the litter layer and become visible (Evans 1982). Thus, when the stalk is not yet formed, only a very careful examination of the litter layer can reveal the true extent of these infections. Evans (1989) notes that the formation of the actual fruit body by the fungus is actually a slow process that may take 2–3 months to complete. Samson et al. (1981) observed that the source of an epizootic of *Stilbella* in the weaver ant *Oecophylla longinoda* was infected queens from the previous season. Hence, dispersing queens could be an efficient mechanism for the spread of the fungal diseases. In addition, whether or not these fungal infections can be observed in focus areas or in a more dispersed fashion depends on the foraging system of the host species. The latter is more likely with individually searching ants such as *Pachycondyla*. Typically too, infected ants abandon their daily routines and may venture outside their foraging ranges or territory.

The systematic study of these relationships is seriously hampered by an unclear taxonomy of the host species, mostly in the *Camponotus* complex, as well as in *Cordyceps* itself where some convenience taxa (*C. myrmecophila*) lump together quite different forms and others have anamorphs (i.e., representing different or incomplete stages such as *Hirsutella* belonging to *Cordyceps*). Hence, the question of host specificity must remain open but seems to be much more complex than previously thought (Evans 1982). For example, in the Amazonian rain forest, the form *Cordyceps lloydii* var. *llodyii* seems associated with large-sized *Camponotus* species, while *C. lloydii* var. *bianatis* is associated with small ant species from the forest understory. *Cordyceps unilateralis* seems specific on the Camponotinii, while *Hirsutella sporodochialis* is probably an anamorph of *Cordyceps unilateralis* specific on *Polyrhachis.* Such evidence, together with the observation that different fungal species typically utilize different parts of the body of ants, led Evans (1989) to suggest an ancient history of coevolution of ants with their fungal pathogens.

Unclear taxonomy and the tendency to classify the parasite after its host are also difficulties that hamper progress with infections by Entomophthorales (Tanada and Kaya 1993), the most important fungal group for temperate-zone ants (Evans 1989). Generally, many Entomophthorales cause serious epizootics in insect populations. Typical infections produce symptoms only late in the process. Then, infected hosts often move to elevated sites, something that has been termed "summit disease." Good examples are the infection by *Entomophthora* in the wood ant *Formica pratensis* (Loos-Frank and Zimmermann 1976), *Alternaria tenuis* in *Formica rufa* (Marikovsky 1962), and *Erynia formicae* in another *Formica* species (Balazy and Sokolowski 1977; see also Oi and Pereira 1993). In these cases, the infected workers tend to climb grasses in the evening. At the same time, tiny fungal threads appear from the body wall which fix the worker irreversibly to the surface. Now the ant also locks itself with mandibles to the grass and dies shortly afterwards. Two days later, fungal mycelia appear and produce conidia, particularly under wet weather conditions. Apparently the fungus then takes control over its host, forcing it into a behavior that benefits the spread of fungal spores from an elevated point. The ants die overnight and the fungus starts to produce spores that are dispersed by air. Many hyphae grow in the vicinity of nerve ganglia, i.e., near suboesophagal ganglia and the protocerebrum. This suggests that close spatial proximity of the fungal hyphae with nerve cells is essential for the behavioral changes. The colony is reported to show hygienic behavior by removing the infected ants from grass (Marikovsky 1962). Marikovsky (1962) also found that the fungal infections are particularly obvious in late summer to early autumn; almost nothing is observed in early summer.

The classical entomopathogenic fungus is *Beauveria bassiana.* It occurs worldwide and probably has one of the broadest host ranges of all known fungi; it even infects vertebrates. *B. bassiana* is a prime infection of soil-dwelling immature insects. It also occurs in the soil as a saprophagous species (Tanada and Kaya 1993). In the fire ant *Solenopsis richteri* infection occurs partly through the alimentary tract and partly directly through the integument. The hyphae enter the body cavity some 72 hours after ingestion of spores. Presumably, the pathogenic effects are due to the production of toxins by the fungus (Broome et al. 1976). *B. bassiana* has also been investigated in *Solenopsis invicta* and *Myrmica* (Oi and Pereira 1993), in the primitive ant genus *Myrmecia,* the wasp *Vespula* (Glare et al. 1993), and in bumblebees (Leatherdale 1970). Similarly widespread is *Metarhizium anisopliae* (appendix 2.3). It is in fact one of the few fungal pathogens reported from termites (Kaschef and Abou-Zeid 1987). Also, *Aegeritella* is found on a variety of ant hosts (appendix 2.3).

In a survey of entomopathogenic fungi of arthropods in Britain, Leatherdale (1970) documented a number of social insect host–parasite associations involving *Cordyceps* and *Paecilomyces* (also *Beauveria*) on social bees (*Apis, Bombus*) and wasps (*Vespa, Vespula*). However, in contrast to Evans (1989), Leatherdale (1970) concluded that these fungi have no special affinity to certain hosts but may indeed occur in many nonsocial hosts. It is tempting to conclude that this discrepancy relates to a general difference between tropical vs. temperate habitats rather than to the specific fungi and social insect species involved. Even if we know little on this subject, it seems likely that fungi are more of a problem in tropical habitats, with their generally warm and humid conditions. At least for the solitary nesting bees, Roubik (1989, p. 231) concludes that mold is the most significant natural enemy and cites figures of over 50% mortality inflicted by fungal infections. Michener (1974, p. 220) agrees but suggests that fungi are less important to the highly eusocial species of social bees due to their ability to control infections. This statement may be inflated by the relative scarcity of threatening fungal infections in the honeybee.

Hölldobler and Wilson (1990, p. 555, based on Benjamin 1971; also Bequaert 1922; Smith 1946) cite ectoparasitic fungi of the order Laboulbeniales as being very host-specific but not creating any pathogenic effects. A rather enigmatic parasite, most likely a fungus, is *Myrmicinosporidium durum,* which has a wide range of hosts in the Northern Hemisphere (Sanchez-Pena et al. 1993; appendix 2.3). Infected ants live even longer than noninfected ones but they become darker brown due to the accumulation of fungal spores. Otherwise, their behavior seems not affected. Sanchez-Pena et al. (1993) note that effects are more typical of chronic infections by protozoa rather than of the usually virulent fungi. So far, nothing is know about its life cycle and transmission.

2.4 Protozoa

Protozoans are a rather diverse group of microorganisms. According to Brusca and Brusca (1990), they are grouped into the phyla Sarcomastigophora (comprising the Mastigophora = Flagellata, Sarcodina, Opalinata), Labyrinthomorpha, Ciliophora (Infusioria), and the former Sporozoa that have now been split into Apicomplexa, Microspora, Ascetospora, and Myxozoa. The classification used by other authors is not always in line with this scheme. Lipa (in Steinhaus 1963) estimated that some 1200 species of protozoa are associated with insects in general.

Normally, protozoa infect insect hosts via the alimentary tract, rarely directly through the integument. Some groups remain in the gut lumen and attach themselves to the gut wall, e.g., many ciliates, flagellates, and gregarines. Others penetrate into the hemocoel and remain there in the hemolymph or within cells of various organs and tissues, e.g., many Apicomplexa and Microsporidia. Microsporidia readily cause pathogenic effects. Some protozoa exhibit tissue tropisms, i.e., they only infect certain types of cells but not others (Tanada and Kaya 1993). In addition to these direct infection routes, protozoa can also be transmitted transovarially to offspring. This is mainly found in the Microsporidia, e.g., by *Nosema kingi* in *Drosophila* (Armstrong 1976). Some are spread by vectors such as mites.

Flagellates

The Zoomastigophora (Flagellates) as a group are mostly parasitic. Those inhabiting the digestive tract of insects are often considered to be mutualistic or commensalistic. From this group, the family Trypanosomatidea (with the genera *Crithidia, Blastocrithidia, Leptomonas, Rhynchoidomonas*), about four hundred species are associated with insects. Typically, flagellates multiply by longitudinal fission followed by growth of the daughter cells. Also, cysts can be formed that are able to survive harsh conditions. Hosts become infected mostly by ingestion of infective cells from contaminated material. Trypansomatidae are thus easily transmitted and appear to have little host specificity and not very harmful effects (e.g., Schaub 1994). They infect the gut at characteristic sites, e.g., the mid- or hindgut, Malpighian tubules, and may occur inside or external to the peritrophic membrane. Those flagellates that invade the hemocoel or glands often may cause more severe disease.

In the honeybee, an area of the pylorus of adult bees (workers and queens) is often colonized by *Crithidia* (*Leptomonas*) *apis* (= *C. mellificae*), although it can be observed in the rectum too. Apparently, the infection does not occur in bees younger than 6 days old. The infections cause little or no harm (Bailey and Ball 1991). In our own studies on the subject, *Crithida bombi* was found to be a widespread inhabitant of the gut of different species of bumblebees (fig. 2.6). In fact, prevalences in the order of 10–30% are quite common. Also, *C. bombi* is a rather mild pathogen. The effects include a reduction of ovary size and slower growth of colonies early in the season (Shykoff and Schmid-Hempel 1992). Some results also indicate increased mortality rates if the infection is by a source from a distant location (Wu 1994; Imhoof and Schmid-Hempel 1998).

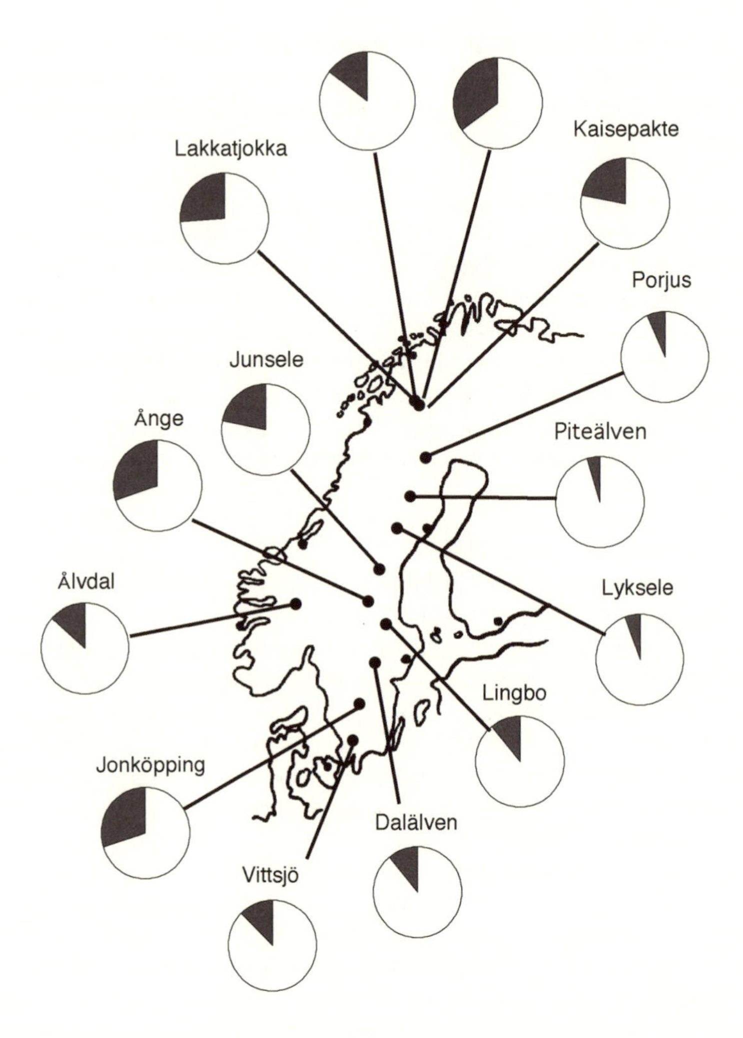

Figure 2.6 The prevalence (dark areas in pie diagrams) of the trypanosome parasite *Crithidia bombi* in workers of *Bombus* species (all species pooled) in Scandinavia. The survey was conducted at several sites in summer 1992. A total of $N=674$ workers were investigated. Note the absence of a latitudinal gradient.

Amoebae and Ciliates

Two other major groups of insect-associated protozoa are the amoebae (Rhizopoda; Sarcomastigophora), which seem to be generally uncommon, and the ciliates (e.g., *Tetrahymena*). Most amoebae are commensals in the digestive tract of their hosts, and few are pathogenic. The life cycle is simple with cysts and trophozoites, the latter being the actual amoeba as we know it. In the honeybee, *Malpighamoeba mellificae* infects the Malpighian tubules of adult bees. It ultimately forms cysts that are passed out in the feces, 22–24 days after infection (Prell 1926; Bailey 1955a; Bailey and Ball 1991). The effects on the individual host appear to be harmless, and the effect for the entire colony is uncertain, despite some earlier reports of fatal infections (see Bailey and Ball 1991). Hassanein (1952) observed that colonies dwindle and eventually perish under severe infestation (i.e., with more than half of the workers infected). The epidemic pattern is similar to some other honeybee diseases as peak infection is observed in early summer. Later it disappears, presumably because the short-lived summer bees and rapid expansion of the brood comb outrun the production and dispersal of cysts. The time from infection to encystation is approximately 21 days, after which cysts pass in the feces (Hassanein 1952). This time period is close to the average life expectancy of a honeybee worker after eclosion (e.g., Seeley 1985). In terms of cyst production, *Malpighamoeba* produces only a fraction of the spores generated by a single infection of, for example, *Nosema apis*. Nevertheless, the two parasites are often associated, perhaps because they are transmitted in the same way—via fecal contamination of the combs (Bailey 1968; B. Ball, pers. comm.).

In the ciliates, a number of species are mutualistic inhabitants of cockroaches and termites. The pathogenic forms, in contrast, usually invade and multiply in the hemocoel and often kill the host in this process. Although important in other groups of (mainly aquatic) insects (e.g., in blackflies and mosquitoes), there seems to be no report of ciliates in social insects, with the well-known exception of termite intestinal symbionts.

"Sporozoa"

A variety of "sporozoa" (a heterogeneous assembly of different phyla) have been found in social insects. In fact, this group comprises the majority of entomopathogenic species. Many infect the digestive tract, fat body, and Malpighian tubules (appendix 2.4). An important group, the Apicomplexa comprise the gregarines and coccidia. Gregarines are characterized by mature forms that are large, extracellular, and found in the digestive tract and body cavity of insects. Gregarines are host specialists as well as more widely occurring forms. Some forms (Eugregarinidae) are usually of low virulence, whereas the Neogregarinidae can cause serious infections (Tanada and Kaya 1993, p. 416). The life cycle of gregarines can be quite complex. Usually it starts with the ingestion of mature oocysts. Several sporozoites then emerge and infect the midgut epithelial cells or stay attached to them externally, or penetrate the gut wall to become trophozoites. These eventually mature, produce gametes, fuse, and form zygotes and new (haploid) oocysts (Brusca and Brusca 1990).

Various gregarines (*Leidyana* and others) infect honeybees where they live in the lumen of the midgut. Because similar parasites have been found in other insects, they are probably not specific to *Apis.* Transfer to honeybees could occur through occasional inhabitants of the nests. Stejskal (1955) found gregarines in honeybees in the tropics (Venezuela). The effects on the host colony are unclear: in Stejskal's observations, only half of the bees that were sick also had gregarines, suggesting that they probably infected already weakened colonies (see also Hitchcock 1948). Gregarines in honeybees are apparently widespread but locally not very common (in North America: Fantham et al. 1941; Hitchcock 1948; Oertel 1965; in the Neotropics: Stejskal 1955; in Europe: Morgenthaler 1926).

Following the general pattern, Neogregarinidae typically infect a host by ingestion, then emerge as sporozoites and reach the hemocoel and target tissue. In adult social insects, this is often the fat body. At this stage, some sporozoites develop into motile merozoites (merogony) that spread throughout the hemocoel to infect other parts of the host body. Eventually, merozoites become gamonts and associate in pairs to become gametocysts with a cyst wall. Each gamont yields four gametes which fuse pairwise to form zygotes that become protected by a spore wall and form oocysts. Usually, oocysts are excreted in the feces, but they may also be spread when the host dies and decomposes. The number of steps (merogonies) in this process as well as the number of oocysts per gametocyst can vary considerably among different gregarines (see Cheng 1986; Brusca and Brusca 1990; Tanada and Kaya 1993). *Mattesia,* for example, is found in several social insects (in bees renamed *Apicystis* by Lipa and Triggiani 1996). The genus is characterized by development through cycles of micro- and macronuclear merogony and by gametogony resulting in the formation of two spores within a gametocyst. It has been found to infect larvae and to have high prevalence in the fire ants *Solenopsis geminata, S. invicta,* and *S. richteri (M. geminata:* Jouvenaz and Anthony 1979; Jouvenaz 1983).

The infection of *Mattesia geminata* in *S. geminata* is located in the larval hypodermis (Jouvenaz and Anthony 1979). It causes disruption of eye development and of the melanization of the cuticle, eventually leading to the death of the pupa. In bumblebees, *Apicystis (Mattesia) bombi* infects the adults, and its presence can be detected by the large spores in feces. The effects are quite serious. Infected workers show a disintegrated fat body (Durrer and Schmid-Hempel 1995; R. Schmid-Hempel, pers. obs.), and infected colonies often have little chance to grow and reproduce. Several other gregarines of honeybees in the tropics also infect adults (Stejskal 1955). In fact, gregarines may be more widespread than hitherto known, as they have been found in quite distant taxa. For example, a yet unidentified gregarine infects larvae of the primitive Australian ant *Myrmecia pilosula* and other *Myrmecia* species. Infected individuals eclose with a brown cuticle, rather than the normal black. They contain the large (13 μm) lemon-shaped spores of the gregarines in their hemocoel (Crosland 1988). Just as in the case of helminthic infections (see section 2.6), the aberrant color makes the affected individuals rather conspicuous. However, as there is no intermediate host known for these parasites, it is unclear what the functional role, if any, of the color aberration is. Two out of five colonies investigated by Crosland were found infected in winter, but none in spring and early summer.

Coccidia seem not to be known from any social insect. This is quite different for the third major phylum in the "sporozoa," the Microsporidia. They are among the smallest eukaryotes. Microsporidia are obligatory intracellular and cause damage to many economically important insects. They are also known from all major vertebrate groups and seem to occur in almost all invertebrate phyla; some are even parasitic on gregarines (Sprague 1977; Canning 1990; Tanada and Kaya 1993). Generally, microsporidia seem sensitive to elevated temperatures, which would explain why warm-blooded vertebrates are not the major hosts (Canning 1990). This aspect gains an additional significance for social insects that can regulate their nest temperature, or for social bees that can maintain high body temperatures by their flight muscles (Heinrich 1979). Some microsporidia depend on intermediate hosts; e.g., *Amblyospora,* a parasite of the mosquito *Culex annulirostris,* also uses a crustacean host (Sweeney et al. 1985).

Besides oral transmission, for example by contaminated feces, microsporidia may also be acquired via the cuticle, e.g., from a parasitoid, or transovarially from the mother. Infection is sometimes in very specific tissues, but the most frequent target seems to be the fat body, and this is often associated with chronic infections. As their name suggests, spores are a distinctive feature of microsporidia. Spores are the only means by which the parasites can survive outside their host. Although small (2–6 μm), spores are probably the sign most easily seen in the infection of an insect. Spores possess a characteristic and unique apparatus by which a polar filament can be extruded that is able to penetrate a host cell. Through the forming tube the sporoblast enters and infects the host cell. After the invasion of the host cell, the sporoblast grows and develops into a meront. In the process of merogony, meronts further divide and can produce more meronts. Eventually, sporogony sets in and new sporonts are produced. This is completed by the formation of walls to produce new infective spores. The whole cycle is very variable among different species, different hosts, or even different tissues, and may be as short as 24 hours. Some microsporidia produce two or more different morphs of sporonts (spores) in the same or different tissue (Tanada and Kaya 1993). In almost all cases, the function of these different morphs is not understood.

Nosema apis is a rather well-known microsporidium occurring in the honeybee. It enters by os and develops in the epithelial cells of the midgut. In bees kept at 30°C, new spores are formed after about 5 days. They pass out in the feces either as free spores or still packed within their host cell that is shed as the gut wall is regenerated. An infected bee can contain 30–50 million spores (Bailey and Ball 1991). Temperature and condition of the host affect the cycle of *N. apis.* The sporonts produced in summer differ from those in winter bees (Steche

1965; Youssef and Hammond 1971). Weiser (1961) suggested that diapause in the host insect also arrests the development of the microsporidia. This would be particularly relevant for social insects with annual life cycles and for those living in strongly seasonal habitats. Recent studies claim that some spores of *N. apis* spread in different ways (Fries et al. 1992; De Graaf et al. 1994).

N. apis seems unable to infect the larvae, hence, newly emerged bees are normally free of the disease (Hassanein 1951; Bailey 1955b). Spores detected in other organs, e.g., in the fat body and hypopharyngeal glands, are probably not *N. apis* (Bailey 1981). Nevertheless, other species of microsporidia have been found to kill the pupae by infection (Clark 1980). Bees infected by *Nosema apis* show no signs of the disease, but are shorter lived (Beutler and Opfiger 1949; Bailey 1958a) and have less well developed, vacuolated hypopharyngeal glands (Wang and Moeller 1969) and fat bodies (Bailey and Ball 1991). They show an accelerated age-related schedule of task attendance, accompanied by a reduction in pollen-collecting behavior (Wang and Moeller 1970; Anderson and Giacon 1992), actually quite similar to the effect of sacbrood virus. The reduction in life span can at least be partly compensated by pollen feeding (Hirschfelder 1964). Moreover, infected bees become dysenteric, which may serve the spread of the disease within the colony (normally bees defecate in flight outside their nest). Queens infected by *Nosema* cease egg-laying and die after a few weeks, often outside the nest (L'Arrivée 1965b). They probably acquire the infection when fed by infected workers.

Naturally infected colonies can recover when infected early in the year, but when infected in autumn they perish during hibernation (Morgenthaler 1941; Jeffree 1955; Jeffree and Allen 1956). In fact, infected colonies lose a roughly constant number of bees per unit time regardless of their size. One reason for this is the lower amount of honey storage that accompanies an infection (down by 20–30% within a week from infection) (Fries et al. 1984; Rinderer and Elliott 1977; Anderson and Giacon 1992). It is furthermore likely that some of the pathological effects are due to the viruses that normally are associated with a *Nosema* infection (see appendix 2.1). *Nosema* epidemics are favored by crowded conditions in the hive. Consequently, an infection of *N. apis* usually disappears in the course of the season, as the bees are released from their winter confinement and are able to defecate outside. In addition, infections in the following year are elevated if the preceding summer was characterized by cold and inclement weather that forced the bees to stay in the hive unusually often (Lotmar 1943). The prevalence of *Nosema* is variable among colonies (Doull and Cellier 1961). Varis et al. (1992) screened their samples of honeybees for various protozoa microscopically. In both Britain and Finland, about one-third of all apiaries were infected, as shown by the presence of spores. The prevalence in the colder climate was not lower; in fact, the figures were quite similar. Note that the absence of a north-south gradient is also true for the trypanosome *Crithidia bombi* in bumblebees (fig. 2.6).

Microsporidia are also known from other social bees. In our own studies, we found that *Nosema bombi* regularly infects a variety of bumblebee species with prevalences of a few percent. It invariably occurs in the midgut and can seriously damage a whole colony. Similar to *N. apis, N. bombi* also seems to build up slowly and to reach peak parasitemia perhaps only 2–3 weeks postinfection (Schmid-Hempel and Loosli, unpubl. data). In a survey in Denmark, Skou et al. (1963) found 18 of 99 queens from five species of *Bombus* infected by *N. bombi.* They also noted that infected queens go into hibernation late and come out early. High levels of nosematosis occurred in autumn. Fisher (1989) observed 10% of spring queens to be infected in New Zealand, but 61% of the colonies were infected at the end of the season, which is an unusually high figure. His experimental manipulations showed no effect of *Nosema* on the weight gain of the colonies. Also, the number of workers produced in infected colonies was 417 ± 74 (S.E.) and not different from the 585 ± 121 workers ($t = 1.22$, N.S.) in noninfected colonies. Unfortunately, the results are inconclusive because the colonies were naturally and not experimentally infected. On the other hand, Fisher's (1989) data suggest that *Nosema* could be a prime cause of overwinter mortality of queens.

Showers et al. (1967) claimed that *Nosema* from both honeybees and bumblebees are cross-infecting. Inocula prepared from honeybees led to symptoms in workers of *Bombus fervidus* after 19 days (e.g., reduced activity) (see also Fantham and Porter 1913). Spores were present in the ventriculus after 21 days, and all bees had died after 24 days. Also, inocula prepared from bumble bees lead to infection of *Apis.* Yet, more recent reports suggest the opposite and find significant biological differences between *N. apis* and *N. bombi* (Van den Eijnde and Vette 1993; McIvor and Melone 1995). In addition, sequence analysis of mtDNA has shown that *N. bombi* is more closely related to *Nosema* species occurring in grasshoppers than to *N. apis* of the honeybee (McIvor and Melone 1995; and Melone, pers. comm.). Furthermore, the susceptible stages of the host seem different for the two forms, with *N. bombi* being especially infective for the larvae rather than the adults (Schmid-Hempel and Loosli, unpubl. data).

The fire ant, *Solenopsis,* is the other well-investigated social insect with respect to the presence of sporozoan parasites (appendix 2.4). The biology of *Burenella dimorpha* infecting *Solenopsis geminata* is particularly interesting (Jouvenaz et al. 1981; Jouvenaz 1986) (see also fig. 3.1). Only fourth instar larvae of the ant become infected. The parasite develops two different kinds of spores: the binucleate, nonmembrane-bound type develops in the hypodermis of the larval host, thus destroying it. The resulting host pupa is characterized by clear areas of head, petiole, and gaster, and by a ruptured skin. The nurse workers usually destroy and cannibalize such aberrant pupae, and the pellet is later fed back to the larvae, completing the parasite's life cycle. The uninucleate membrane-bound octospores develop in the fat body and are expelled via the meconium, i.e., the larval gut that is shed at pupation (larvae do not defecate).

They are not infective when fed to the larvae. Moreover, the developmental rates of the two spore types depend on temperature. The noninfective, membrane-bound octospores, for example, develop in a restricted range of 25°–28°C while the infective binucleate spores develop within a broader range of 20°–32°C (Jouvenaz and Lofgren 1984). In some areas of Florida, where *Solenopsis* is introduced, Jouvenaz et al. (1977) observed up to 40% of all colonies infected by this (or perhaps a similar) parasite.

In a detailed study of biocontrol strategies against fire ants introduced to North America, Wojcik et al. (1987) screened several plots in Argentina for the presence of the microsporidium *Thelohania solenopsae,* a primarily larval infection that may lead to host death at the pupal stage (*T. solenopsae* also infects adults). *Thelohania* apparently causes smaller than normal colony sizes, loss of vigor, and reduced defense behavior upon disturbance. It infects adipose tissue of adults, e.g., fat bodies of workers and sexuals, and the ovaries of queens (Jouvenaz 1986; Patterson and Briano 1990). The infection debilitates the individual host. *Thelohania* is dimorphic and different spores are produced in the same type of tissue (Jouvenaz 1983). The more common spore type is uninucleate, bound in octets (octospores); the other type is binucleate. Nothing seems known about the role of these two spore types. In fact, infection by os seems difficult. Feeding spore suspensions or placing infected brood in a nest does not produce an infection. It is also possible that *Thelohania* is in fact a complex of species (Jouvenaz 1986).

Also in other parts of South America (Uruguay, Argentina), several species of fire ants (*S. richteri, S. invicta, S. blumi, S. quinquecuspis*) harbored infections by *Thelohania* (Allen and Silveira-Guido 1974) (see also reviews in Jouvenaz 1984). For example, Briano et al. (1995a) found *Thelohania solenopsae* in 25% of 185 sites and in 8% of the 1836 screened colonies, with occurrences of up to 40–80% of colonies in certain areas. In a study in Brazil, every colony investigated by Allen and Buren (1974) showed infection by *Thelohania.* In Wojcik et al.'s (1987) survey, half of the six screened plots had parasite occurrence of 35% in all colonies (average density: 198 colonies/ha), while the remainder had 3% and went disease-free within the 6 months of the study. Similar prevalences are reported for the same host-parasite pair in Brazil (2.2% of colonies: Wojcik et al. 1987). Although no further details are known, these authors suspect a role of these microsporidia in the regulation of ant populations (see also Leston 1973). In fact, infected colonies do not live as long as uninfected ones. In contrast, Patterson and Briano (1990) think that *T. solenopsae* has little detrimental effect on the colonies. Again, more detailed and systematic studies are needed to clarify this issue. An interesting observation is that *Thelohania* is transmitted transovarially, i.e., vertically to offspring (Knell et al. 1977).

Wojcik et al. (1987) and Briano et al. (1995a) also found colonies of *S. invicta* and *S. richteri,* respectively, to be infected by a further microsporidium, *Vairio-*

morpha invictae (Jouvenaz and Ellis 1986). In Wojcik et al.'s (1987) sample, 757 of 1000 checked colonies in Brazil contained at least one species of pathogen and/or a myrmecophile (e.g., the scarab beetle *Martinezia,* millipedes, thysanura, etc.; see appendix 2.11). Briano et al. (1995a) report 5% of 185 sites and 1% of 1836 colonies infected. Interestingly, no double infections of *Vairiomorpha* and *Thelohania* were observed in the field, although it occurred in stressed laboratory colonies (Briano et al. 1995a). Similar polymorphic spores as in the other microsporidia are also found in *Vairiomorpha,* but their functions are unknown, too. Binucleate nonbound spores typically account for more than 95% of the total count, while the rest is made up of uninucleate spores (bound in octospores) and a few tetranucleate ones. These types develop in the same kind of tissue, although sequentially, with binucleate spores formed early and uninucleates late in host pupal development. Mature octospores are found in adult ants (Jouvenaz and Ellis 1986). Again, temperature may affect this developmental sequence (Fowler and Reeves 1974). *Vairiomorpha* develops in the cells of the head, the petiole, and the gaster.

Kalavati (1976) screened four termite species in Andar Pradesh (India) for the presence of microsporidia and found infection in all of them. *Gurleya spraguei* was described as a new species that infects adipose tissue (as does *Burenella*) of *Macrotermes estherae,* where it develops into four binucleate spores. If heavily infected, the spores are released into the body cavity. Because of the site of infection, the parasites are unlikely to be transmitted via the feces. Spore dispersal is more likely to occur when the infected body disintegrates or is cannibalized by others. From examining some 4500 specimens, a prevalence of 0–8% was estimated. Similarly, the microsporidium *Pleistophora weiseri* infects the epithelial cells of the foregut in *Coptotermes heimi,* as does *Stempellia* in *Odontotermes* (Kalavati 1976). The effects of such infections are unclear but could be quite serious for the worker and thus eventually for the colony. Seasonal variation is also observed, quite similar to the cycles of *Nosema* in bees.

From the epidemiological point of view, the survival of spores in the environment is an important parameter. Brooks (1988) lists the results of a large number of studies. In situations that approach a natural transmission situation (i.e., spores in cadavers, on surfaces at low temperatures), spores of *Nosema, Thelohania, Vairiomorpha,* and *Mattesia* are found to remain infective for several months. This period is doubled to 6 months or 2–3 years when the spores are kept in aqueous media. *N. apis* was found to remain infective for up to 7 years when cooled. However, sunlight, with its substantial UV component, reduces spore survival drastically to a period of several hours. Hence, in the open these protozoan spores must be transmitted more or less directly, otherwise they degrade quickly. Long-term transmission in contrast could be possible in sheltered places, for example, in an abandoned and reused nest. This has obvious potential consequences for the infection of microsporidia in soil- or cavity-nesting species.

2.5 Nematodes

Nematode taxonomy is mainly based on sexual structures. Therefore, immature stages—the normal stage infecting social insect hosts—are usually difficult to identify. The group as a whole constitutes an important insect enemy, but the ways of life of different species are very diverse indeed. The association ranges from obligatory parasitic to commensal. A number of insect species serve as vectors for nematodes; for example, mosquitoes transmit *Wuchereria bancrofti,* the cause of human filariasis (Cheng 1986). A useful classification (Tanada and Kaya 1993) distinguishes the nematode orders Stichosomidae (Mermithidae, Tetradonematidae), Rhabditida (Rhabditidae, Steinernematidae, Heterorhabditidae, Oxyuridae), Diplogasterida (Diplogasteridae), Tylenchida (Allantonematidae, Sphaerularidae), and Aphelenchida (Aphelenchidae, Entaphelenchidae). The description by Reaumur in 1742 of the bumblebee nematode *Sphaerularia bombi* was one of the first accounts of an insect parasite in general. Not long afterwards, Gould (1747) described the effects of mermithid nematodes in ants, *Lasius flavus* (Tanada and Kaya 1993). In one of the most recent reviews, Poinar (1975) lists over three thousand associations in nineteen insect orders.

The actual life cycle of nematodes is straightforward: egg, juvenile stage (usually four larval instars), and adult. Some species can form a lasting stage ("Dauerstadium", common in Rhabditida), which is normally the third instar juvenile. Nematodes are sexual and the genders are normally separated. Hence, mating is required to produce offspring. But some nematodes have complicated life cycles where sexual and parthenogenetic generations alternate (Cheng 1986; Tanada and Kaya 1993). Poinar (1975) grouped nematodes according to whether they are phoretic, facultative, or obligatory parasitic. In the first group, the juvenile nematodes are carried by the insect to a new host. They are externally attached, e.g., under the elytra of beetles, on legs, etc., or internally in the genital chamber, trachea, head glands, or Malpighian tubules. Normally, little harm is done; this association seems not known in social insects. The facultative parasitic species are able to complete their life cycle outside the host. In insects, they can be found in the body cavity, intestinal tract, pharyngeal and other glands, or in the trachea. Obligate parasitic nematodes cannot live without a host and have no free-living stages. Generally, these species are found in the insect's body cavity; some may be found in the intestinal tract and reproductive system. Life cycles with intermediate hosts may also be found. Tanada and Kaya (1993) observe that obligatory parasitic nematodes tend to have few host species.

A nematode infection occurs when the infective stage invades the host either directly by penetrating the cuticle or through natural openings (mouth, anus, tracheal openings). Some species (Tylenchida, Mermithida) have specialized stylets (spears) used to pierce the host's body wall. Infection via food is certainly also common. The ingested egg then hatches in the gut and may again use a stylet or spear to penetrate the gut wall and enter the hemocoel. It is possible that in many cases of nematode infection that occur via the alimentary tract, a binding system based on sugar receptors in the peritrophic membrane is involved (Stoffolano 1986, p. 123). With active host finding, the nematodes use cues like chemical gradients or host feces (Tanada and Kaya 1993). The pathological effects are manifold but often result in reduced fecundity, sterilization, or host death.

In the honeybee, nematodes have not often been found. This is somewhat surprising given the amount of attention given to honeybee parasites. Bedding (1985) reports only two mermithid nematodes, *Mermis subnigriscens* and *Agamomermis;* normally, the levels of infections are low but substantial prevalences of 61% (in $N=3254$ workers) seem to occur in colonies living on flood plains and river banks (Vasiliadia and Ataksihsiev, cited in Bedding 1985). Poinar (1975) adds another two species (see appendix 2.5). As yet unidentified mermithid nematodes are also occasionally found in bumblebee workers (pers. obs.).

Sphaerularia bombi (Tylenchidae) is a remarkable nematode that exclusively infects overwintering queens of various bumblebee species (see Alford 1975) (fig. 2.7). The usual habit of nematodes to live and mate in the soil is certainly a prerequisite that leads to the opportunity of infecting queens that stay in the soil for several months. The fertilized *Sphaerularia* female invades the hibernating bumblebee queen presumably by os (Madel 1973). Bedding (1985) suspects that juveniles may be deposited on flowers and picked up by others, but this is rather speculative. When the queen emerges from hibernation next spring, the female nematode is found in the hemocoel of the queen. At that stage, the huge and everted uterus is visible and easily mistaken for the actual female that is a tiny appendix by comparison. It is quite common to find more than one *Sphaerularia* in an infected queen: Lubbock (1861, cited in Bedding 1985) found 5–8, Alford (1969) counted usually less than 6, Poinar and van der Laan (1972) found mostly one but recovered up to 72 per host bee. In our own studies, we found an average of 12% of queens infected, with a seasonal trend where early emerging queens rarely have nematodes. Prevalence may rise to over 50% at later times, though (i.e., April to May in the Northern Hemisphere) (Schmid-Hempel et al. 1990 and pers. obs.). The gravid female releases eggs over 1–2 weeks and the larvae hatch 4–7 days later (Madel 1966). She may easily produce 50,000–100,000 newborns. Since the first two juvenile stages are completed within the eggshell, all the freely moving juveniles found in the hemocoel and intestines of infected queens in spring are of the third stage (Poinar and van der Laan 1972). The juveniles pass from the hemocoel to the rectum and to a lesser extent to the oviduct (Pouvreau 1964). They are then passed out in the feces, which are deposited in the soil as the infected queen inspects potential overwintering sites rather than starting a colony on her own

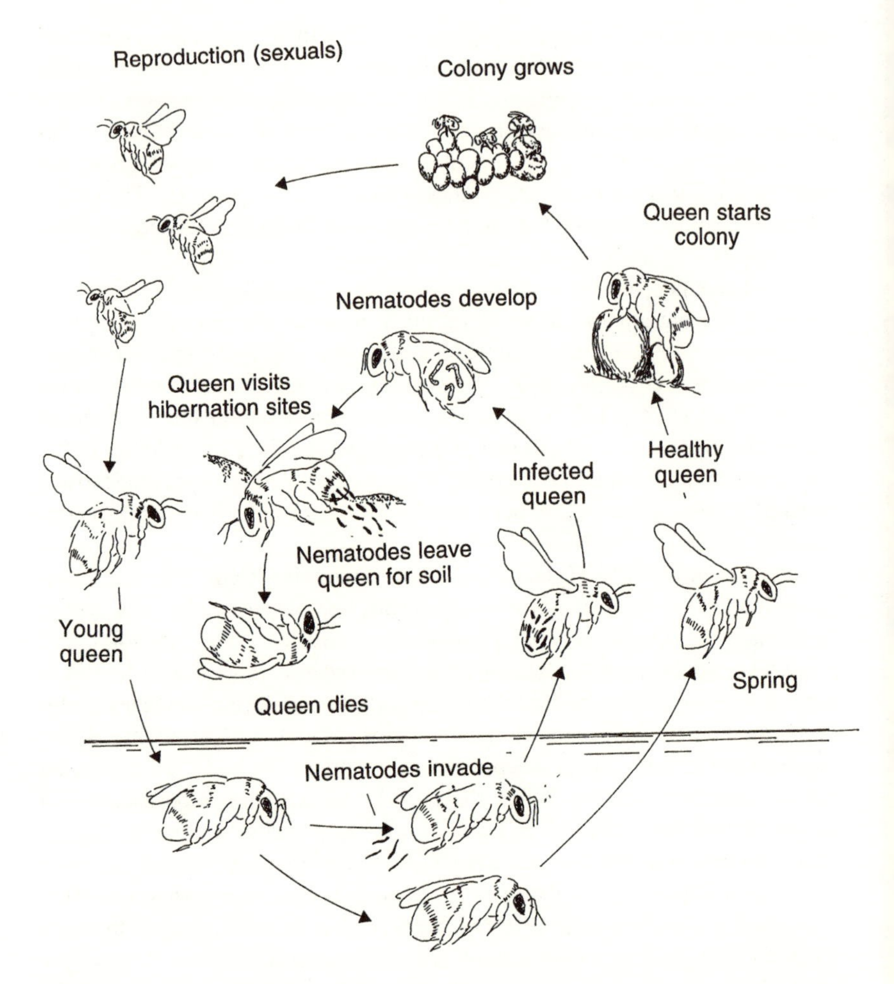

Figure 2.7 The life cycle of the nematode *Sphaerularia bombi* in queens of *Bombus* (after Alford 1975, modified). *Sphaerularia* invades overwintering queens in the soil and develops in the abdomen. Instead of founding a colony, infected queens look for potential hibernation sites and thus deposit larval nematodes that can infect autumn queens going into hibernation at the same site.

(which explains the late-season increase in prevalence among queens). This behavioral change is associated with nematode infection and clearly serves to disperse the parasite's offspring. It is perhaps triggered by the reduced corpora allata that is typical for infected queens. Additional effects include degenerated ovaries and sterilization (Palm 1948). Bedding (1985) suggests that *Sphaerularia* is one of the few really harmful insect nematodes.

The winter survival of infected queens must obviously be important for the prevalence of *Sphaerularia.* In fact, after warm winters and mild springs, the observed prevalence of *Sphaerularia* is higher than average (Hattingen 1956; Stein and Lohmar 1972). Because development from leaving the host to a new infective female is about 10 weeks during summer, young gynes in late summer may sometimes become infected (e.g., *B. hypnorum:* Poinar and van der Laan 1972). Interestingly, Kaya et al. (1979) report a phenomenon that is in fact very similar to the situation in *Sphaerularia.* Flies (*Musca autumnalis*), become infected by the nematode *Heterotylenchus autumnalis,* which eventually invades the ovaries. After this time, the flies no longer visit the feces of cattle as usual but visit only fresh cattle dung, where they "oviposit," i.e., deposit nematodes that are able to infect the offspring laid by other females.

Sphaerularia was inadvertently introduced to New Zealand around 1883–85 together with imported bumblebee queens and released near Canterbury and Waikato. In the 1970s the nematode was found near Canterbury, Christchurch, and Lincoln (MacFarlane and Griffin 1990), which suggests a low rate of spread. According to Hopkins (1914, cited in MacFarlane and Griffin 1990), healthy queens in New Zealand disperse up to 140 km in a single year, and since their introduction they have colonized twenty-three islands, 2–30 km off shore. At the time, queens infected with *Sphaerularia* were found only 30–40 km from their site of original introduction which yields a dispersal of approximately 1 km per year. *Sphaerularia* has been repeatedly studied (Cumber 1949; Pouvreau 1962; Madel 1966; Alford 1969; Lundberg and Svensson 1975) but many facets of its interesting life cycle still remain hidden.

Several rhabditid nematodes have been reported from ants. Typically, they parasitize the pharyngeal glands in the head of their host. Examples are *Caenorhabditis dolichura,* as described by Janet in 1893 and in the following years also, by Bedding (1985) in wood ants (*Formica fusca*); by Nickle and Ayre (1966) in *Camponotus herculaneus* and *Acanthomyops claviger;* and by Wahab (1962) in several other genera of ants (appendix 2.5). While rhabditid nematodes are found in both ants and termites, they are conspicuously absent in the flying social insects (wasps and bees). As usual, this may reflect the lack of studies, but it could also indicate that susceptibility to rhabditids depends on the mode of life, for example, soil nesting.

Mermithids infecting ants are a classical case. These cases are among the best documented for any insect nematode (Bedding 1985; appendix 2.5). In general, mermithid nematodes are found in the ants from early summer to autumn. They leave their host, usually via the anus, in late autumn and midwinter, where they burrow into the soil of the host nest (Vandel 1934). There, mating and oviposition occur. The virgin females can survive for 2–3 years in the soil, in contrast to the mated females that survive for only six months. The infection of the host is very likely by direct penetration of the host's cuticle (Poinar 1975), and can be experimentally established by adding infected soil to an ant colony (Vandel 1934). Passera (1976) demonstrated that third-instar larvae of *Pheidole pallidula* can become infected. This would fit the idea that penetration is only possible in the not yet fully sclerotized larvae but not in adults.

Typically, therefore, mermithids infect the larvae and pupae. If the mermithid enters a juvenile ant after the process of morphological differentiation has occurred, obviously little morphological change results. But sometimes they induce morphological aberrations in the eclosing worker (see examples in Crawley and Baylis 1921; Wülker 1975; Kaiser 1986; Jouvenaz and Wojcik 1990). The resulting morphologically aberrant forms have received a number of different names, e.g., "mermithergate" (Wheeler 1907), "mermithogynes" (Crawley and Baylis 1921), and "mermithostratiotes" (Vandel 1927), which are often intercaste morphs that are produced by an infection in the early third instar. Crawley and Baylis (1921) also found that the mermithogynes had atrophied ovaries, no fat body, and enlarged tracheae. In other cases, infected males and females had shorter wings (Wheeler 1928; Taylor 1933 in Bedding 1985). In several cases, the mermithid-parasitized specimens have been described as variants or new species of ants, since their morphology was quite aberrant (e.g., Wasmann 1909 for *Pheidole symbiotica,* which is in fact *P. pallidula;* or *microcephalus* variants described for *Odontomachus haematoda* by Emery 1890; see Bedding 1985). Mermithid parasitism also affects the behavior of the infected ants, e.g., some become more negatively phototropic, and others more active.

In a detailed study, Kaiser (1986) investigated the relationship between the nematode *Pheromermis villosa* (Mermithidae) and its ant hosts, two *Lasius* species (probably the same relationship as described by Crawley and Baylis 1921). This account illustrates the general patterns mentioned above. *Pheromermis* infects ant larvae when they are fed with prey (oligochaetes, the intermediate hosts) that contain parasite eggs. However, nematode development is only induced as the host pupates. The nematode develops in musculature, both in the intermediate host and in the ant. If sexuals are infected, nematode infection leads to a reduction in flight musculature and shorter wings of females (about 20% of females have been found infected in the study). The nematode hatches from its adult host in summer (July–August) and becomes sexually mature in November. The eggs of nematodes will then be ingested by oligochaetes that are in turn prey of the ants. Nematode infection also changes host behavior, so that infected ants approach water where these intermediate hosts of *P. villosa* live (Kaiser 1986). It is also possible that infected ants may simply become more active in search of food. Interestingly, infected workers of *Lasius niger* seem no longer able to produce the poison normally used in defense of the colony (Gösswald 1930). Due to their life cycle, therefore, nematodes (other than eggs or infective stages) are only present in the adult ants for a few weeks per year (mid-July–August) while they are present in the intermediate host year-round. The eggs are said to stay infective in the soil for 2–3 years. *P. villosa* is a protandrous hermaphrodite that can produce viable offspring on its own.

There are also a number of reports of unidentified nematodes in ants. Examples include a nematode in *Solenopsis richteri* (Jouvenaz and Wojcik 1990) that causes heads and thorax of infected majors to resemble those of minors. The host dies at emergence of the nematode. In *S. geminata,* a large unknown nematode, visible by the enlarged gasters, was found in appreciable numbers (3.3% of 270 workers: Mitchell and Jouvenaz 1985). Some studies have established that ants can be infected by nematodes that have never been observed in this association in the field, e.g., *Steinernema* spp. in *Solenopsis invicta* (Drees et al. 1992; Jouvenaz and Martin 1992).

From detailed investigations on nematode parasitism in different ant families, involving a total of ca. 14,000 host specimens, Wahab (1962) found nematodes in the pharyngeal glands of twenty-five species from six genera. Camponotine ants were more often parasitized than either formicines or myrmicines. Because in the two latter species experimental infection is possible, the author concluded that the feeding habits of these different groups are responsible for the differences. Camponotinae in fact often scavenge and thus come in contact with contaminated food. Most of the affected ants may also have in common the fact that they nest underground in damp soil. Long-lasting stages can be picked up from this reservoir and thus help to sustain the parasite over successive generations of colonies. Also Stein and Lohmar (1972) attributed the differences in the prevalence of *Sphaerularia bombi* in different *Bombus* species to their overwintering habits; the heavily infected species are those normally overwintering in humid-mossy soils. According to these observations, nematodes are not prevalent generally but common locally and over restricted periods of time during the year: they may also be common only during some years but persist in the soil as long-lasting stages.

Only few reports exist for wasps (appendix 2.5). One mermithid species, *Agamomermis pachysoma,* has been found in some common wasps in Europe. In contrast to ants, no morphological change seems to be associated with the infection. The mode of infection is unclear but perhaps happens through food (insect prey) that the wasps use. The nematodes themselves can grow to enormous lengths (several centimeters). Kloft (1951) dissected a parasitized queen of *Vespula (Paravespula) germanica* and found that flight muscles, fat body, ovaries, and the tissue surrounding the thoracic ganglia were all reduced. It is likely that these nematodes leave the dead host and burrow into the soil. Fox-Wilson (1946) cites evidence that almost all male wasps checked were parasitized; the nematode apparently leading to sterility. Cobbold (1888, cited in Poinar 1975) claimed to have found *Sphaerularia bombi* in *Vespula vulgaris* and *V. rufa.* To the extent that other social insect queens overwinter at the same sites as *Bombus* queens, and that nematodes with free-living stages may not be so host-specific, such cross-infections seem at least possible.

A number of reports on nematodes infecting termites, many of them belonging to the rhabtidids, have been summarized by Poinar (1975). Little is known about their biology. Perhaps noteworthy is *Harteria gallinarum,* which uses the termite *Macrorhodotermes mossambicus* as an intermediate host, while the final host is a bird (chicken) (Theiler 1918, in Poinar 1975).

2.6 Helminths

This ad hoc group of parasites belongs to the phylum Platyhelminthes. Of special interest to social insects are the classes Trematoda (the digenetic flukes) and Cestoda (the tapeworms). These are all parasitic species; the digenetic flukes have two or three hosts to complete their life cycle. Social insects are playing the role of one of the intermediate host(s).

Trematoda

So far, only ants have been reported to act as intermediate hosts to helminth parasites. The classical case certainly is the liver fluke *Dicrocoelium dendriticum* infecting workers in the ants *(Serviformica) Formica rufibarbis* and *F. cunicularis.* Inside the ant, the metacercaria develop into cercaria that are finally eaten by the next host, a grazing vertebrate (Hohorst and Graefe 1961; Schneider and Hohorst 1971). In the social insect host, the metacercaria penetrate the pharynx wall after infection and enter the hemocoel. After a brief period of inactivity, the parasites migrate toward the head of the ant, where eventually one "brain worm" settles into the subesophagial ganglion (a part of the ant's "brain"). Within 1–2 days after infection, all other cercaria migrate back to the gaster, where they encyst. By some as yet unknown mechanisms, the brain worm seems to manipulate the worker ant's behavior so that she climbs grasses and fastens herself there with the mandibles. In this way, the ant is more vulnerable to a predator, i.e., the next host of the parasite. It is striking how closely this phenomenon resembles the one observed in ants infected by fungi, as illustrated earlier (Loos-Frank and Zimmermann 1976; chap. 2.3). In both cases, the parasite establishes itself in the vicinity of presumably critical neural tissue of the host. This indicates that seeking the neighborhood for specific host nerve cells is probably a general, active, and adaptive process by which the parasite can attain control over its host's behavior.

Also, the biology of *Brachylecithum mosquensis* infecting *Camponotus pennsylvanicus* and *C. herculaneus* is quite similar. The ants acquire the infection from the intermediate host, a land snail (*Allogona ptychophora*) by way of excreted slime balls containing infective cercaria (Carney 1969). The ingested cercaria penetrate the ant's crop and appear in the hemocoel within one hour. After a cyst wall has formed around the cercaria, the cercaria develop into infective metacercaria within 120 days. Most of these can be found in the ant's gaster, but one or two are also found around the supraesophagal ganglion (the "brain"), quite similar to the situation in *Dicrocoelium.* The brain cercaria never develop such a rigid cell wall as the others do. This suggests a necessity to interact with neural tissue of the host through release of substances. In fact, just as in the liver fluke case, systematic dissections show that the supraesophagal ganglion may be the primary site of cercarial metamorphosis from where all except one or two return to the gaster. Infected ants prefer to stay in open spaces or on tips of grasses where they incur higher predation risks. They also display a sluggish behavior that should make it even more likely that it is eaten by the next host of the parasite. In this case, it is known that the American Robin (*Turdus migratorius*) acts as a final host. The parasite is shed in bird feces that are consumed by the land snails to complete the cycle (Carney 1969). Only *Dicrocoelium* and *Brachylecithum* seem to utilize ants as their intermediate hosts (Svadzhyan and Frolkova 1966) and then only in the Formicinae.

Cestoda

Far from being rare, cestode infections in ants have been reported from the myrmicine genera *Crematogaster, Harpagoxenus, Monomorium, Myrmica, Pheidole, Solenopsis, Tetramorium,* the formicine *Prenolepis,* the ponerine *Euponera,* and *Iridomyrmex* (Dolichoderinae)(Stuart and Alloway 1988). So far, no report seems to exist for bees, wasps, or termites. It is close at hand to suspect that this absence has to do with the complicated life cycle of some of the cestodes and trematodes that often need large, vertebrate final hosts to develop and mature. Bees and wasps often have few vertebrate enemies (apart from nest robbers—the implications of which would be an interesting issue by itself). But they may also be unlikely to encounter the potential hosts or consume appropriate food—e.g., snails, excretions, and so forth—that could harbor infective stages able to infect them in the first place. It would be interesting to know whether carnivorous bee species of the genus *Trigona,* which scavenge on vertebrate carcasses in tropical forests and thus replace pollen with meat as a source of proteins (Roubik 1982), do have helminths that can sustain their life cycle through social bees and a vertebrate host. Termites, in contrast, are well protected in their nests and galleys and may thus not often offer the opportunity for a parasite to pass on to a vertebrate host (again, neglecting nest robbers such as armadillos or pangolins). Moreover, being specialists on plant material, termites will not likely consume an animal host that contains parasites that are able to infect them. On the other hand, mature cestodes, which for example reside in birds, release eggs or shed entire segments of the body (the proglottids) that contain eggs. These segments are mobile and can move around on bird feces. They are therefore presumably attractive for predators such as ants that carry them back to the nest as prey for larvae. The ingested parasite eggs then develop and the oncospheres pierce the gut wall and give rise to many (usually several dozens) worms within the newly hatched ant's hemocoel. The infected ants can live for a long time until they are eaten by a predator that serves as the final host of the parasite.

Case and Ackert (1940) give a list of ant species (*Solenopsis molesta, Crematogaster lineolata cerasi, Iridomyrmex pruinosus, Pheidole dentata, Crematogaster* sp., *Prenolepis* sp.) that can potentially be intermediate hosts of the cestode *Raillietina tetragona* that infects fowls as final hosts. Ants of these species were seen carrying gravid segments (proglottids) of the cestode back to their nest, but no direct evidence for the infection is given. *Raillietina* has also been reported from a variety of other ants (appendix 2.6).

Cestode infection often leads to aberrant coloration and altered behaviors of the infected adult ants. Infected workers of *Leptothorax nylanderi,* for example, are conspicuously yellow as compared to their normal reddish brown healthy nestmates. There are also more subtle morphological differences in the size of the petiole (the segment joining abdomen with thorax). Very similar characteristics have also been observed in several other species of *Leptothorax* and *Harpagoxenus* (Buschinger 1973). Infected ants are more sluggish and have a difficult time leaving the nest after disturbance. They barely engage in brood tending and seem to be constantly hungry, soliciting food from their nestmates. These modifications probably make the ant more conspicuous and easier prey for a predator like a bird, the final host of the parasite. In addition, the infected ants are virtually sterile. Young gynes that are infected apparently dealate in the nest and do not leave the nest for mating flights (Buschinger 1973). In the *Leptothorax* complex, cestodes seem to have low degrees of host specificity (Stuart and Alloway 1988). Stuart and Alloway also found a yet unidentified cestode (probably *Choanotaenia* or *Anomotaenia*) that infects workers of a number of *Leptothorax* ants. The cestode cysticercoides are found in the gaster of adults. Infection affects coloration of the ants, they become yellowish in color rather than dark brown, and thus should be more susceptible to predation by the parasite's final host, probably a bird. Only a few of the investigated nests were infected, however (e.g., 6 out of 1295 nests of *L. muscorum,* 12 out of 2462 nests of *L. curvispinosus*). In infected nests, 5–20% of the workers were found to be parasitized. Stuart and Alloway suspect that some of the described species within the *Leptothorax* group may perhaps be simple color variants due to parasitism.

2.7 Mites

Mites, together with ticks and chiggers, comprise some 30,000 species (perhaps up to one million; Radford 1950 in Evans 1992) and are classified in the order Acari within the class Arachnida. Some classifications group Acari into the three large groups: Opilioacariformes (Notostigmata), Parasitiformes (Mesostigmata, Holothyrida, Ixodida), and Acariformes (Astigmata, Oribatida, Prostigmata) (Brusca and Brusca 1990). Most of the literature that is relevant here uses a grouping into Astigmata, Prostigmata, and Mesostigmata (e.g., Eickwort 1994). This scheme also counts Notostigmata, Holothyrida (Tetrastigmata), Oribatida (Cryptostigmata), and Ixodida (Metastigmata, the ticks) (Cheng 1986).

Mites have a tremendous variety of lifestyles. They are ectoparasitic but are also found in natural body openings (e.g., tracheae) and can serve as vectors for microparasites (see appendix 2.1). Because of this variety and large number of associations, appendixes 2.7 and 2.8 list only the genera, and a few additional notes. Another recurrent problem in mite-host associations is whether or not to classify the mites as being truly parasitic. Apart from a few clear-cut cases (like *Varroa* in honeybees), the effects often appear to be subtle rather than drastic. They may nevertheless be harmful to a colony, although this has rarely been shown in any particular case.

Mites of Social Bees

From the available evidence, it is obvious that bees are prime hosts for mites. Many studies have been devoted to the analysis of their associations, effects, and coevolution. Overall, only a minority of mites seem truly parasitic, most of them belonging to the Mesostigmata (e.g., *Varroa, Parasitellus,* and allies). In fact, in his review on mites in social as well as solitary bees, Eickwort (1994) counted that 53% of all mite genera are saprophagous, i.e., they feed on debris in the bee's nest, 23% are predatory on other mites (this relationship is actually disproportionally frequent in social bees), and only 6% are truly parasitic. All of the parasitic genera were found to be associated with social bees. Some mites may in fact be beneficial to their hosts. For example, Schousboe (1987) noted a negative relationship between mite infections (*Parasitellus*) and the presence of the nematode *Sphaerularia,* and concluded that this indicates a beneficial effect of the mite. *Anoetus* is reported to remove bacteria from its halictid host, which offers an acarinarium in turn, i.e., a pocket on the host's surface that can be inhabited by the mite. *Proctotydeus therapeutikos* is said to deter fungal infections in stingless bees (Flechtmann and de Camargo 1979 in Eickwort 1994; see also Roubik 1989, p. 245). Wilson and Knollenberg (1987) present tentative evidence for a beneficial association with (subsocial) burying beetles (see also Kinn 1984 for bark beetles). However, there is hardly any direct experimental evidence for the benefit gained by these relationships.

Considering morphology, lifestyle, and behavior, Eickwort (1994) concluded that the hosts (the bees) in general have not adapted in particular ways to the mites (with the few exceptions where acarinaria have evolved), but that the mites have adapted strongly to the bees. Nevertheless, transfer of mites among host lineages seems common. Few bee mites have associations beyond the subfamily level, but very few have only one host when there is a congeneric host in the same habitat. Some groups are in fact associated with ants or termites but frequently enter nests of social bees (e.g., of honeybees and stingless bees: *Forcellinia, Oplitis, Triplogynium, Urodiscella, Urozercon*). But around 71% of the groups seem obligatory to bees, although mite phylogeny is in general only loosely related to bee phylogeny. Nevertheless, the social lifestyle is clearly an important element in mite biology. This can be deduced from the data presented by Eickwort (1994): 13 mite genera are known to be associated with the partially social Halictids (comprising 5000 species), 10 genera with the primitively eusocial Bombini (200 species), but 29 genera with the highly eusocial Meliponini (probably 500 species, with 450 as a very conservative estimate; D. Roubik, pers. comm.) and 14 with the Apini (8 species). This yields a measure of 0.003 mite genera per halictid species, and 0.05, 0.06 and 1.75 for the other groups, respectively. This suggests higher mite diversities in social bees. A higher mite diversity is also generally found in wood-dwelling, cavity-nesting bees.

Particularly well investigated are honeybee mites (e.g., De Jong et al. 1982). Bailey and Ball (1991) list *Acarapis woodi* (in *Apis mellifera, A. cerana,* perhaps *A. dorsata*), *A. externus* (*cerana, mellifera*), *A. dorsalis* (*mellifera*), *Varroa jacobsoni* (*mellifera, cerana*), *V. underwoodi* (*cerana*), *Euvarroa sinhai* (*florea*), *Tropilaelaps clareae* (*mellifera, cerana, dorsata, florea, laboriosa*), *T. koenigerum* (*dorsata, laboriosa*), and a *Leptus* sp. (*mellifera*). For example, *Tropilaelaps clareae* (Laelapidae) infests brood cells (Burgett et al. 1983). Usually 1–4 females occupy a single host cell. They are locked inside as the cell becomes capped. The mites live externally and suck hemolymph from the larvae. In infected colonies, 10–90% of the worker cells, and even more of the drone cells, can be affected. *Tropilaelaps* usually invades worker cells much more readily than *Varroa.* Both extract hemolymph from the larvae and can be serious pests. In *A. mellifera,* development of *Tropilaelaps* in larval cells is 8–9 days from egg to adult mite. Adult hosts emerging from infected cells sometimes have malformations (Ritter and Schneider-Ritter 1988) but otherwise the damage seems to be small. More detailed studies (Burgett et al. 1990) have revealed that the distribution of the mites over host cells is aggregated, presumably an adaptation to increase the chances of sexual encounters (mating of mites takes place within host cells). At the same time, this inflicts less damage to the host nest than if distribution were more even. Because *Tropilaelaps clareae* cannot survive outside the brood cell for more than a few days, and since no brood is produced during winter in *Apis mellifera,* this mite seems not well adapted to such a temperate climate host life cycle. However, in Southeast Asia it is quite prevalent and infects not only *A. mellifera* but also *A. dorsata* (Burgett et al. 1983; Nixon 1983).

Infestation by *Varroa jacobsoni* (Varroidae) in *A. mellifera* is quite similar to *Tropilaelaps.* The females are presumably being attracted by volatiles and invade brood cells. After the cell is sealed, the female lays eggs and the developing parasite brood feeds on the bee larva and then pupates. If there is not enough food, the mites are literally trapped and will die. The mite later pierces the host larva or pupa to extract hemolymph. The adult female lays the first (male) egg 60 hours after cell capping and adds female eggs afterwards at approximately 30-hour intervals (Rehm and Ritter 1989). Because these mites also possess a haplo-diploid sex determination mechanism similar to their host, such fine-tuning of offspring sex is readily achieved. Typically, the mite lays 1–

12 eggs in cells, preferentially in drone cells. Overall development takes about 9–11 days in workers cells, 10–14 days in drone cells. Females attain sexual maturity inside the capped host cell 24 hours after the last molt. Since males mature somewhat earlier, mating takes place inside the cell when the females emerge; the males die afterwards (Ramirez and Otis 1986; Donzé and Guerin 1994). Obviously, if host developmental time were short, female mites would not have enough time to develop to maturity, which seems to occur in the Cape Bee (*A. mellifera capensis,* Ramirez and Otis 1986); a reduction in the larval period also normally occurs during times of nectar flow. In fact, small differences in the time available to the mite's development can lead to substantial differences in the number of female mites emerging (Camazine 1988).

The effects of *Varroa* mite parasitization includes a reduction in honey yield (De Jong et al. 1982). In addition, a number of direct effects on the individual host that nevertheless manage to pupate and emerge seem to occur, for example, weight loss, shortened life span, reduced size of glands, reduced flight activity, reduced insecticide tolerance, and smaller sperm loads in drones (De Jong and De Jong 1983; Schneider and Drescher 1988). In fact, the average life span is 27.6 days for adults emerging from healthy larvae, but only 8.9 days if more than two mites were present in the brood cell; the reduced longevity is especially stringent in periods of dearth and during autumn and winter (De Jong and De Jong 1983; Kovac and Vrailsheim 1988). Untreated, the mites may kill a colony within a few years (De Jong et al. 1982), although the available evidence for such a large-scale effect is not overwhelming. As already mentioned before, *Varroa* seems to act both as a vector and an activator of acute bee paralysis virus (Ball 1989). *Euvarroa sinhai* infests drone cells of *A. florea,* but occurrences within colonies are generally low (< 1% of individuals) even though very many colonies may be affected (Mossadegh 1991). For unknown reasons, the typical strategy of *Varroa,* where first the host (larva) is parasitized and mites become phoretic later, has only evolved in the Apinae, but not in the Meliponini (stingless bees) (Eickwort 1994).

Acarapis woodi (Tarsonemidae) infests mainly the tracheae of the first pair of thoracic spiracles of the adult honeybee where it pierces the body wall and feeds on hemolymph (Bailey and Ball 1991). Their presence has an effect on colony survival, particularly during harsh winters. Spring brood area is negatively correlated with infestation intensity (Otis and Scott-Dupree 1992). On average, 41% of the colonies in the sample studied by these authors were infected (with more than 20% of the individuals), and a logistic regression related mite-infestation intensity to colony mortality risk (with 50% colony mortality reached at 50% mite infection). Infestations are particularly high when each colony has a large worker population and foraging is restricted so that crowded conditions prevail, as well as when colony density is high. From 1925 to 1980 the number of colonies in Britain steadily declined, as did the prevalence of *A. woodi* (and paralysis virus) (Bailey 1985). Severe effects on colony survival were also reported by Royce and Rossignol (1990b). Prevalence values of 60% of colonies infected were observed by Eischen et al. (1990) in Mexico, where a particularly rapid spread was observed in the 1980s. The life cycle of *Acarapis* takes only 2 weeks to complete, so a rapid buildup does indeed take place.

Chmielewski (1991) found that 99% of all 328 investigated honeybee hives were infested with mites that are known to be pests of human stored products (a total of fifteen species if the genera *Acarus, Acotyledon, Caloglyphus, Carpoglyphus, Glycyphagus, Lasioacarus, Michaelopus, Thyreophagus,* and *Tyrophagus* are included). Usually, the infestations were in great numbers. Feeding experiments suggested that the mites could be consuming stored honey and pollen as well as comb material and thus be a parasite of nest stores. In addition, the mites may serve as vectors for other diseases. *Tyrophagus putrescentiae* sometimes infests the tracheal system of the bee similar to *Acarpis woodi* and could therefore cause serious damage. In turn, the honeybee colonies would constitute a natural reservoir of mites that are able to contaminate human food stores.

Bumblebees are also infested by a number of mite species. Schousboe (1987) screened for *Parasitellus* and found four species on Danish species of *Bombus,* with 1–12% of the specimen infected. Usually one or two mites per worker bee were found and 0–32 per overwintered queen (see also Pape 1986). In addition, the tracheal mite *Locustacarus buchneri* (see also Skou et al. 1963) and *Scutacarus* (Schousboe 1986) were also found. Schwarz et al. (1996) found ten species of mesostigmatic mites associated with bumblebees in Central Europe. Together with the reports of Richards and Richards (1976) on further *Parasitellus* and of Hunter and Husband (1973) on *Hypoaspis (Pneumolaelaps)* in North America and the Arctic, this suggests that the diversity of bumblebee mites may in fact be quite high.

It is not always clear what the mites are actually doing. Some bumblebee mites are phoretic in the sense that they disembark from their carrier, the spring queen, when she has established a new colony. The mites will then live on pollen or other food available in the nest. Skou et al. (1963) reported that up to 106 nymphs of *Parasitellus fucorum* were found on a single queen of *Bombus terrestris* or *B. lapidarius* but that the effects are probably benign. Goldblatt (1984) compared mite infestations in North America (mainly for *B. bimaculatus* and *B. vagans*) and Europe (*B. terrestris, B. lapidarius, B. pascuorum*) and found rather similar prevalences, at least at a crude level. For example, *Kuzinia* was found on average on 39% (12–70%; $N=339$) of the investigated specimens in North America, which compares with reported rates of 17–83% for European bumblebee populations (Skou et al. 1963 in Goldblatt 1984; and, with some reservations, Chmielewski 1971). Schwarz et al. (1996) found up to 49% of queens infected by mites, but mostly occurrences were around a few percent. Smaller occurrences in North America were also found for *Parasitellus* mites (14% of—mostly early—emerging queens in North America vs. 23% in Europe: Chmielewski 1971) and for *Locustacarus buchneri* (4%–46% vs. 100% in

Europe: Cumber 1949 in Goldblatt 1984) in newly emerged queens. Although these figures are difficult to compare, since they involve the pooling of different mite species and host categories, they at least show that mites are widely distributed and regular elements of bumblebee ecology. Much less detail is known from other groups of social bees. Nevertheless, mites are extremely common in tropical bees (Roubik 1989, p. 244). Probably most of them feed on fungi, pollen, and debris in the nest and have small effects. Stingless bees, however, are said to be almost completely free of phoretic mites (Eickwort 1994).

Mites of Ants

Ant mites have been of interest in the context of phoresy and as guests (Wojcik 1989; Hölldobler and Wilson 1990). Many seem in fact to be purely phoretic; some live socially parasitic by soliciting food from the workers, e.g., *Antennophorus* in *Lasius mixtus*. But others may harm the host in more substantial ways, like *Cilibano comato* that lives on the ant's gaster and sucks hemolymph from its host (Janet 1897, cited in Wojcik 1989). Such a close association is usually correlated with higher levels of host specificity. Hölldobler and Wilson (1990) give an overview: 14 out of 18 cases of mite-ant associations were considered to be phoretic and/or commensal, while only 3 cases were parasitic, on adults and larvae. Most of the latter cases come from a detailed study of the army ant, *Eciton,* where they comprise 95% of all Neotropical army ant ectoparasites (Kistner 1982). As a curiosity, *Oolaelaps* (Laelapidae) lives among the ants' eggs, where it feeds on the secretions that line the eggs and which are deposited by the nurse workers (Kistner 1982). Interestingly, Bowman and Ferguson (1985) report that the mite *Forcellinia galleriella* occurs in the honeybee although it is normally a myrmecophile living in ant nests on debris or dead workers.

In the fire ant *Solenopsis (saevissima) richteri,* Collins and Markin (1971) discovered mites in several colonies, the most common ones belonging to the family Uropodidae. They are attached to alate females but rarely to males or workers. The ants seemed to tolerate the mites. Bruce and LeCato (1980) found that experimental infestation with the mite *Pyemotis tritici* in *Solenopsis invicta* is possible and leads to the decline of the colony. The mite apparently injects a neurotoxin into the ant. The queen then becomes inactive. It is not known whether this is also a natural host-parasite association. Finally, ant-mite associations noted by Eickwort (1990) and O'Connor (1994) include many species of the Acaridae (*Cosmoglyphus, Forcellinia, Froriepia, Lasioacarus, Ocellacarus, Rettacarus, Sancassania, Tyrophagus*), as well as *Lemaniella* (Suidasiidae) and *Histiostoma* (Histiostomatidae).

Mites of Social Wasps

O'Connor (1994) points out that the highest diversity of astigmatid mites are associated with nests of solitary and social Hymenoptera. In particular, the (solitary, partly subsocial) eumenid wasps are associated with mites of the genera *Ensliniella, Kennethiella, Macrophara, Monobiacarus, Vespacarus, Zethacarus, Zethovidia,* and many undescribed genera. *Enslienella, Kennethiella,* and *Vespacarus* feed on hemolymph of host larvae and are therefore truly parasitic, while *Monobiacarus* is cleptoparasitic and feeds on the nest's provisions.

Generally, few mites have been reported for truly social wasps (appendix 2.8). *Polistes* wasps are associated with *Sphexicozela* (Winterschmidtiidae), *Vespula* with *Medeus* (Acaridae) and *Tortonia* (Suidasiidae) (Eickwort 1990; O'Connor 1994). An early report is parasitization of *Dolichovespula sylvestris* by *Acarus* that attacks the larvae and extracts hemolymph until their death (Stone 1865 in Spradbery 1973, p. 265). Nelson (1968) found *Pyemotes* sp. on two *Polistes* species. According to Krombein (1967), the mites attach themselves to wasp larvae and eventually kill them. They are said to be a potentially serious pest. However, they may actually be parasites of moths that are in turn associated with wasps (Nelson 1968). It is likely that as in other taxa, most mites of social wasps are saprophagous or cleptoparasitic, living on debris or nest provisions.

Mites of Termites

Still less is known of mites associated with termites. Eickwort (1990) and O'Connor (1994) include many species of the Acaridae (*Cosmoglyphus, Forcellinia, Froriepia, Lasioacarus, Ocellacarus, Rettacarus, Sancassania, Tyrophagus*) in this group, as well as *Lemaniella* (Suidasiidae) and *Histiostoma* (Histiostomatidae). It is not immediately obvious why termites should not have a similarly rich mite fauna as the other major groups of social insects. Hence, a wide-open field probably remains to be discovered here.

2.8 Parasitic Insects

There are very many insects that parasitize social insects, either brood or adults, live cleptoparasitically on provisions, or are commensalistic associates (such as many beetles). In Godfray's (1994) review of parasitoids, those of social insects are poorly covered. This is perhaps because parasitoids have mostly been considered in the context of controlling pest species, e.g., of various moths. The groups that are particularly relevant for social insects are the Hymenoptera and Diptera. By comparison, Lepidoptera and Coleoptera are not very numerous as parasites, although they are very common associates of social insects. The biology of most of these associates is covered in Kistner (1982). In line with the general scope of this treatise, I will not discuss cases such as robber bees (perhaps as many as ten species: Michener et al. 1994), which are significant enemies of social bees in the tropics, or of socially parasitic ants and wasps.

Parasitic Hymenoptera

A wide variety of hymenopteran parasitoids attack social insects, most of them being brood parasites. A quick look at appendix 2.9 shows that a large number of hymenopteran families are actually parasitic on social insects. The detail of knowledge varies considerably, however. A special mention is deserved for the Eucharitidae—a group of parasitic wasps with very similar biologies and life-styles. Although only studied in a small number of species, all known Eucharitidae seem to be obligatory parasites of ants. This narrow host range is in contrast to the wider ranges known from related chalcidoid families (Johnson 1988). Eucharitidae are mostly ectoparasitic on ant larvae, sometimes also endoparasitic. Some species attack the cocoons. Development continues into the host pupal stage, which is sometimes killed, but in other cases only a deformation of the host's wings can be observed. In some cases, e.g., *Lasius niger* infected by *Pseudometagea schwarzii,* or *Camponotus* by *Stilbula (= Schizaspidia) tenuicornis,* prevalences can be extremely high, with over 90% and 50% of brood, respectively (Ayre 1962; Johnson 1988). Hence, eucharitid wasps are actually not uncommon. As appendix 2.9 shows, they are known in very many ant genera.

The brood of social wasps, particularly those species with combs in the open like *Polistes or Microstigmus,* is attacked by a wide variety of parasitoids, including pyralid and cosmopterigid moths, wasps (Ichneumonidae, Eulophidae), or sarcophagid flies (Nelson 1968; Miyano 1980; Strassmann 1981a,b; MacFarlane and Palma 1987; Matthews 1991). Gadagkar (1991) lists a large number of parasitoids that attack the brood of social wasps in the genera *Belonogaster, Mischocyttarus, Parapolybia,* and *Ropalidia.* Apart from the parasites listed above, he also mentions parasitic flies (Tachinidae, Phoridae) and parasitic wasps (Ichneumonidae, Torymidae, Trigonalidae). In several of these cases, the parasites are suspected to be an important mortality factor for the colony. Nest envelopes seem to provide some defense against such parasites (Jeanne 1991). In fact, it seems obvious that nest coverages or subterranean nesting is an enormous advantage in this respect. Retinue behavior of workers of tropical honeybees having exposed combs may serve the same purpose (e.g., Seeley 1985).

In addition to the species identified and described, many as yet unknown hymenopteran parasitoids attack social insects. For example, Wojcik (1990) gives a list of insect groups that are associated with fire ants but where the effects and biology remain unknown. These include the parasitic families Bethylidae, Ceraphronidae, Diapriidae, and Ichneumonidae. Nelson (1968) lists a number of species that are probably parasitic associates of *Polistes* nests but where the effect is unknown, e.g., *Monodontomerus minor* (Torymidae). Several hymenopteran parasitoids that are associated with social insects are parasites of other insect associates. For example, *Apanteles carpatus* (Braconidae) parasitizes the larvae of moths that are themselves associated with wasp nests. Other such secondarily parasitic species include *Perilampus chrysopae* (Perilampidae), *Dibrachys cavus,* and probably *Pteromalus puparum* (Pteromalidae) that parasitize larvae and pupae of moths and other Hymenoptera (Nelson 1968). Hence, there are many hymenoptera that live on primary nest associates and thus, in the first instance, do not count as parasites of the social insects themselves. For example, Hölldobler and Wilson (1990, their table 13.1) list a number of species that are mainly parasitic or cleptoparasitic on ant associates. These species are perhaps even beneficial as they remove true parasitoids or associates that are otherwise detrimental to the colony. Appendixes 2.9–2.11 suggest that parasitoids are not always very host-specific and may attack quite different groups of social insects.

Dipteran Parasitoids

The Diptera are a very diverse and successful order of insects. Generally, we know much less from dipteran parasitoids than we do from Hymenoptera (see Godfray 1994). Nevertheless, it is evident from the material presented in appendix 2.10 that flies from many different families are parasitic on social insects. Not mentioned are the very many nest associates that are primarily detrivorous or commensal (e.g., many syrphid flies) (see Kistner 1982; Hölldobler and Wilson 1990). At the same time, a note of caution should again be added, because there are almost no studies that have systematically looked at the effects of parasitoids on colony success (but see Müller and Schmid-Hempel 1992b, 1993a). The major relevant families of Diptera include the Conopidae, Sarcophagidae, Phoridae, and Tachinidae.

Phorid flies are among the most common ant associates. They are scavengers, predators, and commensals but also truly parasitic (Wojcik 1989). At least seventeen species of phorid flies are known to be associated with the fire ant species complex (*Solenopsis* spp.: Pesquero et al. 1993). New phorid associations are constantly discovered, sometimes revealing highly aberrant forms, such as a recent discovery of a very odd-looking fly in nests of ants and termites in Sulawesi (Disney and Kistner 1988). The flies also seem to mimic the ant's own pheromonal arsenal (Weissflog et al. 1995). Phorid flies often congregate at mating flights of the ants or attempt to oviposit onto workers nearby. Some species specialize on parasitizing other insects, e.g., orthopterans, that are frightened away by advancing army ant columns rather than on the ants themselves (e.g., Rettenmeyer and Akre 1968). Normally, phorid flies will hover over their hosts, over foraging columns or nest entrances, then dive and place an egg. They are deposited on the prothorax of the host, at the junction of dorsal and ventral sclerites but also in or near the head capsule, and on the gaster. In those cases where the head capsule is invaded by the parasitic larvae, the ant's head will typically just fall off after the parasite has completed its development. In polymorphic host species, the flies preferentially attack majors. Feener (1987) gives a detailed account of how female *Pseudaceton crawfordii* lay eggs on fire ants.

According to Wojcik (1990), only one instance of phorid attacks on sexuals (in *Solenopsis*) is known. The sheer presence of the flies can interfere with the competitive relationships between host and other ant species (Feener 1981). Among social wasps, Nelson's (1968) report of the phorid *Megaselia aletiae* in nests of the wasp *Polistes exclamans,* and the cecidomyiid *Phacnolauthia* sp. in *P. annularis* and *P. metricus* nests are noteworthy too. In all of these cases, no signs of parasitism were detected, and thus the trophic status remains unclear.

Some phorid flies, e.g., *Melaloncha,* are said to deposit larvae onto tropical stingless bees while in flight (Roubik 1989, p. 242). Larviparous parasitism by flies is in fact found in several taxa, mainly in the Sarcophagidae, where this is a group characteristic. For example, the female of the sarcophagid *Senotainia* is said to wait near the hive entrance of the honeybee to swoop down and larviposit (Bailey and Ball 1991, p. 108). The parasitic larva is usually deposited between the thorax and head of the host worker but then develops within the hemocoel of its abdomen. Only after the host bee has died does the larva feed on the thoracic muscles, the abdominal tissues, and then leaves the host to pupate. In tachinid flies both deposition of eggs and larvae is known.

Another group, the Conopidae, has been investigated in some detail in their association with bumblebees. The prevalences can be quite high, with an average of 20–30% of workers being parasitized in most areas, and with seasonal peaks of over 70% of all workers in a given area (Schmid-Hempel et al. 1990; Schmid-Hempel 1994b). Like phorids, the flies attack foraging workers from the air, when they are in flight or on a flower. The female injects an egg into the abdomen of the bee. The morphology of the last abdominal segments of conopids is highly specialized for the task. Given the fact that conopid flies are, due to the body size difference, always slower fliers than their hosts and have to oviposit into the abdomen, it seems evident that attacks are most successful when the host bee is in slow, maneuvering flights in the foraging area. In addition, attacks occur when the wings are spread, since otherwise the female conopid cannot reach the appropriate places (Schmid-Hempel 1994b). Interestingly, even in the same host, eggs of *Physocephala* are mostly inserted through the intersegmental membranes of the tergites (i.e., dorsal), while those of *Sicus* are inserted ventrally. Conopid larvae develop and kill their host within 10–12 days (Schmid-Hempel and Schmid-Hempel 1996). In contrast to phorid flies, the larva pupates within the host and therefore has to overwinter wherever the host dies. This has led to the suspicion that the parasitoids may be able to manipulate the behavior of their hosts. In fact, parasitized bees more often bury themselves into the ground before their death than do nonparasitized ones. This increases the chance of successful overwinter survival for the parasitoid pupae (Müller 1994).

According to Spradbery (1973), several species of conopids are also parasitoids of adult wasps in Britain. They are said to aggregate around the nest entrance of wasp colonies to dive on returning foragers, followed by a brief contact. The wasps themselves, albeit carnivorous, do not chase or capture the flies.

In our own studies on the same and similar conopid species parasitizing bumblebees, we were never able to observe this nest-aggregation behavior. However, it is possible that the more conspicuous nests of social wasps attract the flies, whereas the more cryptic ones of bumblebees are not so easy to locate.

Syrphid flies are associates of bees (e.g., *Bombus*) and wasps (*Paravespula*) (e.g., Rupp 1987). Their effects are probably mostly small, because the larvae subsist on pollen or dead brood. Except for a few exceptions (appendix 2.10), their parasitic status remains unclear.

Little is known about parasitism of termites by dipteran flies. Wilson (1971, p. 398) lists several species belonging to the Psychodidae, Sciaridae, Cecidomyiidae, Phoridae, Termitoxeniidae, Thaumatoxenidae, and Anthomyiidae as associated with termites, though the feeding habitats are unknown or poorly known. Given their general biology, some of those, e.g., the phorid flies, are likely to be parasitic, while others (e.g., the Pyschodidae) are more likely to be associates without any harmful effects. Since termites are usually found under bark, in sheltered galleries, or underground, it is perhaps not surprising to realize that so few reports exist. This may indeed reflect a general paucity of dipteran parasitoids in the termites, associated with their specialized cryptic habits, rather than a lack of studies.

Other Parasitic Arthropods

Some other groups of parasitic insects are of significance, mainly in the orders Strepsiptera, Coleoptera, and Lepidoptera. Often, the distinction with "true" parasitism is difficult to make, especially as the feeding habits of most of these species are not known (see review by Kistner 1982).

Strepsiptera

The early report of Ogloblin (1939) has shed light on a rather intriguing group—the Strepsiptera. Ogloblin found that in several ant species (*Pheidole, Pseudomyrma, Solenopsis, Camponotus*), workers are parasitized by Strepsiptera (*Myrmecolax, Caenocholax*). Strepsiptera also parasitize a wide variety of hosts in the orders Orthoptera, Hemiptera, Homoptera, and Hymenoptera (Riek 1970). The life cycle is quite interesting. Only males are free-living and winged. Their forewings are reduced to sausage-like appendages, while the hind wings are large and characteristically fan shaped. Females are wingless and permanently endoparasitic in their hosts. The males seek out hosts that contain females for mating. Strepsiptera are generally larviparous (ovoviviparous). The first instar, the triungulin, actively seeks out a suitable host and burrows into it. After several instars, strepsipteran parasitization is often detected by the "stylopization" of the host, i.e., the characteristic protrusion of the abdominal parts and genital organs of females, or head capsules of the parasitic males. A single host individual can be parasitized by several strepsipteran larvae at the same time (up to thirty larvae per host have been found: Cheng 1986, p. 684), which gives it the characteristic spiky appearance with many extruded parasite parts. In addition to stylopization, parasitization by Strepsiptera can also lead to changes in the external morphology of hosts: the heads of parasitized hymenoptera often become smaller, rounder, and more hairy. In the solitary bee *Andrena,* the pollen-collecting apparatus of female bees is reduced, hind legs become male-like, and even the yellow coloration of males is attained. Hence, strepsipteran parasitization can lead to a morphological sex change (e.g., Smith and Hamm 1914, in Cheng 1986), although no such observations of morphological change have yet been reported in social insects.

The differences between high parasite prevalences among workers sampled outside the nest and the low prevalences among those inside suggested to Ogloblin (1939) that parasitism is associated with a behavioral change: infected workers are more likely to stay outside, to become diurnal, and to climb on grasses and bushes. More recent reports have doubted the validity of these observations (Kathirithamby and Johnston 1992). But in principle, the observations are quite similar to the behavioral changes noted in workers of *Bombus terrestris* parasitized by larvae of conopid flies, which has been documented in our own extensive studies (e.g., Müller and Schmid-Hempel 1992a, 1993b; Schmid-Hempel and Müller 1991). Apparently, strepsipteran parasitism is not uncommon, given the high prevalences in workers, males, and queens of *S. invicta* in the southern United States (Kathirithamby and Johnston 1992), and the reports from other species of ants (e.g., *Camponotus, Pseudomyrmex, Eciton, Pheidole, Crematogaster:* Teson and de Remes Lenicov 1979; Wojcik 1989; Kathirithamby and Johnston 1992) and wasps (Dunkle 1979). Prevalences of up to 70% have been observed (appendix 2.11).

Stylopidae are the largest family of the Strepsiptera. Most are parasites of bees, some of wasps. The strepsipteran family Myrmecolacidae, in contrast, is smaller but quite intriguing, since males (in Formicidae) and females (in Mantodea and Orthoptera) use entirely different hosts. Yet hosts for females have not often been described. Kathirithamby and Johnston (1992) mention that the female host is known for only five species. The situation for males is actually not much better: the male host is known for only 6 of the 87 described species of Myrmecolacidae; both hosts have been established for even fewer species. No convincing explanation for such heteroxenic parasitism has been formulated to date. In honeybees (Bailey and Ball 1991, p. 109), parasitization by Strepsiptera is known but rare and presumably causes little damage. Wasps are also hosts of this group. Pierce (cited in Dunkle 1979) reported that stylopised wasp queens managed to overwinter. This would be one way to ensure parasite propagation into the next season.

Coleoptera

Again, there are a number of reports on beetle associates with social insects where the biology is not always known, and thus it is unclear whether these species are truly parasitic or not. Among them are species belonging to the coleopteran families Pselaphidae, Scarabaeidae, and Staphylinidae (Wojcik 1990). Some of them are certainly predators on ant larvae rather than true parasites (e.g., the scarabid *Martinezia* spp. in *Solenopsis:* Wojcik 1990). *Trogoderma* found in the social wasp *Polistes* and in bee nests may be parasitic but the evidence is ambiguous (Nelson 1968). Certainly parasitic are the rhipiphorid beetles that are endo- and ectoparasites of larvae of wasps and bees (e.g., *Melöe* in *Apis*) (Appendix 2.11). Some beetles are described as predators of social insect broods. For example, *Trichodes ornatus* preys on bumblebee larvae and pupae (Hobbs 1962). Hölldobler and Wilson (1990, p. 480) list a number of species, e.g., pselaphid and scarabid beetles, as predators on a number of ant species. In some sense, they are parasites of the colony but are perhaps better considered a predator in our context and thus not discussed further.

LEPIDOPTERA

Equally effective as brood parasites are Lepidoptera, where a prime example is given by the various species of wax moths that attack the brood of honeybees or bumblebees (Alford 1975; Bailey and Ball 1991) (Appendix 2.11: *Galleria, Aphomia*). Such attacks can be fatal, especially for small colonies. After the female moth has gained access to the nest and successfully oviposited, the brood is quickly destroyed; the destructive power of such an attack is certainly one of the most impressive (and revolting) experiences that an observer of social insects can have. According to Nelson (1968), the action of parasitic moths may have caused the disappearance of some social wasps (*Polistes*) from islands in the West Indies; given the massive destruction, this seems quite plausible. *Galleria* and perhaps other pyralid moths have a normal host in the Anthophorid bee *Centris* and seem to have readily transferred to *Apis* in the New World (D. Roubik, pers. comm.).

Whitfield and Cameron (1993) list several species of lepidopteran nest associates (*Nemapogon, Plodia, Vitula*) of bumblebees that are mostly commensals—wax consumers and the like—unlikely to be true and harmful parasites. Similar relationships are known from moths attacking social wasps, e.g., *Tinea carnariella* feeding on nest material of *Polistes annularis* (Nelson 1968). Some species are thought to prey on larvae of their hosts, e.g., *Batrachedra* in the ant *Polyrhachis* and many lycaenid butterflies (Hölldobler and Wilson 1990), but are not ranked here as truly parasitic. Wilson (1971) also lists a number of taxa as predators of social insects but their actual trophic relationships are still unknown.

2.9 Summary

Social insects are host to a large number of parasites. All major groups of hosts and parasites have been described in the literature (summarized in appendix 2). However, the existing knowledge is bound to be a massive underestimation, since the true abundance and distribution of parasites remain to be discovered. Among the major parasitic groups, viruses have been studied mostly in the honeybee. Prominent examples are sacbrood and paralysis viruses. Viruses are sometimes associated with other parasitic organisms such as mites or microsporidia. They invade a variety of tissues, e.g., various glands. Mollicutes (Spiroplasmataceae) are bacteria-like organisms found in the host's hemocoel, known from a few bee hosts. From the available knowledge it appears that bacterial infections are relatively scarce in social insects, despite some important cases such as foulbrood in the honeybee. It is likely that a number of bacterial parasites produce opportunistic infections in social insects but otherwise reside in the soil, on plant surfaces, or in other species. Many bacteria in the alimentary tract or in glands are either proven or likely mutualistic organisms that help in the digestion of food or in warding off other microorganisms. Rickettsias, such as *Wolbachia* (a sex-distorting cytoplasmic element), have so far not been investigated but are probably much more common than currently known.

Entomopathogenic fungi are an important and diverse group of social insect parasites. Infection can occur in a variety of ways: through the cuticle, body openings, or by ingestion. Fungal infections often invade the hemoceol and can produce morphological and behavioral changes in the host, as is known from ants. Typically, infected workers climb exposed sites from where the fungal spores are spread. Fungal parasites may perhaps be important for population regulation of ants in some tropical ecosystems. Contrary to some suggestions, there is little difference in the richness of fungal species that attack ants, bees, wasps, or termites, although the terrestrial groups (ants, termites) appear to have slightly fewer fungal parasites than the aerial groups (bees, wasps). Most fungal parasites are not very host-specific and may thus be contracted from reservoirs in other, also nonsocial species. Protozoa comprise some of the most important parasites of social insects. They are mostly acquired by ingestion or via transovarial infection. Among the major groups, flagellates (such as *Crithidia, Leptomonas*) and "sporozoa" (particularly the microsporidia and gregarines) are the most prominent ones, while only one amoeba and no ciliates are known. Several microsporidia (e.g., *Vairiomorpha, Burenella*) are polymorphic as they produce more than one spore type with as yet mostly unknown functions. Spores are generally sensitive to light but can remain infective for longer periods.

Nematodes are important macroparasites. Often they were probably soil dwellers that have managed to infect social insects that nest in damp ground. Nematodes infect by ingestion but often directly penetrate the host's cuticle, some with specialized organs. Rhabditid and mermithid nematodes are the two major groups known from social insects. Mermithids are particularly well known from ants, where they cause morphological aberrations known as mermithergates as well as behavioral changes. "Helminths" (trematoda and cestoda) are only known from ants. Among the trematoda, spectacular cases such as *Dicrocoelium,* which infects wood ants and produces behavioral changes similar to fungal infections ("topping behavior"), have become widely known. Spatial proximity between host nerve ganglia and fungal hyphae or trematode cercaria seems to be necessary for these modifications. Trematode metacercaria can be contracted by scavenging behavior. Many cestodes have also been described. Some infections lead to behavioral and morphological aberrations in the host. The predatory habits of many ants favors cestode infections. In all cases known, the ants serve as intermediate hosts, while the final host is typically a bird or mammal.

Mites are a very large and diverse group of social insect associates, mutualists, commensals, cleptoparasites, or ecto- and endoparasites, with their particular status often unknown. Several groups are predatory on primary associates of social insects. Only a minority seem to be truly parasitic. Mites of social bees have been particularly well studied, a high percentage of them being specialized to bees. *Varroa* infecting *Apis* species is a major concern for beekeepers. Mites show a suite of adaptations to their hosts, but as a whole bees do not seem to have drastically adapted their morphology, lifestyle, or behavior to the mites. Mites are also known from ants, wasps, and termites.

Social insects are hosts to a very large variety of parasitic insects. Among the hymenopteran parasites, the Eucharitidae is specialized on ants. Other important groups include the Braconidae, Diapriidae, Eulophidae, and Ichneumonidae. Hymenopteran parasites are mostly brood parasites, while dipteran parasites attack mainly the adults. Among those, the Phoridae are particularly diverse and prevalent. Strepsitera are found in variety of social insect hosts; the Myremcolacidae are heteroxenic, i.e., males and females use different host species. In addition, Lepidoptera and Coleoptera can also be parasites of social insects. Besides these "true" parasites, a large variety of arthropods are nest associates, guests, or commensals.

3 Breaking into the Fortress

Wilson (1971) and Oster and Wilson (1978) have described social insect colonies as factory fortresses. This is a vivid picture that also describes the relationship of social insects with their parasites. It may in fact not be easy for a parasite to gain access to a colony, but once inside, the opportunities abound. In this chapter parasites are grouped according to where they usually enter the colony. Unfortunately, for most of the cases listed in appendix 2 our knowledge is very limited. The following classification is thus fraught with uncertainties. Nevertheless, it is crucial for both host and parasite where and how infection is established. Therefore, it is necessary to review the pathways of infection and transmission and thus to shed light on some of the examples from chapter 2.

3.1 Infection of Brood

Almost all types of parasites use brood as hosts. But infection and pathways of transmission as well as the effect of parasites vary considerably (table 3.1). Larvae are clearly a major target, but eggs have only a few parasites. This is remarkable as, for example, three wasp families of Hymenoptera consist exclusively of egg parasitoids (the Trichogrammatidae, Mymaridae, and Scelionidae; e.g., Strand 1986). Furthermore, eggs are numerous in nests of social insects and also lack a cellular defense against invading antigens (Salt 1970; Askew 1971). So they should in principle be readily available, easy hosts for parasitoids. On the other hand, eggs and larvae of social Hymenoptera, in particular, are immobile, and in many species (wasps, bees) they are confined to cells. The brood cells or chambers in turn are sheltered and defended against intruders. Indeed, the advantages of communal brood defense against brood parasites have been cited as selective forces toward sociality (e.g., Lin 1964). Parasites must therefore seek ways to find access to these hosts. In parasitoids, for example, females must gain direct access to oviposit, exploit a means of transportation through others, or the parasite larva itself must be mobile to find the host on its own. These requirements are quite demanding, but the parasites have nevertheless achieved remarkable feats.

Table 3.1

Examples of Parasites of Brood

Parasites	*Host*	*Infection, Transmission*	*Remarks*	*References*
VIRUSES				
BQCV (black queen cell virus)	A: *Apis mellifera*	Ingestion by adults when fed together with *Nosema apis*. Infective when injected in pupae (but not adults).	Queen cells become black early in season. A common infection in adults, too. Associated with *Nosema apis*.	Ball and Allen 1988; Bailey and Ball 1991; Ball, pers. comm.
SBV (sacbrood virus) APV (acute bee paralysis virus)	A: *Apis mellifera*	Via salivary glands into food.	SBV-infected workers change behavior. Sometimes found in adults, too.	Bailey and Ball 1991
BACTERIA				
Paenibacillus (Bacillus) larvae Melissococcus pluton	A: *Apis mellifera, A. cerana*	Ingestion of contaminated food. Transported by nurses. Spore deposits.	Agents of American foulbrood and European foulbrood. Spores are extremely long-lived. Queen can infect workers.	Bailey and Ball 1991; Bitner et al. 1972
FUNGI				
Ascosphaera apis	A: *Apis mellifera*	Mechanical contamination of nurses. Direct ingestion.	Agent of chalkbrood. Infection most successful at lower temperatures. Old larvae more susceptible. Also transfer via nurse bees.	Bailey and Ball 1991
Aspergillus flavus	A: *Apis mellifera*	Ingestion of contaminated food. Can also germinate on cuticle and penetrate.	Agent of stonebrood. Infected larvae become hard, most die after cells have been capped. Can also infect adults. Multiple passage of spores increases virulence.	Bailey and Ball 1991

Table 3.1, continued

Examples of Parasites of Brood

Parasites	*Host*	*Infection, Transmission*	*Remarks*	*References*
FUNGI, continued				
Beauveria bassiana	F: *Solenopsis richteri* V: *Vespula germanica*	Ingestion of contaminated food. Also via air in tracheal openings, wounds possible. Spread by mechanical contamination of nurses?	Infection primarily per os. Can cause 80% case mortality.	Broome et al. 1976; Glare et al. 1993
Penicillium cyclopium *Rhodotorula* sp.	A: *Apis mellifera*	Perhaps by os. Spores remain infective in brood intestine.	Unclear significance for bee populations.	Prest et al. 1974
PROTOZOA				
Mattesia (=Apicystis) geminata	F: *Solenopsis geminata, S. invicta, S. richteri*	Per os and contact does not produce infections.	Develops in hypodermis of larvae; sometimes dual infection with *Burenella*.	Jouvenaz and Anthony 1979; Jouvenaz 1983
Burenella dimorpha	F: *Solenopsis geminata*	Ingestion of contaminated pellet given by the nurse worker.	Only infects 4th instar larva. Dimorphic spores, only one type infective.	Jouvenaz et al. 1981
Gregarinidae	F: *Myrmecia pilosula*	?	Leads to color change of eclosing adults.	Crosland 1988
Vairimorpha invictae	F: *Solenopsis invicta*	Per os and contact does not produce infections.	Mature octospores in adults.	Jouvenaz and Ellis 1986

Table 3.1, continued

Examples of Parasites of Brood

Parasites	*Host*	*Infection, Transmission*	*Remarks*	*References*
HELMINTHS, NEMATODES				
Anomotaenia brevis (Cestoda)	F: *Leptothorax* spp.	Ingested with infected food in the larval stage.	Stays in host to adulthood. Produces aberrant color morphs.	Stuart and Alloway 1988
Pheromermis villosa (Nematoda)	F: *Lasius flavus, L. niger*	Acquired via infected food items.		Kaiser 1986
Mermithid nematodes	In many ants	Likely to directly penetrate the body wall of larvae. Also ingestion possible.	Larva has specialized stylets or other piercing organs. Stays in host to adulthood. Leaves host via anus to burrow into soil.	Bedding 1985
MITES, PARASITOIDS				
Varroa jacobsoni *Tropilaelaps clareae* (Acari)	A: *Apis mellifera, A. dorsata*	Mites prefer nurse workers as carriers. Spread by nectar-robbing and drifting.	Feeds on hemolymph of larvae. Removed by nurse workers if recognized.	Bailey and Ball 1991; Fuchs 1992
Eucharitidae (Hymenoptera)	F: Specialized on many ant species	Planidia larva attaches itself to foragers. Burrows into host or stays ectoparasitic.	Attacks larvae. Endo- or ectoparasitic. May kill host pupa. Oviposition onto plants. Dispersal with seeds possible.	Johnson 1988
Various Hymenoptera: Braconidae Chalcididae Diapriidae Eulophidae Ichneumonidae	F: In many ants, wasps, and bees	Usually egg directly deposited in or near nest.	Endoparasites of brood.	Nelson 1968; Strassmann 1981b; Wojcik 1989; MacFarlane et al. 1995; Whitfield and Cameron 1993

Table 3.1, continued

Examples of Parasites of Brood

Parasites	*Host*	*Infection, Transmission*	*Remarks*	*References*
MITES, PARASITOIDS, continued				
Apodicrania termilophila *Apocephalus aridus* (Phoridae)	F: *Solenopsis* spp., *Pheidole dentata*	*A. termilophila*: Fly walks around in nest and oviposits on or near larvae.	Female is ignored by ants. Adult females tended by ants.	Wojcik 1989
Volucella inanis (Syrphidae, Diptera)	V: *Paravespula vulgaris*	Eggs laid on external envelope of wasp nest. First instars seek host actively.	Ectoparasite. Feeds on hemolymph. Destroys pre-pupa, leaves, and buries in soil.	Rupp 1987
Megaselia sp. (Phoridae, Diptera)	V: *Mischocyttarus labiatus*	? Direct oviposition.	Attacks brood and is responsible for a large proportion of nest failures.	Litte 1981
Strepsiptera: Caenocholax fenyesi and other Myrmecolacidae	F: *Solenopsis invicta, Eciton, Camponotus,* and various genera	Triungulin larvae carried back, attached to host or collected with prey. Burrows into host (egg, larva).	In general: only males use ants. Endoparasitic. Emerges from adult host.	Kathirithamby and Johnston 1992
Xenos spp. (Stylopidae, Strepsiptera)	V: *Polistes* spp.	Burrows into host.	Endoparasitic; skewed sex ratios.	Dunkle 1979
Rhipiphoridae (Coleoptera)	V: *Dolichovespula, Paravespula*	Triungulin attaches itself to foraging worker.	Endo- and ectoparasites of wasps.	Clausen 1940a; Askew 1971

NOTES: Parasites that are transmitted by attendant workers are necessarily found in adults, too. The infection pathways refer to the most common pathway, according to current knowledge. Host groups are A: Apidae, F: Formicidae, V: Vespidae. For complete host records, see appendix 2.

Infection by Contact or Exposure

Contact and subsequent penetration is the typical mode of infection by fungal parasites. For example, *Beauveria bassiana* can infect larvae of the ant *Solenopsis richteri* by penetration via the cuticle, the tracheal openings, or wounds. Infection will eventually lead to death of the host (Broome et al. 1976). It is not reported how the parasite is transmitted further. It seems likely that spores released by the killed larva are spread to other larvae by workers or by the contamination of surfaces. Therefore, the major pathway is by contact. Overall, however, there appear to be relatively few parasites that rely exclusively on direct exposure or contact to infect brood.

Infection by Ingestion of Food or Cells

Entomopox viruses usually infect larvae of insects and develop slowly. It is thought that they normally enter per os (e.g., in bumblebees: Clark 1982; but see appendix 2.1). Also, sacbrood virus in the honeybee can be transmitted when contaminated food is ingested. In this case, however, the nurse workers are involved in special ways, as described below.

The most prominent bacterial infections of brood that can be acquired by simple ingestion are those already mentioned for the honeybee. Detailed studies with *Paenibacillus larvae* (American foulbrood) have shown that queen larvae are the most susceptible ones, followed by worker and drone larvae. Spores can also be passed by the queen to the attendant workers (Bitner et al. 1972). Interestingly, secondary infections seem unable to get established, perhaps due to some kind of antibiotics released by *P. larvae* as it sporulates (Holst 1945). Larvae get infected when placed into contaminated comb cells that contain spores, and sometimes also by direct contact with attending workers that are contaminated. In a classical study of behavioral genetics, Rothenbuhler (1964a) demonstrated that the hygienic behavior of bees, consisting of uncapping cells and removing infected larvae, prevents the disease from spreading and that these behavioral elements are heritable. A further defense mechanism is the attendant bee's proventriculus that filters out spores from contaminated nectar (Sturtevant and Revell 1953; see also chap. 7).

Melissococcus pluton (European foulbrood) is also ingested with food (Bailey and Ball 1991). In contrast to *P. larvae,* older larvae are also susceptible. Infected larvae sometime survive and, as the larval gut lining (the meconium) is shed at pupation, a batch of bacteria is discharged that can remain infective in the cell, perhaps for several years. Improper cleaning of cells may lead to new infections in the next larva housed in this cell. When the cells are cleaned by the young workers (the nurses), they are likely to pick up infective material and transmit it to other larvae when the larvae are fed. As workers clean out the brood cells and eject diseased larvae, a balance with the new infections can become established. An infection can thus persist in colonies for many years without major outward symptoms (Bailey and Ball 1991). An outbreak follows when larval ejection and cleaning can no longer keep up with the dissemination of bacteria. This may occur, for example, when many new larvae are produced and workers are needed for foraging rather than nest cleaning—a situation typical for spring conditions.

The spores of the fungus *Ascosphaera apis* (chalkbrood in honeybees) are known to occur on the outer surface of larvae. Infection is by ingestion. Spores are very persistent and can remain infective for up to 15 years (Bailey and Ball 1991). Infections are most effective when the larva is not kept at the usual brood temperature, e.g., when the comb is chilled at 30°C rather than the usual 35°C. This can occur when the colony is small and has many larvae, or in larvae located at the periphery of the brood area (as is typical for drone brood). Hence, the disease is most likely to break out when the colony is growing rapidly and larvae receive less attention (e.g., Heath 1982; Koenig et al. 1987). In contrast to susceptibility against *Paenibacillus larvae,* older larvae are more susceptible toward infection by *Ascosphaera.* The epidemiological consequences of this are unclear. But older larvae need more food and thus are presumably likely to be subject to a larger force of infection (i.e., the chance to get exposed to the parasite and to become infected; see chap. 6). As the larva is eventually killed by the parasite, the infective spores get spread by contamination of the nurse workers that remove the dead larvae (Bailey and Ball 1991, p. 58). As mentioned above, some fungi can penetrate the cuticle of the host, but the normal way of infection is by ingestion of spores. Examples are *Ascosphaera flavus* or *Aspergillus flavus* (stone brood) of the honeybee. The latter can also multiply and develop in adults in similar ways as in larvae (Bailey and Ball 1991).

Similarly, larvae of the ant *Solenopsis richteri* are infected per os by the fungus *Beauveria bassiana.* The tip of the hyphen then penetrates the gut wall within 60–72 hours postingestion and enters the hemocoel. Characteristically, the pH of the host hemolymph drops after penetration, but the significance of this is not really known. Further infection and spread may also occur through ingestion of spores with contaminated food or accidental feeding. *Myrmicinosporidium durum* is known from a variety of ant hosts in the families Formicinae and Myrmicinae (Sanchez-Pena et al. 1993; appendix 2.3). It presumably infects the larva by os and remains present into the adult stage. Contrary to a more typical entomophagous fungus, though, it is nonlethal and chronic, although it leads to a change in cuticle color with unknown consequences. *Myrmicinosporidium* may thus be illustrative for a range of parasites that cause few symptoms and are consequently hard to detect in the first place. Several other unidentified fungi of the fire ant are also contracted by ingestion of contaminated food (e.g., Jouvenaz 1983). In general, grooming of nestmates together with filtering by the infrabuccal cavity and necrophoretic behavior by the workers (i.e., removal of dead nestmates) serve as a means of defense against the spread of such pathogens.

Sporozoa (gregarines) and Microsporidia are protozoan parasites that typically infect young stages of the host, then remain and gain transmission from the adult. But exactly how general this particular life cycle is remains unclear for the social insects (see tables 3.1, 3.2). Furthermore, gregarine infections are likely to occur by ingestion of spores while further transmission is by discharge of infective propagules with the feces, but the evidence is often ambiguous. For example, *Mattesia geminata* infects larvae of *Solenopsis* spp. (appendix 2.4) and leads to aberrant pupae that never mature. But spore feeding or placing diseased pupae in a colony does not lead to novel infections (Jouvenaz 1986). *M. geminata* is restricted to immature hosts.

Two different infection strategies are illustrated by the microsporidium *Nosema* in *Apis* (e.g., Bailey and Ball 1991) and *Bombus* (e.g., Durrer and Schmid-Hempel 1995), respectively. In both cases, the infection results from the eating of spores. But while *N. apis* seems to be infective for adults only, experimental evidence for *N. bombi* shows that this species also infects larvae (Schmid-Hempel and Loosli, unpubl. data, in contrast to Van den Eijnde and Vette 1993). The development of the parasite (up to 2–3 weeks) is often slow in both cases (L'Arrivée 1965a; Schmid-Hempel and Loosli, unpubl. data). This is remarkable, because this time period is comparable to the expected life span of the adult host, which is around 4 weeks for both *Apis* and *Bombus* workers in the summer (Seeley 1985; Goldblatt and Fell 1987). *Nosema* must therefore infect the host as early (i.e., as young) as possible to be able to multiply sufficiently. In the honeybee with its ten-thousands of workers, young adults are numerous and available in high densities, so that transmission to new hosts is very likely. This is not the case in bumblebees, which have small colonies by comparison (a few dozen or hundreds of workers). The parasite will therefore not find young adults as readily. Hence, infecting the immature host may be an alternative that helps to gain time because larval mortality, at least after the initial stage(s), due to other causes is usually low in social insects. Such selective pressures could thus lead to the differences in the life cycles of *Nosema apis* and *N. bombi,* respectively, although this remains speculative for now. Nevertheless, the time until half of the infected bees have died is highly variable among colonies of honeybees, suggesting the potential for coevolutionary change (Sylvester and Rinderer 1978).

As discussed in chapter 2.5, mermithid nematodes typically invade host larvae and are still present in the adults. Their mode of infection is basically simple: the juvenile has to find a host and penetrate the body wall or become ingested when the host feeds on infected prey. The latter seems more likely in predatory ants and wasps (appendix 2.5, table 3.1). In fact, infections by rhabditid nematodes seem more prevalent in ant genera that are predatory and

Table 3.2

Examples of Parasites of Workers

Parasites	*Host*	*Infection, Transmission*	*Remarks*	*References*
VIRUSES				
APV (acute bee paralysis virus) SPV (slow bee paralysis virus) Deformed wing virus	A: *Apis mellifera*, *A. cerana*	Activated from infected tissues when mites pierce integument.	Vector/activator is the mite *Varroa jacobsoni*. Also found in adults. Multiplies better at 35°C than at 30°C.	Ball and Allen 1988; Bailey and Ball 1991; Ball, pers. comm.
BVX (bee virus X)	A: *Apis mellifera*	Ingestion of contaminated food. Not infective if injected.	Only when given in food at 30°C and not at 35°C. BVX associated with *Malpighamoeba mellifica*.	Bailey and Ball 1991
BVY (bee virus Y)	A: *Apis mellifera*	Ingestion of contaminated food. Not infective if injected.	Only when fed at 30°C; BVY associated with *Nosema apis*.	Bailey and Ball 1991
CBPV (chronic bee paralysis virus) CWV (cloudy wing virus) Kashmir bee virus	A: *Apis mellifera*	Airborne, via openings (trachea, wounds). Kashmir virus by direct contact (rubbing). Not transmitted via trophallaxis.	Wounds can occur with broken hairs.	Bailey and Ball 1991
Entomopox-like virus	A: *Bombus impatiens*, *B. pennsylvanicus*, *B. fervidus*	? Perhaps fed to larvae in pollen.	Virus found in salivary glands, sometimes in body wall and hemolymph. Unclear pathology.	Clark 1982

Table 3.2, continued

Examples of Parasites of Workers

Parasites	*Host*	*Infection, Transmission*	*Remarks*	*References*
VIRUSES, continued				
FV (filamentous virus) IV (iridescent virus)	A: *Apis mellifera* A: *Apis cerana*	?	In fat body and ovarian tissues; hemolymph becomes white. Associated with *Nosema apis*. IV in many tissues.	Ball and Allen 1988
BACTERIA				
Pseudomonas apiseptica= *Serratia marcescens?*	A: *Apis mellifera* V: *Vespula germanica*	Via trachea. Perhaps as aerosol.	Perhaps only transitory, opportunistic infection.	Bailey and Ball 1991; Glare et al. 1993
Spiroplasma melliferum *S. apis* (a mollicute)	A: *Apis mellifera, Bombus impatiens, B. pennsylvanicus*	Ingestion of contaminated food.	Found in hemocoel and gut. Infected bees become moribund and eventually die within. May be transmitted via flowers in spring. Optimal growth at 32–35°C in media.	Clark 1977; Clark et al. 1985
FUNGI				
Alternaria tenuis	F: *Formica rufa*	Contact with diseased individuals through hygienic behavior. Feeding on diseased workers.	Workers search and remove diseased nestmates. Behavioral change.	Marikovsky 1962
Beauveria bassiana	A: *Bombus pratorum* F: *Solenopsis richteri* T: *Reticulitermes*	Through integument and also by ingestion.		Leatherdale 1970; Broome et al. 1976; Kramm and West 1982

Table 3.2, continued

Examples of Parasites of Workers

Parasites	*Host*	*Infection, Transmission*	*Remarks*	*References*
FUNGI, continued				
Cordyceps sphecocephala *Paecilomyces farinosus*	V: *Vespa crabro, Vespula sylvestris*	Contact? Through necrophoric behavior.	Fungal parasites with unclear effects and biology. Most seem not to be host-specific and occur in many nonsocial insects. Behavioral change.	Leatherdale 1970
Cordyceps spp. *C. lloydi* *Hirsutella formicarum* *H. sporodochialis* *H. acerosa* *Desmidiospora myrmecophila*	F: *Cephalotes atratus, Camponotus abdominalis, C. sericeiventris, C. unliateralis, C. pennsylvanicus, Polyrhachis* sp., *Dolichoderus attelaboides, Monacis bispinosa, Pachycondyla* spp., and many other tropical ants	? Ingestion, feeding. Contact with diseased individuals as a consequence of necrophoric behavior. Also likely by scavenging (e.g., in *Solenopsis*).	A large complex of ant-fungus interactions. Behavioral change. Length and complexity of spore-carrying trunks correlates with burrowing habit of infected hosts.	Evans and Samson 1982, 1984; Evans 1982; Samson et al. 1982, 1988
Erynia spp. *Entomophthora* spp.	F: *Formica, Serviformica*	Contact. Penetration through integument.	Behavioral change.	Loos-Frank and Zimmermann 1976; Evans 1989
Metarhizium anisopliae	T: *Cryptotermes brevis*	Ingestion.	Kills host rapidly.	Kramm et al. 1982; Kaschef and Abou-Zeid 1987

Table 3.2, continued

Examples of Parasites of Workers

Parasites	*Host*	*Infection, Transmission*	*Remarks*	*References*
PROTOZOA				
Apicystis (= Mattesia) bombi *A. geminata*	A: *Bombus* spp. F: *Solenopsis*	Likely by ingestion of spores. In the larval stage? *A. geminata*: Experimental infections not successful.	Cysts in fat body, spores detectable in feces.	Jouvenaz and Anthony 1979; MacFarlane et al. 1995; R. Schmid-Hempel, pers. obs.
Crithidia bombi (Trypanosomatidae)	A: *Bombus* spp.	Ingestion of spores. From contaminated surfaces.	Widespread infection. Suppresses early colony growth, ovaries, and delays reproduction.	Shykoff and Schmid-Hempel 1992; Schmid-Hempel, pers. obs.
Gregarine sp.	F: *Myrmecia pilosula*	Likely by ingestion of spores in the larval stage.	Present in entire body cavity of adults.	Crosland 1988
Leidyana spp.	A: *Apis mellifera*	? Perhaps by ingestion of spores.	In lumen of midgut attached to epithelium. Usually little harm done.	Bailey and Ball 1991
Malpighamoeba mellifica	A: *Apis mellifera*	Ingestion of cysts.	In Malphigian tubules. Cysts pass out via feces some 3 weeks after infection. Associated with *N. apis*.	Hassanein 1952; Bailey and Ball 1991
Nosema apis (Microsporidia) *N. bombi*	A: *Apis mellifera, Bombus* spp.	Ingestion of spores with contaminated food.	*N. apis* associated with FV (filamentous virus). *N. bombi* also in larvae.	Bailey and Ball 1991; Varis et al. 1992; Schmid-Hempel and Loosli, in prep.

Table 3.2, continued

Examples of Parasites of Workers

Parasites	*Host*	*Infection, Transmission*	*Remarks*	*References*
PROTOZOA, continued				
Thelohania solenopsae	F: *Solenopsis invicta, S. richteri, S. quinquecuspis*	? Not infective by os or contact. Perhaps transovarial transmission.	In adipose tissue of workers and sexuals. In ovaries of queen. Dimorphic spores. Also infects brood. Infected colonies smaller than normal. Workers show less vigor and pursuit when disturbed. Perhaps important for population regulation.	Allen and Buren 1974; Knell et al. 1977; Jouvenaz 1983
HELMINTHS, NEMATODES				
Agamomermis pachysoma (Nematoda)	V: *Vespula vulgaris*	Ingestion of eggs? Perhaps penetration by juveniles?	Eggs, larvae brought into nest by foragers.	Bedding 1985
Brachylecithum mosquensis, Dicrocoelium dendriticum (Trematoda)	F: *Camponotus pennsylvanicus, C. herculaneus, Formica* spp.	Ingestion of cercaria, expelled by intermediate host.	Metacercaria infect gaster. Some migrate to brain and cause behavioral changes.	Hohorst and Graefe 1961; Carney 1969; Schneider and Hohorst 1971
Mermis subnigrescens, Agamomermis sp. (Nematoda)	A: *Apis mellifera*	Perhaps taken up with contaminated water from foliage.	Nematodes deposit eggs on foliage.	Bedding 1985
Rhabditid nematodes	F: In many ants	Attachment of larva to ant? Ingestion of contaminated food?	Often migrate to pharyngeal glands. Can also survive and develop outside ants.	Welch 1965; Bedding 1985

Table 3.2, continued

Examples of Parasites of Workers

Parasites	*Host*	*Infection, Transmission*	*Remarks*	*References*
HELMINTHS, NEMATODES, continued				
Sphaerularia bombi (Nematode)	A: *Bombus* spp.	Direct penetration by juveniles.	Only infects overwintering queens.	Alford 1969
MITES, PARASITOIDS				
Kuzinia, Parasitus (Parasitellus), Scutacarus (and many other mites)	A: *Bombus* spp., and many other bees and ants	Phoretic stage is vectored by adults.	Often with minor effect on host. But high prevalences in some cases. May act as vectors of disease among colonies.	Goldblatt 1984; Eickwort 1990, 1994
Syntretus splendidus (Braconidae)	A: *Bombus* spp.	Oviposition.	Polyembryonic development. One host can contain very many larvae.	Alford 1968; Schmid-Hempel et al. 1990
Conopidae (Diptera)	A: *Bombus* spp., *Apis* spp.	Oviposition.	Egg inserted into abdomen, often in flight.	MacFarlane and Pengelly 1974; Schmid-Hempel et al. 1990
Phoridae (Diptera)	Many ant species	Egg is deposited on host, or inserted.	At least 17 species of phorids known in the fire ant species complex.	Wojcik 1989; Pesquero et al. 1993
Sarcophagidae (Diptera)	A: *Apis* spp., *Bombus* spp.	Deposition of larva on host (larviparous).	Often attacks on flowers.	Pouvreau 1974

NOTES: Host groups are A: Apidae, F: Formicidae, V: Vespidae, T: Termitidae. Some parasites are also found in immature hosts.

scavenging, such as *Camponotus,* which often come in contact with food items on which the nematodes reproduce (Wahab 1962; Bedding 1985). Also, many helminth parasites (i.e., trematodes, cestodes) infect larvae when those are fed with infected prey. For example, *Anomotaenia brevis* (Cestoda) is known to infect the larval stage of the ant *Leptothorax* (Stuart and Alloway 1988). Later, parasitism leads to aberrant coloration of the adult.

Infection Related to Brood Care

An important difference in this context is whether the parasite is transmitted to the brood simply by "passive" contamination of attendant workers, in the handling of food, or through an "active" involvement of nurses (table 3.1). The latter could involve gland secretions or discharge of pellets. In this active case, an accelerated turnover of the worker population or a decrease in task specialization are of consideration and would reduce the force of infection to the brood as the contact rate decreases. This is not the case (at least not to the same degree) when the infection, for example, occurs through contaminated food (see also chap. 4).

Indeed, a major problem of brood microparasites is how they get transmitted to new hosts of the appropriate stage once development is successfully completed. As an example, consider again sacbrood virus in the honeybee (chap. 2.1; appendix 2.1) (Bailey and Ball 1991). It multiplies in tissues of young larvae. The host larva produces huge numbers of new viruses that can potentially infect hundreds of new colonies. Nurse bees routinely remove infected larvae. Therefore, the infection in any one colony usually remains light. At the same time, however, the nurses and young workers inside the nest can also become infected via the ecdysial fluids that the dying larvae release when manipulated. Viruses also accumulate in the hypopharyngeal glands of workers within days without causing symptoms. Then, the gland secretions are fed to other larvae during normal attendance and thus produce new infections (Bailey 1967c, 1969).

Fortunately, workers are not such efficient vectors for sacbrood virus as one might expect, because they change their behavior. Infected honeybees cease to eat pollen and soon stop caring for larvae. Instead, they will start to forage earlier in their life than noninfected bees. But as a further twist to the story, by changing their behavior in this way, honeybee workers will sometimes collect pollen. Then many virus particles find their way into the pollen load as the infected worker adds her secretions (a typical behavior of pollen-collecting bees; Bailey and Fernando 1972). But Clark (1982), interested in the potential cross-infectivity from wild bee species, could not infect larvae of honeybees with entomopox-like viruses collected from bumblebees, because the particles (pollen and the viruses contained inside) were removed efficiently by the nurses.

In the development of holometabolic insects (as in the social Hymenoptera),

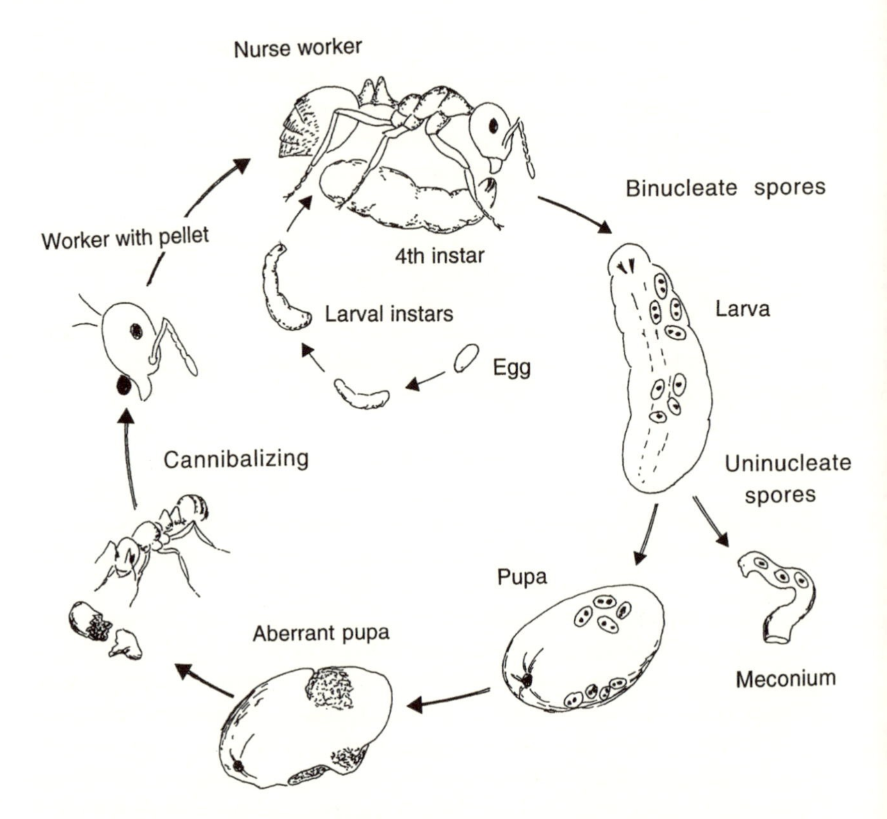

Figure 3.1 The life cycle of the microsporidium *Burenella dimorpha* in its host, the fire ant *Solenopsis* sp. The spores are fed to a fourth instar larva by a nurse worker (*top*). The microsporidia produces two types of spores: binucleate, non-membrane-bound spores, and uninucleate, membrane-bound spores. Non-membrane-bound spores stay in the pupa and accumulate in the proventriculus of a cannibalizing nurse workers from where they are fed as a pellet to the next larva. Membrane-bound spores are shed with the meconium (the larval gut) at pupation of the host. (After Jouvenaz and Hazard 1978.)

the lining of the gut is shed at pupation, while in the hemimetabolic pathway (the termites), a process of self-sterilization occurs before the adult stage (Krieg 1987). In either case, microparasitic infections of the alimentary tract cannot persist from the larva to the adult stage. In these cases, and in particular when immature stages are immobile, the infection chain must be maintained by adults that tend and feed larvae. *Burenella dimorpha,* when infecting fire ants, provides a good example for a parasite strategy that depends explicitly on social brood care (fig. 3.1). As already described in chapter 2.4, *B. dimorpha* develops two different kinds of spores: the binucleate, nonmembrane-bound type develops in the hypodermis of the host (the larva and pupa) and leads to aberrant pupae. The nurse workers cannibalize them and accumulate the spores in their infrabuccal cavity. This pellet is later fed back to the fourth instar larvae, leading to a novel infection. This larval stage is the only one that receives solid food. The uninucleate membrane-bound octospores, in contrast, are shed with the meconium and are not infective when fed to the larvae.

For many pathogenic bacteria in holometabolic (nonsocial) insects, only the larvae are susceptible, while eggs and pupae are refractory. Adults usually tolerate the infection and thus serve as a major reservoir (Krieg 1987). For bacteria this basic pattern is a suitable preadaptation to use social insect hosts. In addition, adults are concentrated in relatively high numbers in the nest and regularly visit their larvae so that transmission of parasites is readily possible. Because adults are tolerant to the parasite, this pattern is an example of an "active" nurse-to-larva transmission because it is not just a simple contamination of the attending workers.

Infection by Actively Seeking Parasite Stages

Many mermithid nematodes have specialized structures to pierce a host's cuticle. Hence, they can infect their target by search and penetration of the body wall. But although nematodes are mobile, their range of movements is naturally quite small: Ishibashi and Kondo (1990) estimate 245 μ/min for *Steinernema carpocapsae*. Therefore, infection requires the host insect to be close. For example, ants that are affected by nematodes have in common that they nest underground in damp soil (Wahab 1962). Unmated female nematodes can in fact survive in such soil for several years. In this way, long-lasting stages will occasionally encounter a new host and thus infect future host generations (Bedding 1985). As a consequence, nematodes are not expected to be prevalent everywhere but are common locally and over restricted periods of time during the year. They may also be common during some years and otherwise persist in the soil as long-lasting stages.

Macroparasites, in general, have the problem that the female, the larva, or any other appropriate stage must actively seek out a host. This is nicely illustrated by the parasitic Strepsiptera. The first larval instars of these insects, the triungulins, possess well-developed eyes and legs. They leave the body of the female (Strepsiptera give "birth" to larvae), seek out and enter a host larva (observations of the entire life cycle are difficult; thus, this pattern is at least strongly suspected as true). The biology of *Caenocholax fenyesi* illustrates a number of these characteristics. Its development occurs via various instars alongside the host's development into an adult. Sexual maturity of the parasite is attained in the adult host worker or in a sexual (drone or queen). In social Hymenoptera, the parasite normally seems to be in its second instar when the host undergoes its final molt (Cheng 1986). Although unknown, it is likely that endocrine synchronization modulates these relationships between host and parasite. As is typical for Strepsiptera, the parasitic males of *C. fenyesi* extrude the head and cephalotheca through the host cuticle of the adult worker or sexual host ant (which leads to "stylopization"). At this time, the larval parasite lives in the adult host's hemocoel and may or may not kill it when the free-living parasite males emerge. In fact, in some cases, before emerging, the males break the larval protruding cap, but the hole is closed again after the parasite has left, perhaps to allow the development of their brothers in superparasitized hosts (Kathirithamby 1991). The emerged parasitic larvae of both sexes, the triungulins, disperse in the neighborhood of the affected ant nest. To find a suitable host later, the triungulins must be carried back by the ants to the nest. Often this is accomplished by foragers of the same nest from where the female came. But the larvae may also be picked up by workers of a neighboring colony and thus become transferred horizontally to another colony. Within the nest, the larva is dispersed by the help of nurse workers. In the heteroxenic species, the emerged female larvae must seek out a different host., e.g., a grasshopper or preying mantis. How this feat is achieved is largely unknown (e.g., Waloff and Jervis 1987). To complete the story, a prey item (e.g., a grasshopper) already stylopized by a female parasite may be collected by ant foragers and brought into the nest. While Strepsiptera normally attack the larval stages of the host, Kathirithamby and Johnston (1992) report that *Caenocholax fenyesi* attacks eggs of *Solenopsis invicta* in Texas.

The triungulins of Strepsiptera illustrate a strategy that is also successfully employed by two other, unrelated groups and therefore must have arisen independently. In fact, the morphology of first instar larvae of Encharitidae (Hymenoptera) and Rhipiphoridae (Coleoptera) shows remarkable convergences with Strepsiptera (Godfray 1994). Similar forms are also known from some Ichneumonidae and Perilampidae (Hymenoptera), and from the dipteran families Acroceridae, Tachinidae, and Sarcophagidae (Johnson 1988). In all of these groups, the first instar larvae are specialized to actively seek out hosts. The predatory habit of ants or wasps, in particular, makes it more likely that triungulins or stylopized insects are recognized as prey items and brought back to the nest, from where they can invade the host brood.

In the eucharitid wasps, a group specialized on parasitizing ants, the first instar is termed a planidia (see Johnson 1988). It hatches from an egg deposited in or on a plant by the female wasp, e.g., on *Plantago major.* In *Schizaspidia (Stilbula) tenuicornis* the eggs hibernate in buds of plants. Clausen (1923) counted up to 7200 eggs per bud. In spring, eggs from open buds fall to the ground and perish. Only when buds remain closed for a while will larvae hatch and develop into planidia. Quite often, the time when these buds open coincides with a time when the plant has many aphids that are tended by ants. This increases the chances of encounter and close contact. Planidia can attach themselves to foraging ants and thus get carried back to the nest. In perennial social insects, the female eucharitid wasp may also overwinter as an internal parasite of prepupae and then complete its development as an ectoparasite at pupation of the host in spring (e.g., *Pseudometageae schwarzii* in the ant *Lasius neoniger;* Ayre 1962). Once inside the nest, the planidia normally invades a host larva and develops internally. In some cases, though, the planidium remains ectoparasitic on the ant larva until it pupates, or it may switch from endo- to ectoparasitism as the host develops (e.g., *Pseudometageae* above, or *Orasema* on *Solenopsis* sp.; Johnson 1988). At the later stages, the parasite starts to consume the host and quickly develops into an adult (Heraty 1985). The host pupa can be killed, or the adult ant emerges with morphological deformations. Because the ants cannot distinguish the planidia from their own brood, Vander Meer et al. (1989) suspect that they are able to chemically "hide" and thus exploit attendance and feedings. In fact, the parasite obviously acquires the colony odor passively but can later also produce its own compounds. The adult parasites mate and look for oviposition sites, some close to their host nest. The female dies within a few days. But the deposited eggs overwinter on a plant and thus infect the next year's generation of hosts (Clausen 1940b). Eucharitid parasitism seems to be local, which is not surprising given this lifestyle.

The biology of rhipiphorid beetles, which are parasites of social wasps, shows remarkable convergence (*Metoecus* spp.; appendix 2.11). Their triungulins also attach themselves to foraging workers and thus gain access to the nest. For example, the triungulins of *Retoecus paradoxus* have long prehensile legs with suckers at the end of the tibiae plus a large paired sucker on the terminal segment of the abdomen. The triungulin "stands up" on the terminal sucker and tries to grasp a passing forager. After arrival in the nest, triangulins drop off, search for a wasp larva, and bury themselves into the host abdomen between the second and third thoracic segment (Spradbery 1973). The beetle larvae are first endoparasitic on host larvae and later become ectoparasitic (Clausen 1940a; Askew 1971). The switch from endo- to ectoparasitic rather than vice versa has therefore evolved repeatedly in independent groups of macroparasites. The general pattern of invasion is also similar in the beetle *Melöe* attacking social bees (*Apis, Bombus*). Bailey and Ball (1991) report that eggs of beetles are oviposited in the soil, from where larvae hatch. These triungulins climb on vegetation and wait for visiting bees, which they grasp. In the bee nest, they burrow themselves into the host near the joint between the thorax and abdomen.

Vectored or Activated Infections

Mites are mobile and possess dispersing stages. These can be the deuteronymphs (such as in *Parasitellus* and *Kuzinia* on honeybees and bumblebees) or the females themselves (as in *Hypoaspis, Scutacarus* on other bee groups; appendix 2.7). These stages are carried by foragers, queens, or males to a new host colony and thus are dispersed by a vector (Eickwort 1990). Often, only the dispersal stage is found on the host while the other stages live in the nest itself. In tracheal mites (e.g., *Acarapis woodi* in honeybees or *Locastacarus buchneri* in bumblebees), all developmental stages are on the host and these mites do not use vectors for dispersal. Dispersal in these groups presumably occurs through larviform females that leave the tracheal system of their host to find another bee in the hive (Pettis et al. 1992). Horizontal transfer is probably achieved through queens or workers that drift to other colonies. In fact, mites are very successful as brood parasites (appendixes 2.7 and 2.8).

Varroa jacobsoni in *Apis mellifera* parasitizes both the worker and drone brood. When the host cell is unsealed, the young mites leave their natal site and disperse within the colony. In choice experiments, the free *Varroa* prefers to attach itself to young workers (less than 14 days old) where they reside under abdominal sternites. On older workers, the mites are found between the thorax and abdomen. Clearly, these preferences facilitate the finding of new brood cells via transfer on (young) nurse workers, while it allows the spread to other parts of the nest or other colonies via old workers (Kraus et al. 1986). As might be expected, the spread of *Varroa jacobsoni* within the colony is facilitated when the colony is prevented from carrying out its normal activities so that the density of individual bees within the nest is high (De Jong et al. 1984). This is characteristic for the spread of other parasites too (Bailey and Ball 1991). Fuchs (1992) has modeled the choice behavior of mites, whether they will infect drone or worker brood cells. He found that this decision depends on mite density. Not surprisingly, the generalist strategy to infest both types is preferred under high infestation intensities.

Interestingly, colonies infested by *Varroa* seem to have more drifting workers, i.e., workers that "erroneously" enter foreign nests (Sakofski 1990). At the causal level, this is likely to be related to the bad condition of infected bees. But in terms of mite transfer between colonies, drifting should favor horizontal transmission of mites to other colonies. Nevertheless, robbing behavior is probably the major quantitative factor for this kind of transmission. Nectar robbing is in fact quite frequent among neighboring colonies of bees (Sakofski and Koeniger 1988). After 2 hours of watching robbing activity, Sakofski (1990) found that 14 ± 6% ($N=7$ colonies) of all mites were transferred; after 24 hours this proportion had increased to about one-third ($N=9$ colonies). High densities of colonies consequently favor the spread of the mites (Ritter 1988). Besides robbing, absconding and swarming have also been invoked as explanations for the rapid spread of mites, such as observed for the tracheal mite *Acarapis woodi* (infecting adults) in Mexican honeybee populations (Eischen et al. 1990). Komeili and Ambrose (1990) found that bees from mite-free colonies in apiaries in North America became infested within 5–30 months via infested source colonies nearby. De Jong et al. (1982) report a natural advance of *Varroa* infestations of 3 km per year, with maxima of 6–11 km within 3 months.

The Southeast Asian honeybee *Apis cerana* is the natural host of *Varroa jacobsoni*. In this host, the mites can only reproduce in drone cells because they are removed by the attending workers from worker cells. In fact, mites are also removed from the bodies of nestmates. In addition, mites experimentally introduced from *A. mellifera* hives are readily recognized and removed within 6 days

(Koeniger et al. 1981; Ritter and De Jong 1984; Rath and Drescher 1990; Tewarson et al. 1992). Therefore, although most colonies are infested, the overall damage is small. Similarly, nurse workers are to some extent capable of recognizing a brood cell infected with *Tropilaelaps clareae,* a mite parasite of *A. mellifera* and *A. dorsata.* Nurses then uncap and at least partly remove the affected pupae. Furthermore, in *A. mellifera* the ability to remove mites from infested cells is known to be variable among bee lines (Boecking and Drescher 1991, 1992), although there is some conflicting evidence as to the significance of race as compared to other factors (e.g., climatic conditions: Moretto et al. 1991). Overall, *A. cerana* can better cope with *Varroa jacobsoni,* while *Tropilaelaps* is better controlled with *A. mellifera* and *A. dorsata* (Burgett et al. 1990).

How mites manage to be brood parasites in taxa other than the honeybee is not well known. Many mites are certainly just phoretic or commensalistic but not true parasites (e.g., Eickwort 1994; appendixes 2.7, 2.8). It seems obvious that two problems have to be solved by the mites: to get transferred to brood and to get transferred between colonies. As we have just seen, robbing behavior aids in the latter. There is also some evidence that bee mites may be transferred to new colonies via flowers (Schwarz et al. 1996; Schwarz and Huck 1997).

Several rare and apparently opportunistic infections of brood that illustrate the "activation principle", are known from the honeybee (Bailey and Ball 1991), i.e., that the parasite develops and becomes established or cause damage only after some external events occur. For example, the bacterium *Hafnia alvei* (= *Bacillus paratyphi alvei*) must be transferred from the contaminated outer surface into the hemolymph. This happens when a mite (*Varroa* in this case) pierces the body wall, thereby transferring the parasite by contamination. Acute bee paralysis virus is also activated and released from infected tissue when the tissue is pierced by a mite (Ball and Allen 1988). Activation is also necessary for some infections in adults (see below).

Direct Oviposition by the Parasite Female

Social insects often have sheltered combs that are protected in cavities and by envelopes. As was just discussed, strepsipteran, eucharitid, and rhipiphorid larvae use sophisticated strategies to break into this fortress. But there are more direct intrusions. As an example, wax moths could be mentioned. The females more or less directly oviposit into or near the nest. Similarly, syrphid flies hover in front of a nest entrance and flip eggs into the colony. The emerging larvae are mostly detrivorous, however, and do not seem to cause real harm.

Although it is simple in principle, the "direct-access" method must involve sophisticated means of coping with outward nest defense, specialized behavior, or chemical communication of hosts. According to the literature, this has been achieved mainly by lepidopteran, hymenopteran, and coleopteran parasites. There are also a number of examples for diptera (appendix 2.10): *Apodicrania* sp. and *Apocephalus aridus* (Phoridae, Diptera) are proven endoparasites of ant larvae in *Solenopsis* and *Pheidole,* respectively (Jouvenaz 1983; Wojcik 1989); *Helosciomyza* (Helosciomyzidae, Diptera) feeds on larvae in the ants *Chelaner* and *Prolasius. Aenigmatis* (Phoridae) eggs are placed on larvae of the wood ant *Formica. Brachicoma* (Sarcophagidae, Diptera) attacks the brood of bumblebees, *Bombus.* The phorid *Megaselia* seems to attack the brood of social wasps (*Mischocyttarus,* Litte 1981), and *Sarcophaga* (Sarcophagidae) that of *Polistes; Koralliomyia* (Tachinidae, Diptera) attacks that of other wasps, such as *Ropalidia, Parischnogaster,* and *Liostenogaster. Volucella inanis* (and other *Volucella* species; Syrphidae, Diptera) are parasites of larval *Paravespula vulgaris.* Despite these examples, the direct access strategy is generally not common for dipteran flies (cf. table 3.3). Obviously, the brood of social wasps, particularly of species with combs in the open like *Polistes or Microstigmus,* is more accessible and seems to be attacked in a direct way by a wider variety of parasitoids (table 3.3; chap. 2.8; Appendixes 2.9–2.11).

Transovarial Infection

A special category of egg parasites are those being transmitted transovarially, i.e., via the ovaria of mothers to her eggs and offspring (a case of 'vertical' transmission). Acute bee paralysis virus could be transmitted this way, but this has not been demonstrated conclusively (Bailey and Gibbs 1964). Given the biology of social insects where typically only a few individuals in a colony actually produce eggs to reproduce, the problem must be a formidable one for any parasite, simply because these opportunities are rare. For the same reasons, therefore, we should expect that vertical, transovarial transmission is more likely to have evolved in host species with polygynous nests or with a large proportion of reproducing workers.

Thelohania solenopsae is a larval infection of fire ants, some of them polygynous, that may lead to host death at the pupal stage (table 3.2; chap. 2.4). It is a dimorphic species that produces different spore morphs with as yet unclear functions. Knell et al. (1977) report that *Thelohania* is transmitted transovarially. Other attempts to infect larvae, e.g., spore feeding, have not been successful. *Vairimorpha invictae* is also a dimorphic microsporidium found in fire ants. The infection persists from the larva into the adult. Mature octospores are found in head, petiole, and gaster. Again, attempts to infect by feeding spores or placing diseased individuals in a colony did not produce novel infections (Jouvenaz and Ellis 1986). Thus, natural infection in these cases may perhaps occur via vertical transmission. But so far, the exact transmission and infection pathways remain enigmatic, although it is likely that ingestion of spores is sometimes necessary. Given the fact that microsporidial diseases are widespread, though not always very common, one nevertheless wonders how these parasites are maintained in a host population.

Table 3.3

A Synopsis of Parasitoid Genera of Brood (Including Pupae) and Adults for Hymenopteran and Dipteran Species

Parasitoids of brood				
Hymenoptera				Diptera
Bethylidae	*Mimopria*	*Rhipipallus*	*Polistophaga*	Heliosciomyzidae
Pseudisobrachium	*Myrmecopria*	*Schizaspidia*	*Sphecophaga*	*Helioscyomyza*
Braconidae	*Phaenopria*	*(Stilbula)*	*(Anomalon)*	Sarcophagidae
Heterospilus[a]	*Plagiopria*	*Tricoryna*	*Toechorychus*	*Brachicoma*
Chalcididae	*Solenopsia?*	Eulophidae	Mutilidae	*Sarcophaga*
Dibrachys	*Tetramopria*	*Alucha*	*Dasymutilla*	Syrphidae
Elasmus	*Trichopria*	*Brachicoma*	*Mutilla*	*Volucella*
Nasonia	Eucharitidae	*Elasmus*	*Pychotilla*	Phoridae
Trichokaleva	*Austeucharis*	*Melittobia*	*Tropidotilla*	*Aenigmatis*
Chrysididae	*Chalcura*	*Pediobius*	Pteromalidae	*Apocephalus*
Chrysis	*Epimetagea*	*Tetrastichus*	*Pheidoloxenus*	*Apodicrania*
Diapriidae	*Eucharis*	Ichneumonidae	Torymidae	*Borophaga*
Ashmeadopria	*Eucharomorpha*	*Arthula*	*Monodontomerus*	*Dicranoptera*[b]
Basalys	*Isomerala*	*Camptotypus*	Trigonalidae	*Megaselia?*
Bruchopria	*Kapala*	*Christelia*	*Bakeronymus*	*Palpiclavina*[b]
Bruesopria?	*Obeza*	*Ephialtus*	*Poecilogonalus*	*Pseudohypocera*
Geodiapria	*Orasema*	*(Hemipimpla)*	*Pseudonomadina*	Tachinidae
Gymnopria	*Pheidoloxeton*	*Latibulus*	*Seminota*	*Anacamptomyia*
Hemilexis	*Pseudochalcura*	*Mesostenus*	Braulidae	*Koralliomyia*
Lepidopria	*Pseudometagea*	*Pachysomoides*	*Braula*	
Loxotropa	*Psilogaster*	*(Sphecophaga)*	*Megabraula*	

Table 3.3, continued

A Synopsis of Parasitoid Genera of Brood (Including Pupae) and Adults for Hymenopteran and Dipteran Species

PARASITOIDS OF ADULTS			
Hymenoptera	Diptera		
Braconidae	Conopidae	*Cataclinusa*	*Pseudaceton*
Elasmosoma	*Conops*	*Ceratoconus?*	*Pseudohypocera?*
Hybrizon	*Dalmannia*	*Cremersia*	*Puliciphora*
Neomeurus	*Leopoldius*	*Dacnophora*	*Rhynchophoromyia*
Syntretus	*Myopa*	*Diocophora*	*Stenoneurellis*
Diapriidae	*Physocephala*	*Diplonerva*	*Styletta?*
Mimopria	*Sicus*	*Gymnoptera*	*Thalloptera*
Eucharitidae	*Thecophora*	*Iridophora*	*Thaumatoxena*
Orasema?	*Zodion*	*Megaselia*	*Triphlesta*
Pheidoloxenus?	Drosophilidae	*Melaloncha*	*Trucidiphora*
Tiphiidae	*Drosophila*	*Menozziola*	Sarcophagidae
Myrmosa?	Ephydridae	*Microselia*	*Boettcharia*
	Rhynchopsilopa	*Misotermes*	*Sarcophaga*
	Phoridae	*Myrmosciarius*	*Senotainia*
	Acanthophorides	*Neodohrniphora*	Tachinidae
	Apocephalus	*Plastophora*	*Rondanioestrus*
	Apterophtora?	*Plastophorides*	*Strongylogaster*
	Auxanommatidia	*Pradea*	*(Tamiclea)*
	Borgmeierphora?	*Procliniella*	

[a]On the sphecid wasp *Microstigmus*.

[b]In nymphs of termites.

Based on references given in appendixes 2.9 and 2.11. Cases where status is uncertain are marked with a question mark. Excluding these cases, the distribution of genera is not random with respect to parasite order and type of host attacked. (For the confirmed cases: $X^2 = 55.66$, $P < 0.0001$, $N = 125$ genera). There are few genera that attack both adults and brood.

Some Conclusions

One can draw some general conclusions from the cases discussed here (cf. table 3.1; appendix 2). For example, with the highly evolved brood care behavior in social insects, the larval stages are usually quite helpless and immobile, dependent on being fed and cared for by attendant workers. This opens the possibility for nurse-to-larva transmission of parasites. The parasite must, for example, be able to infect adults and to reach the organs from where it gets back to the larvae (the pharyngeal glands, as in viruses). Alternatively, the parasite could enter the food chain and become fed to larvae. In this case, infected food should also not be recognized by the feeding workers. These feats have been achieved by a number of parasitic groups.

A naive expectation is that large parasites, i.e., macroparasites such as nematodes or cestodes that generally need more time to develop as compared to microparasites like viruses or protozoa, might have evolved to infect larvae rather than adults. This would offer the advantage of developing over longer time spans along with the host's own development. But even a short look at the evidence suggests that there is no such relationship between the type of parasite and how it is infected/transmitted. Quite to the contrary, macroparasites seem rare in brood. Perhaps it is more difficult for them to pass through the hosts' metamorphosis, or perhaps they simply have better ways of getting from adults to larvae in due time (see also chap. 4). In this context, it is interesting to note that the nematode *Hexamermis,* after leaving its adult ant host, passes through an intermediate host that is then fed to a larva—the final host—as a prey item (Bedding 1985; Kaiser 1986). Hence, the parasite uses another host to grow and complete its development before returning to the ant. In fact, social insects, such as sedentary ants, should be quite suitable hosts for this kind of parasitic strategy because the life cycle of the parasite can be completed in the vicinity of the nest. The intense patrolling and food-gathering activity of ants will furthermore ensure a high probability of contact with the intermediate host of the parasite. Absconding in social bees, in contrast, will leave sedentary parasites behind, often those that infect brood.

3.2 Attacking the Workers

In a nest of social insects, workers are more numerous relative to any other class of individuals. They show wide-ranging activities and therefore have many occasions to have contact with parasites— specially since they routinely leave the sheltered environment of the nest. It is therefore not surprising to find that workers are attacked by many different parasites and that the prevalence and intensity of infections can be very high. But parasites that attack workers also must overcome a number of difficulties. As will be discussed in more detail later (chaps. 4 and 7), such defenses range from behavioral mechanisms to chemical defense and the immune system of insects. Parasites that are acquired at the larval stage but persist to the adult host—quite a large group (table 3.1)— were mentioned in the preceding section (chap. 3.1).

Infection by Contact or Exposure

Prime examples of this route of infection are parasites that can infect a host through body openings. This is likely for chronic bee paralysis virus and cloudy wing virus' which are probably airborne and infect honeybees over short distances through lesions in the cuticle or tracheal openings. Hence, crowded conditions are favorable to the intensive transmission of the virus (Bailey and Ball 1991, p. 13). So far, only one bacterial infection is suspected to be directly infective through tracheal openings: *Pseudomonas apiseptica* (perhaps identical to *Serratia marcescens*). This is somewhat surprising, because bacteria must be everywhere in and around the nest, and it is inevitable that workers become exposed to many bacterial cells in the course of their daily activities.

The microsporidium *Gurleya spraguei* infects the adipose tissue of workers of the termite *Macrotermes estherae* in India (Kalavati 1976). The parasites are not transmitted via the feces and hence presumably not ingested by os. Instead, if the termite is heavily infected, the spores are released into the body cavity. Thus, spore dissemination from the decaying host body and direct contact are a likely kind of transmission, although it is unclear how this works.

Direct contact with spores and subsequent penetration of the cuticle occurs in fungal infections. Workers of *Formica rufa,* for example, are infected by *Alternaria tennuis* (Marikovsky 1962). They show no overt signs of infection until shortly before their death. But, as mentioned in chapter 2.3, at some stage after infection the workers start to climb on grasses, from where the fungal spores are eventually spread from the dead host's body. This suggests that *Alternaria* may be infective as airborne spores that can be acquired by direct contact. For *Cordyceps* (=*Hirsutella*), a widespread fungal parasite of tropical ants, infection is likely to occur by direct contact with diseased individuals, perhaps followed by ingestion of spores. The fungus invades the tissues of the head and thorax only (Nnakumusane 1987). The fruit body develops slowly in ants that have fixed themselves to the substrate. From there, spores pass on to ants that either scavenge on these cadavers or to workers that approach their infected nestmates. *Beauveria bassiana* infects its host by attaching to the surface and penetrating the cuticle, in addition to ingesting spores. Workers of the fire ant *Solenopsis invicta* apparently reduce the spread of fungal spores by highly specialized necrophoretic behavior. They remove dead nestmates and pack soil around them. On the other hand, infected workers tend to move around the nest erratically. This may help to spread the fungal spores. Some authors also consider such behavior to be an instance of adaptive suicide, because the ant will more readily fall prey to a predator or become lost outside. As a consequence, the argument goes, the worker prevents the further spread of the fungus to its relatives within the nest (Smith Trail 1980).

It is indeed striking in how many different systems, in which the parasite depends on contact with a new host for transmission, infected workers will change their behavior. Fungus-infected ants often move around more frequently (Evans 1989; Oi and Pereira 1993) or leave their nests (Allen and Buren 1974). Often, particular sites are preferred by the infected workers. These sites are characterized by properties that ensure transmission of fungal spores to new hosts, either by wind, by rain, or by scavenging ants. A classical example is of course the infection by the metacercariae of the liver fluke, where ants serve as intermediate hosts (e.g., Schneider and Hohorst 1971; see also below). Some examples are given in table 3.4. Further observations can be found in Bequaert (1922), Paulian (1949), Turian and Wuest (1969), Balazy and Sokolowski (1977), Humber (1981), and Samson et al. (1982).

In honeybees, the mite *Acarapis woodi* (Tarsonemidae) infects the tracheae of adult bees (Bailey and Ball 1991). Within the nest, the mites transfer to new hosts directly by walking from hair tip to hair tip of workers that come in contact with one another. Among colonies, however, they are carried by dispersing hosts.

Infection by Ingestion

Adults of social insects are active foragers and consume food. Hence, this route of infection is particularly relevant for this group, in particular when food from nest provisions is consumed. But ingestion of contaminated food at the source, e.g., from flowers or cadavers, is also important.

Some honeybee viruses are infective when ingested but cannot produce an infection when directly inoculated into the hemocoel. This includes bee viruses X and Y. Unfortunately, our knowledge from viruses in species other than the honeybee is rather limited (see appendix 2.1). Entompox-like viruses found in adult bumblebees (Clark 1982) may be infective per os. The virus is, moreover, found in their salivary glands (Goodwin et al. 1991). Because bees usually wet their collected pollen loads with saliva, the virus may be efficiently spread to brood by food and pollen collecting.

Bacterial infections of adult social insects are normally acquired by ingestion of infective cells. For example, as Miller and Brown (1983) have shown experimentally, feeding of a range of bacteria to fire ants produces a severe septicemia that leads to host death. The mollicute *Spiroplasma* sp. has been identified in several bee species (chap. 2.2; appendix 2.2). In this case, infection is by ingestion of contaminated food, e.g., pollen. In tests, the spiroplasm then entered the hemocoel of young adult honeybees some 7–14 days after infection. The affected host soon dies. The pathogens may be deposited on plant surfaces

Table 3.4

Examples of Behavioral Changes Associated with Parasitic Infection

Host	*Parasite*	*Behavioral Change*	*Possible Function, Remarks*	*References*
		DIRECTLY TRANSMITTED PARASITES, PARASITOIDS		
A: *Apis mellifera*	*Nosema apis*	Accelerates individual age-dependent sequence of activities, e.g., earlier foraging career.	Possibly facilitates spread of parasite spores to other colonies in the population.	Wang and Moeller 1970
A: *Bombus* spp.	*Sphaerularia bombi* (Nematode)	Infected queens activate hibernation behavior in a wrong context.	Facilitates deposition of parasites at sites where transmission is possible. Possibly related to changes in hormone titer.	Lundberg and Svensson 1975
A: *Bombus terrestris*	Conopid flies: *Physocephala, Sicus* (Diptera)	Infected workers leave nest, particularly at night.	Places host in cool environment that damages parasite. Increases worker longevity.	Müller and Schmid-Hempel 1993b
Many ants	Mermithid nematodes	Ants become negatively phototropic. In permanent state of hunger.		Bedding 1985
F: *Cephalotes atratus, Paltothyreus*	*Cordyceps* spp. (Fungus)	Workers wander around and avoid nest. Perish at particular places.	Prevents infection of nest. Facilitates dispersal of spores.	Samson et al. 1988

Table 3.4, continued

Examples of Behavioral Changes Associated with Parasitic Infection

Host	*Parasite*	*Behavioral Change*	*Possible Function, Remarks*	*References*
		DIRECTLY TRANSMITTED PARASITES, PARASITOIDS, continued		
F: *Formica pratensis* and many other species	*Entomophthora* sp. (Fungus), and many other fungi	Infected workers climb on top of grasses, fix themselves with mandibles, and die.	Facilitates (windborne) dispersion of fungal spores. Possibly triggered by hypha that grow into vicinity of critical host nervous tissue.	Loos-Frank and Zimmermann 1976; Evans 1989
F: *Solenopsis* sp.	*Beauveria bassiana* (Fungus)	Infected colonies dislocate more often. Infected ants also often start to move around erratically and climb grasses.	Facilitates dispersion of spores. Colony movements possibly reduce infection levels as contaminated nest cavity is left.	Oi and Pereira 1993
		PARASITES WITH MORE THAN ONE HOST		
F: *Camponotus* sp. as intermediate host	*Brachylecithum mosquensis* (Trematoda)	Infected ants (metacercaria) more likely to stay in open spaces, on tips of grasses; sluggish behavior.	Increases predation risk toward final host of parasite (birds?).	Carney 1969
F: *Formica rufibarbis*, *F. cunicularis*	*Dicrocoelium dendriticum* (Trematoda)	Infected ants climb onto grasses and fix themselves there with mandibles.	Increases predation risk toward final host of parasite (a grazing mammal).	Hohorst and Graefe 1961; Schneider and Hohorst 1971

NOTES: Possible functions are mostly speculative and not proven. Host groups are A: Apidae, F: Formicidae.

when bees defecate. They are presumably picked up from there by the next forager (Mouches et al. 1984).

Some fungi infect through the ingestion of their spores. Examples include an unidentified yeast and *Beauveria bassiana* infecting workers of the fire ant *Solenopsis* (Broome et al. 1976) in the laboratory (Jouvenaz 1990), and *Metarhizium* and *Beauveria* infecting termites (in *Reticulitermes:* Kramm and West 1982; in *Cryptotermes brevis:* Kaschef and Abou-Zeid 1987). However, the rate of transmission along this route appears to be generally low. This could be different when scavenging behavior is common, particularly in ants. For example, the fungal genus *Aegeritella* is found in ants that regularly scavenge, such as *Cataglyphis cursor, Lasius flavus, Camponotus sericeiventris,* and various species of *Formica* (summarized in Balazy et al. 1986). In several of those instances, however, the prevalence is fairly low. It is likely that scavenging also increases the risk of fungal infections by *Cordyceps* and *Erynia* species (Evans 1989). Fungal spores are not always infective if ingested because of a filter apparatus established by the infrabuccal cavity and/or the proventriculus just in front of the stomach (Eisner and Happ 1962). Furthermore, as will be discussed in later chapters, social insects, especially ants, also possess an arsenal of chemical substances that ward off fungal infections (Beattie et al. 1985, 1986; Veal et al. 1992). In addition, behavioral mechanisms, such as grooming or avoidance of infected nestmates, help to limit the spread of fungal and other infective parasites (reviewed by Oi and Pereira 1993).

Many protozoan parasites also infect their host via ingestion of cells, cysts, or spores. For example, cysts of the amoebae *Malpighamoeba mellificae,* when ingested by the honeybee, enter the alimentary tract and develop into a flagellated form that migrates into the Malpighian tubules, where they change into a trophic form. Particularly in bees kept at low temperatures, new cysts are soon formed and are excreted in the feces (Schulz-Langner 1958; Bailey and Ball 1991, p. 72). Similarly, the flagellate *Crithidia bombi* is highly infective by ingestion for many species of bumblebees. It can be contracted by visiting flowers and imbibing infected nectar (Durrer and Schmid-Hempel 1994). Honeybees also become infected by ingestion of spores of *Nosema apis* as they come into contact with brood combs that are contaminated with spores. The younger workers that normally clean the cells are more susceptible to infection. Many other microsporidia are not infective by os, as many experimental studies have shown (e.g., in fire ants: *Thelohania,* Knell et al. 1977; *Vairimorpha invicta,* Jouvenaz and Ellis 1986; *Mattesia (Apicystis) geminata,* Jouvenaz and Anthony 1979).

Trematodes often have complex life cycles on two or three hosts. Infection of adults social insects is typically by ingestion of cercaria that later penetrate the gut wall and invade the appropriate tissues of the host body. This pattern is known from the liver flukes *Dicrocoelium* and *Brachylecithum,* which use ants as intermediate hosts. Although not demonstrated, the same mode of infection is likely for as yet unidentified cestodes, e.g., in *Leptothorax* workers (Stuart and Alloway 1988). In the case of cestodes, the predatory habit of ants aids the parasites. For example, as Case and Ackert (1940) have observed, ants of a number of species accept and carry back the gravid segments of cestodes as a prey item. These proglottids are mobile and therefore are likely to attract the attention of predators by their movements.

Among ants, Camponotinae seem to be more often infected by nematodes than Formicinae. This probably relates to their habitats, which ensure contact with food that is suitable for nematode reproduction (Bedding 1985). However, nematodes could also be grouped with parasites that actively seek hosts. For instance, many rhabditid nematodes produce durable stages, usually late-stage instars that can survive adverse conditions. They are capable of erecting their bodies, so with waving movements they can then rapidly attach to passing insects. Similarly, Nickle and Ayre (1966) were not able experimentally to infect these ants with nematodes. But they observed that juvenile nematodes attached themselves to the ants, aided by sticky secretions. Perhaps later in the nest, allo- and self-grooming transfers these juvenile nematodes into the buccal cavity of the cleaning worker and leads to ingestion and infection.

Nematodes that infect pharyngeal glands of adult ants are known from many ant species and involve several genera of rhabditid nematodes (Janet 1893, 1894, 1897, cited in Crawley and Baylis 1921; also Wahab 1962) (appendix 2.5). Again, it is likely that infection occurs by ingestion of eggs or durable stages. Bedding (1985) also lists cases where rhabditid nematodes infect just one of the four pharyngeal glands in the head of ants (*Formica*). Although the significance of this infection pattern is not known, the case is reminiscent of the parasitic mite that usually infects and damages only one ear of some noctuid moths. This seems to ensure that the host remains a safe mite vehicle that can still escape predatory bats (Treat 1975). For ants, the pharyngeal glands are important not only for feeding but also for social interactions (Hölldobler and Wilson 1990). Infection of just one gland out of four is not too debilitating and should ensure continued host survival and acceptance within the nest until parasite development or transmission is completed. It should be possible to test some of these hypotheses in a relatively simple way. Wheeler (1928) found that mermithid-infected ants showed higher levels of negative phototropism. The infected ants did not feed immature ants, however, and appeared to be in a chronic state of hunger.

Infection by ingestion of infective stages (perhaps in addition to direct penetration by the larval nematode) is suspected for *Agamomermis pachysoma* in *Vespula* (Bedding 1985). This nematode infects only adults, which makes ingestion as the primary route the most likely one. Mermithids also infect the honeybee, more so in generally wet habitats. This indicates that workers may become infected when taking up water from foliage (Bedding 1985). It should be mentioned that the nematode families Steinernematidae and Heterorhabditidae have mutualistic associations with bacteria of the genus *Xenorhabdus* (Kaya and Gaugler 1993). The nematodes rely on the bacteria to kill the host as well as on the production of antibiotics to keep secondary infections away. These secondary parasites are likely to be contracted by ingestion. At the same time, host defense against bacteria would probably also inactivate the nematodes.

Vectored and Activated Parasites

Some viruses of adult social insects appear to rely on activation by mites. In these cases, the viruses presumably have resided in tissues without producing symptoms (see also fig. 2.3). When a mite pierces the body wall of the host, virus particles are mechanically transferred to the hemocoel, where they induce a systemic infection. This appears to be the case with acute bee paralysis virus in association with the mite *Varroa jacobsoni*. Moreover, the mite is also known to be a vector of the virus and to transfer to adult bees or to pupae (Ball and Allen 1988; Allen et al. 1986; Bailey and Ball 1991). Adult bees apparently can also directly infect young larvae, most likely via secretions from the glands that contain the infective particles (Ball and Allen 1988). The role of microsporidia (*Nosema*) and ciliates (*Malpighamoeba*) in their association with bee viruses Y and X (BVY, BVX), respectively, is not quite clear. But BVY does increase the pathogenicity of *Nosema;* a similar possibility is discussed for black queen cell virus and filamentous virus in Bailey et al. (1981).

Similar to mites that infect brood, many mites that utilize adult hosts rely on vectors (Eickwort 1994). The choice of a carrier, however, is not passive, but the phoretic stages are well capable of making choices between alternatives (e.g., Kraus et al. 1986; Schwarz et al. 1996). This suggests that mites are well equipped with sensory organs to detect and identify appropriate carriers, hosts, or food sources. Phoresy as a strategy to find new hosts seems to be useful, because nests of social insects are widely scattered in the habitat and could not possibly be reached in any other way. In many species, the phoretic stages possess organs (claws, chelicerae, suckers) that ensure a firm grasp on the carriers (Eickwort 1994). Since males of social Hymenoptera do not normally return to their nest once they have left it, they could serve as vectors to females only during mating, and would thus generally be unefficient carriers. There is some evidence that males are in fact less preferred by the phoretic stages (in bumblebees: Richards and Richards 1976; Schwarz et al. 1996). In contrast, Eickwort (1990) concluded that in most of the studied cases, mites do not discriminate. However, bee mites that use males are often found in the genital chamber (Eickwort 1990, p. 239), which supports the idea that males are only useful as vectors to females. In annual species such as bumblebees, the mites must locate a daughter queen to overwinter. Hunter and Husband (1973), Richards and Richards (1976), and Schousboe (1986, 1987) all found that the phoretic instars of *Pneumolaelaps, Parasitellus,* and *Scutacarus* were indeed more common on queens than on males or workers. In addition, phoretic stages of *Parasitellus* have also been collected from flowers (Richards and Richards 1976). This has also been confirmed in more recent studies with bumblebees in Europe (Schwarz and Huck 1997). It suggests that mites may transfer between colonies in this way. Compared to bees, the mite fauna of ants is not well studied. Many are phoretic, some are cleptoparasitic by soliciting food from the workers, and others are truly parasitic (chap. 2.7; appendix 2.8).

Oviposition by the Parasite Female

There are many parasitoids that directly oviposit into adults, workers, queens, or males of social insects, although actual observations of ovipositions are few. These include many species of hymenopteran or dipteran parasitoids (appendixes 2.9, 2.10). Some bethyloid wasps apparently oviposit into the gaster of ants (Wojcik 1989). Similarly, braconid wasps have been observed to hover over foraging columns of raiding ants and to oviposit into the gaster of the workers (Hölldobler and Wilson 1990, p. 485). Brood of the braconid wasp *Syntretus* is occasionally found in workers, males, and queens of bumblebees (Alford 1968; Goldblatt 1984; Schmid-Hempel et al. 1990; the genus *Syntretus* contains polyembryonic species). Pouvreau (1974) noted that female *Syntretus* attack on flowers rather than in the nest, and that they insert the egg into the host.

Direct ovi- or larvaposition on or into adults is a specialty of dipteran parasitoids. Phorid parasitoids of ants are good examples. For example, females of *Apocephalus* deposit or insert an egg in the host as it swoops down to attack the worker or soldier (Wojcik 1989). Sarcophagid flies are larviparous (ooviviparous) and deposit larvae externally on their hosts (e.g., on *Apis* workers; appendix 2.10). These subsequently burrow into the host and usually float freely in the hemolymph. Quite similarly to phorids, conopid flies also attack their host on the wing, in flight, or on the flower. Within a fraction of a second, the egg is injected into the bee's abdomen (pers. obs.). Multiple parasitism is quite common, and up to nine larvae or eggs have been found in a single bumblebee worker, although only one can successfully develop into a pupa (Schmid-Hempel and Schmid-Hempel 1989).

Table 3.3 shows that Diptera are mostly parasitoids of workers and adults, while Hymenopteran parasitoids primarily attack the brood. Hymenoptera may be better equipped than diptera with the necessary behavioral and pheromonal arsenal to gain access to host nests. They can thus exploit the communications of the host, which is a prerequisite to enter the colony fortress and to have access to the brood. On the other hand, the highly mobile and agile adult workers of social insects are probably difficult targets. It requires high maneuverability in flight to carry out a successful attack. This is where Diptera excel—Hymenoptera are generally more clumsy fliers. This difference is probably more important than the presence of a heavily sclerotized ovipositor typically found in hymenopteran but not in dipteran parasitoids. In fact, some dipteran parasitoid flies, such as the conopids attacking bumblebees, manage to insert eggs directly into the host's body cavity even without a hard ovipositor (Schmid-Hempel and Schmid-Hempel 1996).

3.3 Parasites of Sexuals

In the biology of social insects, sexuals are only a very small fraction of the adult individuals present in any one host population. Considering this simple fact, one sees that parasites do not have as much opportunity to specialize on sexuals as on workers. Hence, we should expect strict associations to occur only when there is a distinct advantage of parasitizing sexuals or when there are distinct pathways of transmission that require sexuals rather than workers. On the other hand, sexuals are usually larger hosts than workers; they disperse or can bridge important temporal gaps such as hibernation periods in temperate climates. Moreover, with the haplo-diploid sex determination system in social Hymenoptera, parasites can attack a haploid male rather than a diploid worker or queen. However, it is not known whether this genetic difference has an effect on host defense in the first place, as one would naively expect when host defense is more effective in heterozygous rather than homozygous or hemizygous (e.g., hymenopteran males) individuals. So it is not surprising that almost all of the parasite groups mentioned before occur on queens as well as males. For example, the tracheal mite *Acarapis woodi* also infects queens of honeybees (Pettis et al. 1989; Burgett et al. 1989), although there are no reports of damage inflicted in the queens. For a wide range of parasitic organisms, therefore, sexuals are just another host in addition to the workers.

Only a few parasites may have specialized on sexuals, although the evidence is ambiguous. For example, the fungus *Metarhizium anisopliae* seems attack to only queens of *Solenopsis* in South America ("queen's disease"). However, as Allen and Buren (1974) pointed out, this may only look like a queen pathogen because infected workers leave the nest and are never found. Also, the rare fungus *Desmidiospora myrmecophila* may be specific only to queens of *Camponotus* ants (appendix 2.3; Evans and Samson 1982; Hölldobler and Wilson 1990). The dipteran parasitoid *Strongylogaster* (Tachinidae) is also reported to be an endoparasite of young queens of the ant *Lasius* (Gösswald 1950). Queen melanosis has been observed in honeybees (Bailey and Ball 1991, p. 62) and bumblebees (Skou and Holm 1980). This syndrome involves melanization (blackening) of epithelial cells in the ovaries, in the poison sac, and in the rectum of queens. The etiologic agent is unknown and perhaps it is not even a parasitic disease at all. Similarly, Skou and Holm (1980) mention several species of fungus that have been found to infect the reproductive organs of honeybee queens. The fungus *Paecilomyces lilacinus* kills queens within 3–5 days in 10% of cases when experimentally infected. But it is not known to be a natural associate of honeybees. The bacterium *Aerobacter cloaca* is found in ovaries of queens in honeybees and bumblebees (appendix 2.2).

One of the most remarkable examples of an exclusive parasite of sexuals is the nematode *Sphaerularia bombi* (Tylenchidae) attacking overwintering queens of various bumblebee species (see Alford 1975) (fig. 2.7). Similarly, Pouvreau (1974) noted that queens are among the most frequent hosts for the braconid wasp *Syntretus splendidus,* which parasitizes bumblebees. But workers may become parasitized, too (Schmid-Hempel et al. 1990). The wasp produces many offspring and substantially harms the queen. The diaprid wasp *Plagiopria passerai* seems to be an exclusive parasitoid of queen cocoons in *Solenopsis*—the first example of this group that is parasitic on ants (Lachaud and Passera 1982).

Sexually transmitted diseases would obviously also be restricted to the sexuals only. Skou and Holm (1980) have discussed this possibility for several fungi (*Candida, Paecilomyces, Saccharomyces,* etc.) and bacteria (*Areobacter*). The fungi can indeed be transmitted experimentally when queens of honeybees are inseminated artificially. But no cases from natural populations seem to be known. In a recent review, Lockhart et al. (1996) suggested that sexually transmitted diseases should be much more common than hitherto known. For a variety of reasons, their effects are expected to be rather benign, and hence sexually transmitted parasites may have escaped attention all too often. In fact, Lockhart et al. (1996) found only two cases of a parasite (a nematode, *Acrosticus;* see appendix 2.5) that is likely to be sexually transmitted in two species of social *Halictus* bees. Note, however, that in social insects sexually transmitted parasites are not necessarily exclusive parasites of sexuals. In fact, a parasite may be transmitted sexually but could still also infect the workers or brood of the colony and multiply before infecting the sexual hosts, the young males and females that will mate and disperse. In addition, one should also distinguish between parasites transmitted sexually in a vertical way, e.g., via transovarial infections, and those transmitted during mating, e.g., by contact between mating partners as is likely for mites.

3.4 Social Parasites

Social parasitism in social insects is a relationship between two species where the parasite benefits in many ways from brood care or from the other socially managed resources of its host. Although in general this kind of exploitation is not really restricted to truly social species—there is, for example, egg-dumping and brood parasitism in birds—social parasitism in social insects is an exceedingly complex phenomenon. It does involve a large number of species and diverse lifestyles. Examples range from workerless obligatory social parasites that have completely lost a worker caste (e.g., the cuckoo bee *Psithyrus* or the "final" ant *Teleutomyrmex schneideri*), to temporary parasitic species (as in some *Tetraponera* and *Aphaenogaster* ants), to slavemakers, dulosis, inquilinism, each type is associated with its own sophisticated method to control or kill the host queen (Buschinger 1986). Tables 3.5 and 3.6 give examples and a synopsis of the different types of social parasitism that are usually recognized in social insects.

In 1909, Emery proposed that social parasitic species are usually derived from species that share an ancestry with their hosts ("Emery's rule"). This refers to the remarkable morphological similarity between many host and parasite species. Typically, though, the parasite has lost or degenerated many of the characteristics that are related to resource collection or nesting. For instance, parasitic bees cannot produce wax that is otherwise needed for nest construction. Emery's rule has since been buttressed by additional evidence, at least as far as true inquilines are concerned. For example, in a recent attempt to revise the group, Williams (1994) grouped the social parasitic cuckoo bees *Psithyrus* with their hosts, *Bombus*, which is also borne out by molecular evidence (Koulianos et al., in prep.). Emery's rule also seems to be a useful approximation for yellow-jacket species, *Vespula* (Greene 1991), and for bees in the Ceratini group (*Allodape*, *Allodapula*, *Macrogalea*, *Braunsapis*) (Michener 1974). As Roubik (1989) notes, in over 80% of fifty-two analyzed parasitic bee genera that occur in the tropics, both host and parasite are classified in the same family (most of them are solitary species). Also, no case is known where dulosis has evolved that connects ants from different tribes or subfamilies (Hölldobler and Wilson 1990, p. 464). It is not surprising then to find that there are no cases known where social parasitism is between the major taxa, e.g., a wasp parasitizing a social bee or an ant parasitizing a termite, although at first sight this would seem entirely possible.

The problem underlying Emery's rule is one of how social parasitism can evolve. One possible way is through "classical" allopatric speciation, which leads to two closely related lines that may come into contact again, and then leads to parasitism of one by the other. An expectation from this scenario is that social parasitism should be frequently observed in specious and proliferating lineages. This seems to be approximately the case in certain groups of ants (*Pheidole*, *Leptothorax*, *Plagiolepsis*, *Lasius*, *Formica*: Hölldobler and Wilson 1990), and bees (Roubik 1989). In contrast, Buschinger (1986; see also Bourke and Franks 1991; Buschinger 1990) proposed sympatric speciation as the most important mechanism. This process involves divergence in mating patterns

Table 3.5

The Different Kinds of Social Parasitism

Type	*Definition*	*Examples, Remarks*
Plesiobiosis	Very simple association. Different species nest close by but have little interaction.	
Cleptobiosis	Some smaller species nest near larger species and may rob their foragers or use debris.	*Crematogaster* using *Holcomyrmex*. Many bee species (Wheeler 1910, cited in Hölldobler and Wilson 1990).
Lestobiosis	Small species stay inside nest of larger species or enter their chambers stealthily to steal food or prey on brood.	The "thief" ant *Solenopsis fugax*.
Parabiosis	Species live in same nest and may use same foraging trails. Keep and tend brood separately. No particular host-parasite relationship obvious.	Probably some species of *Crematogaster, Monacis,* and *Dolichoderus* in South America.
Xenobiosis	Colonies of one species live in nests of another, dominant species. "Parasite" moves freely and obtains food but keeps and tends brood separately.	*Formicoxenus nitidulus* in *Formica rufa*.
Temporary social parasitism	Newly fertilized queen seeks host and gets adopted. As her own brood is assisted by the host, the host queen is killed and slowly the nest is taken over by the parasite.	In many groups of ants, e.g., *Formica difficilis* = autoparasitism after Bolton 1986.
Dulosis (slavery)	Parasite dependent on host species, which are raided. Parasite foragers only raid, slave workers become fully functional colony members, tend and raise parasite brood.	Also facultative dulosis known.
Inquilinism	Permanent parasitism. Entire life cycle is within host colony. Parasite workers usually absent, or highly atrophied. Host queen often killed or "forced" to cooperate.	*Teleutomyrmex schneideri* (Stumper and Kutter 1950).

SOURCE: After Hölldobler and Wilson 1990.

Table 3.6

Examples of Social Parasites

The Social Parasite	*The Social Host*	*Remarks*	*References*
ANTS			
Teleutomyrmex schneideri	*Tetramorium caespitum*	Among the most advanced inquilines known. Only occurs in host nests. Parasitic queens stay and ride on back of host queen. Reduced mandibles, sting and poison sac.	Stumper and Kutter 1950; Kutter 1969
Epimyrma spp.	*Leptothorax* spp.	Inquilinism. A whole complex of species. Have diverse methods to eliminate host queen, from strangling to mutilating.	Buschinger 1989
Harpagoxenus americanus	*Leptothorax longispinosus*, *L. curvispinosus*	Dulotic relationship.	Creighton 1950
BEES			
Psithyrus spp.	*Bombus* spp.	Usurp nest; host queen may or may not be tolerated.	Fisher 1988
Lestrimelitta, *Cleptotrigona* spp.	Other social bees	Raid nest of other bees and rob food and nest material. Neotropics, Africa. Can also usurp nests.	Michener 1970
Tetragonisca angustula	Stingless bees, *Plebeia*	Usurp nests of host bees.	Roubik 1989
Inquilina spp.	*Exoneura* spp.	Obligately parasitic. No overt aggression observed, but host brood is absent. Australia.	Michener 1965, 1970

Table 3.6, continued

Examples of Social Parasites

The Social Parasite	*The Social Host*	*Remarks*	*References*
WASPS			
Polistes sulcifer	*Polistes gallicus*	True inquiline.	Scheuven 1958, cited in Wilson 1971
Vespula squamosa	Many *Vespula* species	Usurp nests.	Greene 1991
Vespula austriaca, Dolichovespula arctica	Many *Vespula* species	True inquilines. Have short intense oviposition period, physically dominate host colony members.	Greene 1991
TERMITES			
Ahamitermes, Incolitermes, Termes spp.	Mound-building termites	Nest parasites, close to xenobiosis. Live in other termites' nests. Some live on material of nest structure.	Wilson 1971
Termes insitivus	*Nasutitermes magnus*	Nymphs and alates live in nest galleries of host. Perhaps a true inquiline.	Hill 1942, cited in Wilson 1971

within the population that eventually leads to reproductive isolation, for example, triggered by mating at different times or at different places. If this is also associated with loss of polygyny in potential host species, then, as Buschinger (1986) suggested, the evolution of forceful host queen elimination may become possible and likely.

With different mechanisms, true inquilinism can evolve through different routes. Most likely, this is either via the dulosis route or the temporary parasitism route. In the former case, dulosis may result evolutionarily from habitual predation on other ants, from territorial disputes, from transport in polydomous colonies, or from any combination of these. Raiding behavior in the context of territorial disputes is considered to be a very likely starting point, later followed by facultative raiding and an ever-increasing complex suite of behaviors and pheromonal arsenals to increase raiding efficiency (Hölldobler and Wilson 1990). In the temporary parasitism route, the process is likely to start in species where young queens are often adopted by other colonies in the population, as is the case in many ant species. From there, evolution could lead toward adopted queens taking over the foster colony. If this process starts to involve different species, a route to true social parasitism is open. This pattern is well demonstrated by the wood ants, *Formica* spp. There is also a startling repertoire of behaviors by which the parasite queen, or the host workers "seduced" by her, eliminates the host queen (their mother). This can involve bold actions such as decapitating the host queen (in the ant *Bothriomyrmex* decapitating *Tapinoma*), slowly strangling her (*Epimyrma* strangling *Leptothorax*), or piecemeal mutilating the antennae or legs (*Epimyrma*).

Obviously, similar life histories, similar communication arsenals, and similar morphology facilitate evolution along these routes. However, the simple fact that the same type of nesting sites are used should also facilitate social parasitism. Consequently, interspecific or paraspecific (i.e., by a closely related species) parasitism, for example among true *Bombus* species, is a much more specialized relationship, whereas the "true" cuckoo bees, *Psithyrus*, utilize a phylogenetically more variable host spectrum (Alford 1975; Sakagami 1976). In addition, the rarity of nest sites or environmental harshness, e.g., short seasons and small chances of renesting later in the season, must be environmental constraints that favor the evolution of social parasitism. This is suggested by Wcislo (1987), who found that the percentage of bee taxa known to be parasitic within a given group increases with latitude.

True inquilinism seems absent in termites (Wilson 1971), although a large number of guests are known to occur in their nests. It is possible that several characteristics of termites act against the evolution of social parasitism in its incipient stages. Indeed, new queens and kings of termites rarely reenter existing colonies. Instead, new colonies are founded independently as a rule. Note that fertilized queens also return to a nest in the stingless bees and the honeybees. However, in those cases it is their own parental nest and thus parasitization would occur among sisters or close relatives, which, by kin selection arguments, is not favored to the same degree. In addition, territorial disputes may be rare among termites, although it is admittedly difficult to study this issue in subterranean species, where conflicts may occur only when galleries of neighboring colonies meet.

3.5 Summary

Colonies of social insects are profitable targets for parasites because of the locally high density and similar genotypes of hosts in a nest. On the other hand, invasion of this "factory fortress" is difficult. Eggs of social insects appear to have few parasites, but larvae and pupae are prime hosts. Infection of brood by direct contact or exposure of brood is not common, but typically it occurs by ingestion as parasites are spread by contamination of food, nest material, or attending workers. Bacteria and spores of fungi can remain infective for years, or produce infections with few outward symptoms. Some parasites, e.g., microsporidia, have exploited brood care of social insects and can thus infect a particular larval stage of the host. Their transmission is ensured via the attending workers, e.g., when nurses feed larvae with secretions or pellets. Macroparasites from several taxa actively find their hosts. The larval stages of Strepsiptera, Encharitidae, Rhipiphoridae, of some Ichneumonidae, Perilampidae, and several dipteran families, all of which parasitize brood of social insects, show specializations such as mobility as well as sensory and attachment organs. Often, these parasites actively attach themselves to passing foragers of the host nest and thus gain entrance to the colony. The intensive patrolling and food gathering activities of social insects aids this strategy. Some parasites, e.g., the *Varroa* mite, are vectored by workers drifting to other colonies. In turn, some viruses and bacteria are vectored and/or activated by the mites when they pierce the hosts's cuticle. Female parasites can also oviposit directly on the brood inside the nest. This strategy is relatively less common, because brood is defended against invaders. Transovarial infection seems possible in viruses and microsporidia.

Workers of all castes are numerous in social insect nests and actively move around. A number of fungal parasites infect by direct contact or exposure and by subsequent penetration of the cuticle. The infection by several parasites that depend on dispersal of spores or on finding a suitable next host is associated with changed behaviors of the host worker, showing remarkable convergences, for example between fungal and trematode infections. Infection of workers by ingestion is important for all microparasites. Scavenging and predation are particularly prone to contracting such parasites. Macroparasites, such as nematodes or helminths, can be acquired by feeding on contaminated prey items. As in brood, some viruses are activated/released by mites when the particles are transferred into the host hemolymph. Direct ovi- or larviposition on adult hosts is common in dipteran but less so in hymenopteran parasitoids. A prime example is the rich variety of phorid flies that parasitize ants. The difference may relate to the high maneuverability of flies as compared to parasitic wasps. Many parasites infect the brood but persist into the adulthood of the host. This includes not only microparasites but also macroparasites such as Strepsiptera or nematodes.

Almost any group of parasites that attack brood or workers can also attack the sexuals (gynes and drones). But in social insect populations sexuals are rare. Therefore, few parasites seem to have specialized on this class of individuals. A notable exception is the nematode *Sphaerularia*, which is specialized on overwintering queens of bumblebees. Only very few cases of sexually transmitted diseases have been reported in social insects, although they should be common. A brief look at social parasites shows that similar diversity of associations, invasion strategies, and effects exist.

4 Parasites and the Organization of the Colony

In this chapter I will first give a brief sketch of the internal organization and structure of social insect colonies, and then discuss the relationship between colony organization and parasites. Few concepts have been set forth on this topic as well as the ones following in later chapters. Therefore, an important aim of this and the following chapters is to clarify these issues, to develop new concepts, and to formulate expectations that can eventually be tested. In the discussions, infectious microparasites serve as the model, but the arguments could readily be extended to other categories as well.

4.1 The Basics of Colony Organization

Colonies of social insects are structured in subtle ways, particularly in their division of labor systems, which are based on morphological castes (i.e., workers with different sizes and shapes) and age-related polyethism (i.e., a systematic change of behavior over an individual's lifetime). By many accounts, such systems are efficient in terms of increasing a colony's performance (Hölldobler and Wilson 1990; Schmid-Hempel 1991a), although the evidence comes more from investigations in the social Hymenoptera than in termites. For example, the observed allocation of different worker sizes to different tasks in the leafcutter ant, *Atta*, matches the expectation derived from an allocation that maximizes the energetic returns on investment and running costs of operations in the colony (Wilson 1980a,b). In addition, in almost all cases, a large worker number has been demonstrated to be of overriding importance for colony success (see Schmid-Hempel et al. 1993).

On the other hand, phenotypic variation among colonies in morphological caste proportions has been reported in several independent studies (e.g., Talbot 1957; Davidson 1978; Herbers 1980; Johnston and Wilson 1985; Fowler 1986). But there is little evidence that such variation correlates with relevant environmental characteristics, for example with the presence of predators and competitors, or the availability of food in space and time (e.g., in the ants *Camponotus impressus*: Walker and Stamps 1986, and *Pheidole dentata*: Calabi and Traniello 1989; but see Passera et al. 1996 for *Pheidole pallidula*). In other cases, an experimental change in morphological caste structure was found to have little effect on eventual reproductive output (e.g., as with soldiers in the termite *Reticulitermes*: Haverty 1979; Haverty and Howard 1981; see also Kolmes 1986; Schmid-Hempel 1991b). In addition, if such colony structure is to be explained in terms of microevolutionary processes, this trait should possess heritable variation. This has been shown at least for caste-specific behaviors that underlie the division of labor in the honeybee (e.g., Robinson and Page 1989a). Nevertheless, our knowledge on this point is quite limited. Several factors tend to reduce the significance of caste structures as a universal adaptation for efficient design of a colony. Foremost among them is the existence of substantial behavioral flexibility (Oster and Wilson 1978). In addition, physiological and developmental constraints of caste determination further narrow the scope for morphological caste evolution (Wheeler 1986).

Wilson and Hölldobler (1988) suggested that insect societies are organized in "dense heterarchies." The concept is not well investigated and it is unclear whether it applies generally to social insects. Here, it only serves as a useful metaphor. According to this view, the societies are "dense" in the sense that almost all individuals interact with one another. Obviously, not all can really interact, if only because of the huge number of possible pairwise contacts in large colonies. Societies are "heterarchic" in the sense that there is no strict top-down flow of signals, but rather each individual reacts to many others within the colony. In the spread of an infectious disease within the colony, this metaphor of social organization makes it clear that any member of the colony could in principle contract an infectious parasite that is passed on along the pathways of interactions. In a somewhat different view, Oster and Wilson (1978) have noted that an important element of social insect biology is that operations are performed by many workers in parallel arrangements, and that compartments of work may thus exist. This system is well known from technical applications. Every airplane, for instance, has at least two or three copies of an important system to increase the reliability of the whole system by way of redundancy. Oster and Wilson (1978) have consequently used the results of reliability theory to analyze the ergonomics of work in social insects (see box 4.1). As will be discussed, such organizational arrangements are likely to have effects on the parasitization of the colony.

Another remarkable feature is that social insect colonies are capable of resisting large fluctuations in their external environment. For instance, morphological caste ratios have a homeostatic property in that they are often restored into previous proportions after disturbance, as a result of a "passive" regeneration process, determined by the intrinsic birth and mortality rates in the colony (e.g., Gentry 1974; Herbers 1980; Fowler 1985; Johnston and Wilson 1985; Wilson 1983). Thus, transient external disturbance has often surprisingly little effect on the eventual reproductive output of the colony. Only if perturbations

BOX 4.1 RELIABILITY AND SOCIAL INSECTS

In a social insect colony, work is organized along one or several parallel chains. For example, in the honeybee, food has to be collected outside, passed on to an appropriate receiver bee and finally stored in the comb. Later, the food is retrieved and used to feed larvae or adults. This process occurs independently and concurrently in several chains involving many workers each. Only when all steps in the sequence have been successfully completed will the colony and its members have success, for example, manage to raise brood. The whole arrangement is reminiscent of a technical device that ensures reliable performance. Oster and Wilson (1978) have applied the concepts of reliability theory to analyze the economic design of work in social insects ("ergonomics").

a) Serial arrangement

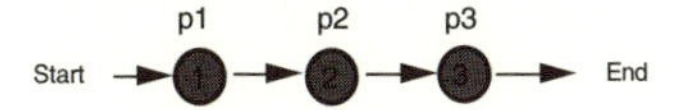

b) Parallel-Series

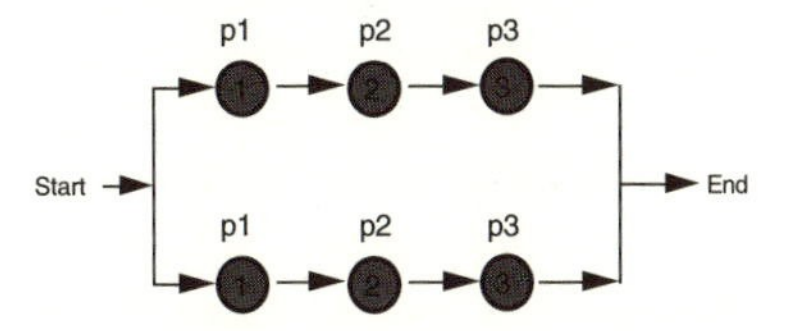

c) Series-Parallel

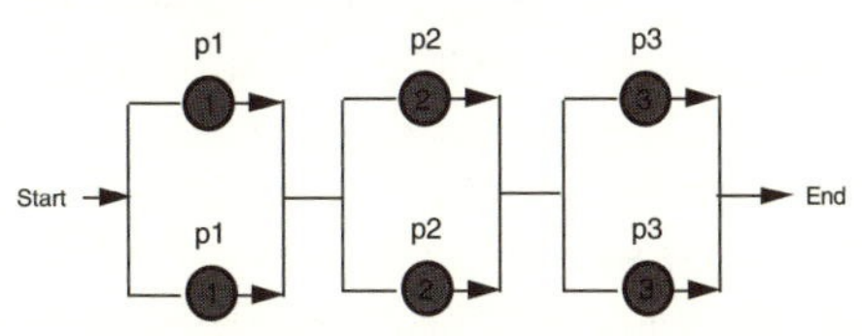

With work organizations such as those shown in the accompanying graph, the total system reliability, P_s, i.e., the probability that the process will be completed from start to end in at least one of the branches of the network, can be calculated as follows:

With a serial arrangement: $$P_s = \prod_{i=1}^{n} p_i \qquad \text{(1a)}$$

With a parallel-serial arrangement: $$P_s = 1 - \prod_{j=1}^{m}\left(1 - \prod_{i=1}^{n} p_{ij}\right) \qquad \text{(1b)}$$

With a series-parallel arrangement: $$P_s = \prod_{i=1}^{n}\left(1 - \prod_{j=1}^{m}(1 - p_{ij})\right), \qquad \text{(1c)}$$

where p_{ij} is the probability that an individual in chain j will perform task i successfully (p_i for the individual in a single chain), and where m individuals and n tasks

BOX 4.1 CONT.

are involved. As the equations show, the overall success strongly depends on the kind of organization. For example, with $n=4$ and $m=10$, and $p_i=p_{ij}=0.2$, one finds that the serial arrangement is worst with P_s= close to zero, followed by the parallel-series arrangement with $P_s=0.02$ and best with series-parallel $P_s=0.63$. Redundant arrangement of work therefore buffers the colony against failure of individual components, due to "errors" committed by individual workers. Oster and Wilson (1978) have concluded that this is a major reason why individual members of social insect colonies need not be as competent as their solitary counterparts.

Strictly parallel-serial or serial-parallel schedules may not be very common in nature, however. An alternative is to model the situation by assuming that the colony needs to meet just a minimum number k out of a total of n contingencies in order to survive and reproduce. Alternatively, we could think of a minimum viable colony size for survival and reproduction. If the tasks are equivalent and in random order, then the probability of success in this system is given by a binomial distribution (e.g., Oster and Wilson 1978):

$$P_s=\Sigma_{i=k}^{n}\binom{n}{i}p^i(1-p)^{n-i}.$$

are much more severe or long-lasting will negative effects on survival and reproduction become evident (e.g., Kolmes and Winston 1988; Sutcliffe and Plowright 1988) (see table 4.1). Resiliency against external perturbations is also ensured by short-term behavioral flexibility. If the work allocation of the entire colony is disturbed by an external event, individuals can shift their individual work profiles to restore the former activity profile of the colony, for instance by switching from idleness to foraging (e.g., Gordon 1986, 1989, 1991).

The general principles underlying the organization of social insect colonies are well known and have been repeatedly described (see the reviews by Oster and Wilson 1978; Hölldobler and Wilson 1990; or Bourke and Franks 1995). However, their functional significance in terms of contributions toward colony survival and reproduction is much harder to demonstrate. Unfortunately, we also know very little about the effects of parasitism on caste structure. But given the above considerations, the stage for caste evolution in response to parasitism seems set. So, how could parasitism be connected with the evolution of caste structure? For the parasites themselves, colony organization and caste structure determine the environment with which they have to cope. Hence, the significance of these elements for parasite success is potentially much more obvious. For example, the pathways of parasite transmission should depend on caste structure and organization of work, especially for those that are transmitted by contact or via food. Several such examples have been described in chapter 3. If

Table 4.1

Effects of Experimental Stress on Established Colonies

Species, Reference	*Experimental Manipulation*	*R Counted?*	*Effects*
Pogonomyrmex badius Gentry 1974	(F) Removal of 10 or 50 ants/day over 23 weeks.	Yes	Fewer flights in "50." Smaller, fewer workers. Activity decrease.
Pheidole dentata Johnston and Wilson 1985	(L) Predator challenge over 19 weeks.	No	No effect on major production. Ratios soon restored (specifically).
Atta cephalotes Wilson 1983	(L) Media workers removed once.	No	No effect on foraging rate. No effect on production of media workers.
Formica obscuripes Herbers 1980	(F) Exclusion from food. Removal of 200–300 ants/week.	No	No effect on average caste ratio. Less variability in stressed colonies.
Apis mellifera Kolmes and Winston 1988	(F) Moderate removal of young or old workers.	No	Little effect on age polyethism. Increase in brood care.
Bombus terricola Sutcliffe 1987	(L) Pollen access restricted to 8, 14, or 24 hr after 2d brood.	Yes	Smaller gynes, males.
Bombus lucorum Schmid-Hempel and Heeb 1991	(F) Doubling worker mortality rate after 2d brood.	Yes	Marginal effects on colony size and number of gynes, males.
Bombus lucorum Müller and Schmid-Hempel 1992b	(L) Doubling worker mortality early or late in colony cycle.	Yes	Stress early in cycle more effective than later. Small effects on colony size and gynes. Strong effect on male production.
Polistes metricus Dew and Michener 1981	(F) Removal of older workers.	No	Idle workers more active, more efficient.
Polybia occidentalis Jeanne 1986	(L) Colony size reduced.	No	Small colonies require more time to complete construction cycle.
Polistes bellicosus Strassmann et al. 1988	(F) Nest removal once.	Yes	Smaller young. Less-developed ovaries. Production ca. halved. No effect when > 4 queens present.

NOTES: Reproductives (R) have not been counted in all studies. F = field, L = laboratory study.

the flow of interactions or food among individuals is highly structured, as with a sophisticated caste system, then transmission from start to end of the work chain becomes less likely (see Akre and Reed 1983 for a case of limited contacts between queen and workers in wasps). This is the argument intrinsic to box 4.1 and will be detailed below. There is, however, an additional element. Morphological variation among workers of the colony will often relate to variation in the differences in the life span of the various morphs. For example, large workers are typically longer-lived than smaller ones (e.g., Garofalo 1978; Calabi and Porter 1989). Also, body size usually affects at what age a worker will attend a given task (e.g., Jeanne et al. 1988). Both properties are important for parasite establishment, success, and transmission. Caste structure therefore does affect the temporal dynamics of the demography of the colony and thus how parasites can establish themselves and become transmitted among colony members. A central element in all of these scenarios is the nature of interactions among individuals.

4.2 Interactions among Individuals

Colony organization requires that individuals interact with one another in various ways. Whenever these interactions occur, it can affect the parasites and, vice versa, the hosts can be affected by their parasites. For example, when one worker grooms another, spores of a pathogen may not only be removed from the body of the groomed individual but also be transferred to a new host—the grooming nestmate. But food transfer, nest construction, and the spatial distribution of colony members are also important for the success of parasites of social insects.

Trophallaxis and Provisioning of Larvae

The meaning of the term "trophallaxis" has had a troubled history, at times essentially encompassing all communication that occurs in a colony. Today, it refers to exchange of food among colony members (Wilson 1971). This is how the term is used here. Trophallaxis exists in many taxa of social insects but it is not universal. In ants, the degree of trophallaxis is highly variable and reflects both phylogenetic position and feeding habits. For example, in *Amblyopone* or other primitive ponerine ants (see Haskins and Whelden 1954), it is absent or very poorly developed. In most species of the higher Myrmicinae, in contrast, it is present. The same pertains to bees and wasps where nesting and feeding habits, particularly the construction and management of cells and brood, necessitates interactions among individuals. In fact, liquid food, such as nectar or fluid extracts of meat, is difficult to transfer to another individual without direct feeding, especially to larvae that are more or less confined to a particular location inside the nest. In the primitively eusocial bumblebees and the eusocial halictid bees, trophallaxis is nevertheless absent or very rarely observed. It is present in allodapine bees and in the honeybee, where food sharing is a sophisticated activity and where all workers exchange food with one another (Roubik 1989; Seeley 1985). In social wasps, adults may engage in larval feeding by regurgitating food (e.g., in the Stenogastrinae) (Turillazzi 1991). Also in the lower termites (Kalotermitidae, Rhinotermitidae), the exchange of liquid food is found quite universally (Wilson 1971). Colony members feed one another with either stomodeal food that originates from the salivary glands and the crop, or with proctodeal (anal) food from the hindgut. Stomodeal secretions are used to feed the royal pair and the larvae, while proctodeal excretions are rich in the symbiotic flagellates on which the lower termites depend for the digestion of cellulose. Proctodeal food is routinely given to colony members because termites lose their symbionts during each molting (where the gut lining together with the intestinal fauna is shed; see also fig. 1.2). The higher termites (Termitidae) do not depend on symbiotic flagellates and, consequently, the habit of proctodeal trophallaxis has disappeared.

Where it occurs in social insects, trophallaxis is among adult workers, from adults to larvae and also from larvae to adults. Trophallaxis seems to be an open system. Any one individual does give food to a large number of others as demonstrated by radioactive tracers in honeybees (Nixon and Ribbands 1952) and myrmicine ants (Bhatkar and Kloft 1977). Discrimination according to relatedness may perhaps occur, i.e., a more frequent food exchange among related workers, although the available evidence (Frumhoff and Schneider 1987) is now considered to be unconvincing. The transmission of food from larvae to adults has long remained enigmatic. It seems now that larvae have secretions that may be important for the adults' nutritional balance, especially in terms of amino acids (e.g., for a review in social wasps: Hunt 1991). An interesting case is provided in some primitively social bees, where the returning forager feeds the guards at the entrance in order to gain access to the nest (Michener 1985). This should be particularly prime for transmission of pathogens that are contracted on flowers (Durrer and Schmid-Hempel 1994).

So, how might trophallaxis be involved in parasite transmission? As for so many areas discussed in this book, we do not know exactly. Consider the case of the lower termites where symbiotic flagellates inhabit the digestive tract of the workers. When individuals molt they lose their flagellates but regain them from others by proctodeal feeding. These protozoans are considered to be mutualists with beneficial effects on their carriers. However, such food sharing would allow a parasitic habit to evolve in any protozoan that is transmitted this way. For example, the strains of flagellates that would grow faster in any one individual would be more likely to be transmitted to a new colony member when it receives its postmolt inoculation by proctodeal feeding. Hence, the need to pass on the flagellates relatively often (each individual termite will sooner or later die or molt) would open the door for the selfish interest of flagellates to multiply and become transmitted. To the extent that a high growth rate of a microorganism will eventually affect the carrier termite adversely, this would increase the "virulence" of the flagellate. We would therefore expect (for the arguments, see chap. 7) that parasitic habits are more likely to evolve in termite species that have either a high turnover of workers or else complicated caste systems, leading to high frequencies of flagellate transmission by proctodeal feeding, or in species that have a high diversity of flagellates leading to high levels of multiple "infection" of the same host. Clearly, this scenario is frequency-dependent because not all flagellates can evolve a parasitic habit, or else the host, depending on the microorganisms' beneficial services, would evolve countermeasures in turn. However, the point is that from the first principles of microevolution we would expect that such parasitic strains or variants occur among termite flagellates. Few studies have explicitly considered such a possibility, but the evolutionary perspective adopted here aids in identifying promising candidates. Frank (1996) discusses this problem also with respect to host-symbiont conflicts over virulence and multiple infection.

For trophallaxis to be an effective pathway of parasite transmission, some conditions have to be met. First of all, the parasite has to be able to be transferred in the food. This often requires that the parasite is able to survive in glandular secretions, in the physical-chemical environment of the digestive tract, and in the food itself, e.g., under high concentrations of sugar. If trophallaxis is restricted to the sharing of food via mouth (as with stomodeal feeding in termites), then the parasite must also be infective by os for the recipient, while its transmissible stage must also leave the host body via the mouth opening and not via the hindgut. This poses restrictions on the kind of parasites that can be spread by this kind of trophallactic interaction, particularly among adults. Bailey and Gibbs (1964) describe that chronic bee paralysis virus of adults is spread in honeybee colonies via trophallaxis among the workers and characteristically occurs in large, crowded colonies. Chronic bee paralysis virus is also found in the honey stomach of workers and in pollen prepared with the addition of honey. The virus is presumably secreted by the glands of the bees. Although this is a case in point, this virus is more effectively spread through injuries of the cuticle (Bailey et al. 1983b). Jouvenaz (1986) assigns an important role to the demonstrated food exchange among workers (Summerlin et al. 1975) for the transmission of pathogens in colonies of fire ants. In addition, some rhabditid nematodes that reside in the pharyngeal glands of ants may become transmitted by trophallaxis. As chapter 3 showed, many other parasites are contracted by feeding on contaminated food. Sometimes at least this transmission must involve trophallaxis, but this speculation clearly awaits further studies.

"Mouth-to-mouth" parasite transmission via trophallaxis should also be important in interactions between adults and larvae. A parasite that manages to live in the larva and to become transmitted to new hosts via the nursing workers exploits this niche. In chapter 3 I listed some of these cases. A typical example is American foulbrood in the honeybee (the bacterium *Paenibacillus larvae*) (see chaps. 2, 3). Bee larvae become infected when imbibing contaminated food. Spores can be passed among workers (Bitner et al. 1972) and an infected worker will shed infective cells for up to two months (Wilson 1972). Transmission is most effective when workers clean the cells of previously infected larvae and then move on to the next cell, where they attend another larva. Even spores from dead larvae can be transmitted in this way. Transmission is just as effective if the nurse bees are from a resistant line than if they are from a susceptible line, demonstrating that the resistance mechanism is not a behavioral one. This could be the case, for example, if cell-cleaning workers would not attend and feed live larvae (Rose and Briggs 1969). Because only young larvae are susceptible (Bamrick and Rothenbuhler 1961; see also Hitchcock et al. 1979), this also indicates that separation of brood by age class may help to decrease the rate of transmission within the colony. Such a separation is to some degree realized in honeybees, where the queen lays many eggs in one comb at a time. Hence, cohorts of broods are spatially separated.

Not only bacteria, but also other parasites, e.g., microsporidia in some social insects (Canning 1982), are transmitted in larval feeding (e.g., *Burenella dimorpha* in the fire ant; chap. 3). In several myrmicine species of the genera *Crematogaster, Leptothorax, Monomorium, Myrmica, Pheidole, Pheidologeton, Solenopsis,* and *Tetramorium,* the eggs of cestodes are fed by the attending adults to larvae. Other subfamilies of ants are also known to be affected, e.g., *Prenolepis* (Formicinae), *Euponera* (Ponerinae), and *Iridomyrmex* (Dolichoderinae).

There seem to be no known cases of parasites being transmitted specifically via hindgut secretions, although parasites could exploit this pathway. Such transmission is different in kind from transmission via the feces, as it necessitates special adaptations for the parasite to enter the respective glands or to get into the fluids that are discharged on the occasions. In contrast, transmission via feces is of course common, as the examples of chapters 2 and 3 have shown (e.g., *Nosema apis* in the honeybee).

The host is not without countermeasures. A structure called the proventriculus sits between the crop (or "Sozialmagen," the "social stomach") and the midgut, and thus regulates which food is consumed by the individual worker and which is regurgitated to be shared by trophallaxis with others. In bees, this device is also important in eliminating foreign particles. In ants, the infrabuccal pocket, a cavity just beneath the "tongue" of workers, also functions to filter out and collect most of the particles that might otherwise clog the narrow channels of the proventriculus (Eisner and Happ 1962; Hölldobler and Wilson 1990). Generally, the accumulated waste material is disgorged and removed from the nest in form of a pellet. In some species, for example in the Pseudomyrmecinae (Wheeler and Bailey 1920; Wilson 1971), the pellet is fed to the larvae.

Another example is the fire ant *Solenopsis,* where the microsporidian parasite *Burenella dimorpha* is maintained in this way (see fig. 3.1) (Jouvenaz et al. 1981). Strictly speaking, in this case, transmission does not depend on adults feeding or being fed by larvae. Rather, the symptoms of disease, i.e., that infected pupae develop a different morphology, elicits a hygienic behavior by the nurse workers that leads to parasite transmission. Where the pellet is removed rather than fed to larvae, the infrabuccal cavity acts as an efficient barrier to further transmission. The particle sizes that are removed are in the size range of spores (e.g., Eisner and Happ 1962) (cf. chap. 7). If the infrabuccal cavity was shaped by selection through parasite pressure, one would expect to see sophisticated structures in those species that are especially prone to infestation. Examples would be species with high rates of trophallactic interactions, or those with large colony size if it promotes the maintenance of parasites. The available data do not permit a check of these expectations.

In some cases, trophallaxis occurs between different species of social insects. This situation arises when colonies are composed of different species and individuals engage in food exchange (e.g., in ants: Bhatkar 1983). This mechanism could lead to parasite transmission across species boundaries. Similarly, trophallaxis sometimes occurs among different colonies in the population (Bhatkar 1979a,b). This could lead to the horizontal transfer of parasites to different colonies within the population. However, these two phenomena seem rare and are therefore not expected to be a major route by which parasites are shared among different colonies or species of social insects in a given area. It is more likely that the shared use of resources (see Durrer and Schmid-Hempel 1994) is the major pathway that links different species and colonies. But there is clearly a need to explore these possibilities more systematically.

Grooming and Hygienic Behavior

Workers routinely remove particles from the surface of their own or a nestmate's body. As with trophallaxis, such grooming behavior occurs in all major taxa, but it is not universal. The removed particles may include a foreign object. In particular, it can include parasite spores and hyphae of fungi, or eggs and juveniles of nematodes. Thus, grooming can function to prevent infections by parasites (Farish 1972). However, there are two sides to grooming. Whereas the process removes potentially dangerous parasites from a nestmate, the grooming worker could at the same time become infected, e.g., as with the entomophagous fungus *Metarhizium anisopliae* in the termite *Reticulitermes,* where infected workers are more attractive to groomers. However, individuals that have already been killed by the fungus are normally avoided and are not cannibalized or groomed by healthy workers. At this point, the disease is thus not spread further (Kramm et al. 1982). Honeybees infected by the chronic bee paralysis virus elicit a "grooming" behavior, i.e., nestmates "attack," lick, and chew the body of the infected worker, resulting in the diseased bee's characteristic hairless appearance ("hairless black syndrome"). Behaviorally, these attacks resemble those against submissive intruders or against workers that develop ovaries (Waddington and Rothenbuhler 1976). In cage tests with virus-infected workers, diseased individuals also elicit more aggressive responses than healthy nestmates or even foreign intruders (Drum and Rothenbuhler 1985). Although these attacks may eventually result in isolating diseased individuals, such behavior should also benefit the parasite by increasing the chances of transmission, since this virus can be spread to body openings (chaps. 2, 3). It may indicate that the parasite has won in this evolutionary race. Entomophagous fungi are typical cases of parasites being able to infect the host in the ways described, though bacteria or nematodes could also qualify. Host defenses that aim at eliminating parasites from this transmission route should therefore include specific patterns of grooming. In particular, diseased individuals would probably not be groomed if they pose a risk of spreading the disease. A possible example could be infection by chronic bee paralysis virus, if aggression is elicited but no licking and chewing.

Transmission by grooming has an additional dimension when grooming among nestmates is not random. Especially if close kin were more often groomed than less related individuals. Unfortunately, the study by Frumhoff and Schneider (1987), suggesting such an effect, is ambiguous. With more frequent interactions among kin, parasites would more often be transmitted to another host with a similar genotype. This should make it even more likely that it can establish itself on the next worker. In fact, Shykoff and Schmid-Hempel (1991a) found that sibs in mixed groups of workers of the bumblebee *B. terrestris* were interacting with one another more often than non-sibs. However, in this case, this difference had no relation to the probability of transmission of the trypanosome *C. bombi*. Nevertheless, nepotism could act to increase net transmission rates and thus the likelihood that a parasite can be maintained in the population. On the other hand, if grooming relationships are structured in this manner, the effective number of available susceptible hosts becomes smaller, as does the potential for parasite maintenance (see chaps. 6 and 7 for such epidemiological considerations). Chapter 3 gives an overview of further instances where hygienic behavior by the workers leads to the transmission of parasites. In particular, parasites that infect adipose tissue (e.g., the microsporidian *Burenella* in ants and *Gurleya* in termites) depend on hygienic behavior (e.g., cannibalization, necrophoretic behavior) for their spread because they do not simply pass out in feces of the host.

It is not known to what extent these processes affect parasites and their ways. However, for a successfully established parasite it would be beneficial if its host preferentially interacted with close kin and thus with susceptibles that are more prone to becoming infected by the same parasite. This scenario leads to a number of expectations if the parasite could influence these interactions.

For example, for parasites that are transmitted by social interactions, be it grooming, trophallaxis, or even simple spatial proximity, the degree of nepotism should increase with infection levels. The opposite should be the case when the host controls the infection by modification of social interactions. This could mean interacting less or interacting less frequently with close kin. Therefore, in species where colonies consist of different patri- or matrilines, the analysis of social contacts and how these interactions are distributed among kin has an intriguing additional dimension—the parasites.

A very crude test of these ideas is to compare the frequencies of social interactions, especially grooming, in relation to parasite load. This special behavior is readily recognized and quantified. A useful additional assumption is that larger colonies suffer more from parasites (see below, fig. 4.10). Hence, if the host is in control and everything else is the same, self-grooming as an act of hygienic behavior should become more frequent as colony size increases, while allogrooming, being a potential risk, should be reduced. Data for a number of ant species suggest the opposite, i.e., that the frequency of self-grooming decreases with the typical colony size of the species (Schmid-Hempel 1990), while a reanalysis of this data suggests that allogrooming activity increases with colony size (fig. 4.1). The same analysis suggests that grooming activity declines with larger colony size, but general activity increases. Workers of ant species with larger colonies are therefore typically more active in allogrooming but less in self-grooming. Hence, if large colony size is a correlate of increased parasite pressure, then workers seem to be less hygienic for themselves but more often engage in activities that promote transmission. Ideally though, these expectations should be analyzed for experimentally manipulated or infected colonies. More general theoretical expectations can be formulated by analyzing the consequences of the host-parasite dynamics within the colony.

4.3 Parasite Dynamics within the Colony

Division of Labor

Caste and division of labor generate smaller units within the colony, for example as a result of canalized interactions or different spatial domains of individual activities. A turnover problem might arise too, because individuals of the different groups may have different life spans or developmental rates. They are thus available as hosts for different lengths of time. How does this affect the pathways of transmission and the dynamics of infections?

Sophisticated division of labor represents a more heterogeneous environment for the parasite, particularly in the process of transmission. Consider the case of a clearly delineated age-related division of labor, where young workers would tend the brood and then switch to foraging at an older age, with little

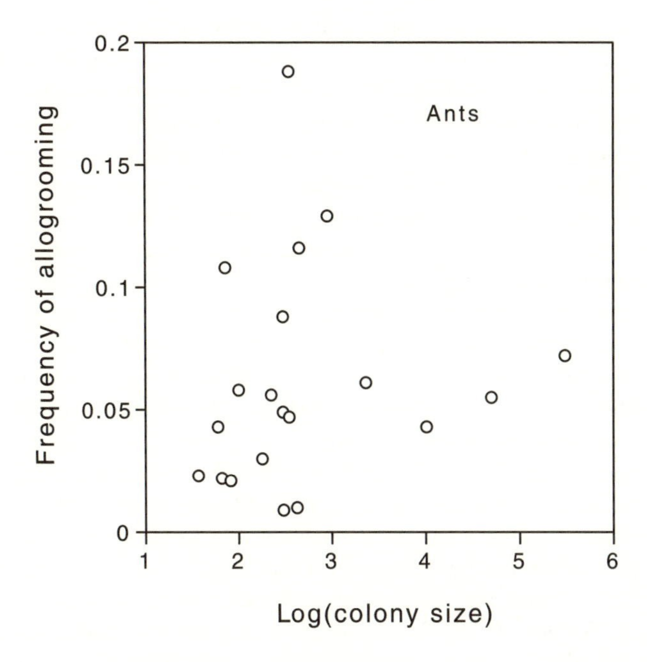

Figure 4.1 The frequency of allogrooming by workers (proportion of this activity out of total activity budget) plotted against the typical colony size (number of workers) of the species. Each dot is a different species of ants. (Data from Schmid-Hempel 1990 and Kaspari and Vargo 1995.) The correlation is significant (Spearman's $z=2.253$, $P=0.024$, $N=20$).

overlap between these two activities. Such a scenario approaches that observed in honeybees (Seeley 1982) and in the ant *Pheidole dentata* (Wilson 1976). Consequently, a parasite that is able to infect only the young workers or brood and which does not multiply fast enough will not be transmitted often. This is because the host worker will have become older in the meantime and may already have switched to other activities outside the nest, thus making it less likely that the parasite is transmitted back to the young workers or brood. Parasites that rely on being transmitted to certain classes of individuals within the nest should thus be under considerable selection pressure to adapt their developmental rates and timing of transmission. Schmid-Hempel and Schmid-Hempel (1993) have tested for this possibility with the trypanosome *Crithidia* infecting *Bombus* spp. They found a trend showing such that rapidly multiplying variants of the parasite fare better in infecting other members of the same colony. Slowly multiplying variants, in contrast, are more likely to infect unrelated individuals. This would correspond to transmission within and among colonies. The magnitude of these effects was small, however.

The phenomenon of division of labor forces parasites to adapt their ways not only with respect to host physiology but also with respect to the individual host life histories that unfold against the background of a social life. *Varroa* mites infecting honeybees, for example, attach to different sites of the body on young and old host workers, respectively (Kraus et al. 1986). This arrangement helps to infect brood cells via the young nurses while it facilitates transport to other parts of the hive or other colonies by the older workers and foragers. More sophisticated social organizations should correlate with more specificity in the parasites and, perhaps, with fewer parasite species being present. This cannot be tested for the time being.

The organization of work in redundant chains (box 4.1) makes the system as a whole more reliable. In addition, reproductive success of the colony depends on its size at reproduction (see also chap. 5). Now consider again the problem of reliability as illustrated in box 4.1. Note that the pathways of work are also potential pathways of transmitting an infectious disease, with the circles of box 4.1 being individual hosts. The overall probability of success, P_s, is now the probability that the infection has spread at least once from start to finish of all possible chains and hence to the colony as a whole. Consequently, p is the probability of establishment and successful development of the parasite inside each individual host along the chain, combined with the rate of transmission to the next hosts. In other words, p is the probability that the infection is passed on by one individual to the next as they come into contact during their normal social activities. Considering box 4.1 and figure 4.2, it is clear that work schemes that are reliable in the sense of increased system performance in the ergonomic sense are also those that ensure the highest probability for the spread of an infection. The "series-parallel" scheme is perhaps most relevant for social insects because, typically, many individuals (high values of m) are organized in a few castes (i.e., with a comparatively low value of n). If parasites can spread freely among individuals, but only along the defined work chains, this typical social organization is both reliable and also highly likely to allow an infectious parasite to spread throughout the entire colony.

Assume that individuals within a colony have a chance p to become infected, and if infected their contribution to colony development shall become negligible. In the general case, the colony is supposed to be successful if at least k of the n individuals in the colony escape infection and thus ensure a sufficient amount of work to be done. Under these conditions, colony success is (see box 4.1, eq. 2).

$$P_{Colony} = \Sigma_{i=k}^{n} \binom{n}{i} (1-p)^i p^{n-i}. \qquad (4.1)$$

Colony success is thus related in a sigmoidal fashion to the probability that individuals fail similarly but in an inverse manner to figure 4.2. The probability of infection and transmission (the x-axis of fig. 4.2) can now also be considered to reflect the level of resistance (physiological and behavioral) that individuals mount against the infection. Low values of p mean that the individuals are refractory to the infection. With low p, the infection is likely to grind to a halt

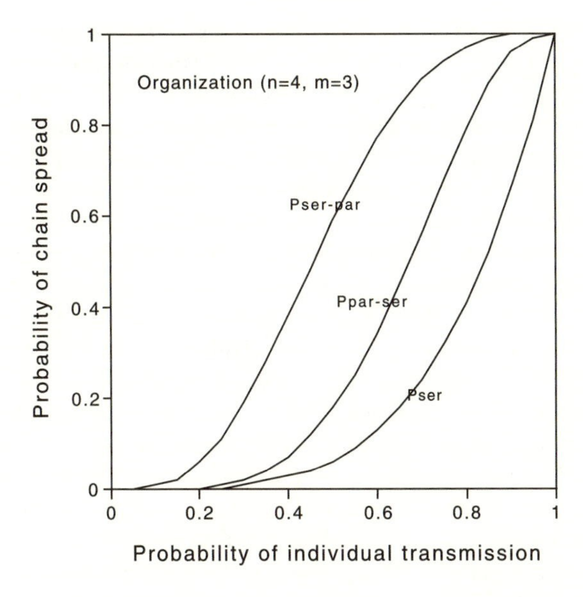

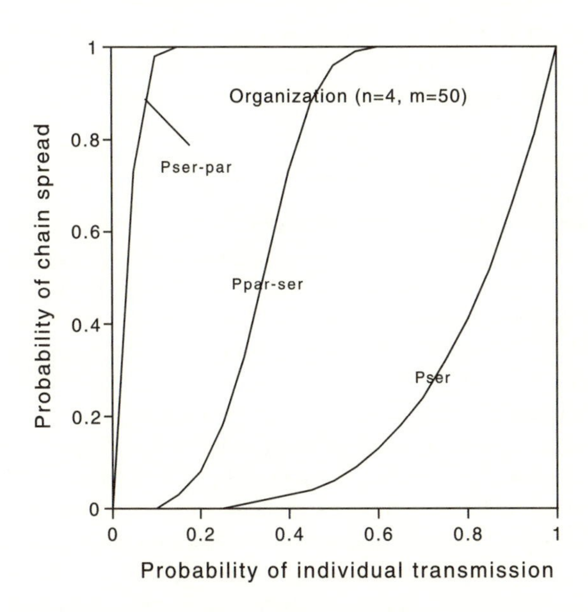

Figure 4.2 The probability of the spread of an infection throughout an entire transmission chain (ordinate) as a function of the probability of establishment and transmission, p, per individual (abscissa). The work chains are organized as: *ser*=serial, *par-ser*=parallel-series, *ser-par*=series-parallel; see box 4.1 for further details. The parameters are (a) $n=4$, $m=3$; (b) $n=4$, $m=50$. n characterizes the number of compartments in a colony (e.g., the distinct castes); m defines the number of individuals within each compartment. Case (b) is typical for social insects since m is much larger than n.

along the transmission chain, depending on social organization. In passing, it should be noted that the same is to be expected when the pathogen is too virulent and kills its host too quickly before transmission is achieved (see also chap. 7). In addition, k would indicate the amount of parasite-induced damage that the colony could tolerate before failing to survive or reproduce. Low levels of k would indicate that the colony is robust and only a few individuals must survive to ensure colony success. Note that k is an absolute number rather than a proportion. Therefore, contrary to intuition, this would suggest that small and more primitive societies are probably more likely to succeed if only a few individuals are not infected than larger and more advanced societies. In the latter, probably a higher threshold number of individuals is required to maintain the colony's functional integrity. Hence, if p reflects resistance, then large and highly advanced societies should be investing more in parasite defenses than small and more primitively organized ones (but see the discussion of resistance in chap. 7). That parasites can actually knock out entire colonies is known for the honeybee. Outbreaks of foulbrood, sacbrood, *Nosema, Varroa,* and other diseases can rapidly kill a colony (Bailey and Ball 1991) and should naturally select for controls of the epidemic. However, these expectations have so far not been tested, and table 4.1, at least, does not support them strongly. But in principle, direct experimental tests of how the organization of work facilitates or impedes parasite infections could be done, for example, with techniques similar to those used by Wilson (1980b) to test economic concepts.

The Dynamics of Infection

So far, I have considered a rather static case of host-parasite interactions. This seems justified as a first glance at how parasitism and colony organization may interact. But of course the dynamics that allow an infection to establish itself and render the hosts unavailable, then clear up and make the hosts susceptible again, have to be considered in determining how division of labor affects parasitism.

A simple model like this is formulated in box 4.2: a colony has two castes—foragers and nest workers—both of which contribute to colony growth in relation to their numbers, and in relation to the ratio of the two castes. Nest workers age and develop to become foragers. As a measure of colony success, worker numbers (colony size) are counted at the end of the cycle (Schmid-Hempel et al. 1993). The model assumes that new infectious parasites are picked up by the foragers outside the nest and at an arbitrary time in the life cycle of the colony. This point marks the beginning of the epidemic process as modeled here. Only the subsequent dynamics are considered in box 4.2. The parasite is then transmitted among the foragers and among the nest workers at the same or different rates. Hence, the division of labor can be modeled by defining a matrix of who acquires the infection from whom (a "WAIFW"-matrix, sensu Anderson and May 1991); its entries are the transmission rates between classes of individuals.

BOX 4.2 DYNAMICS OF PARASITISM WITH DIVISION OF LABOR

A useful basic model of a simple host-parsite dynamics within a social insect colony is the following:

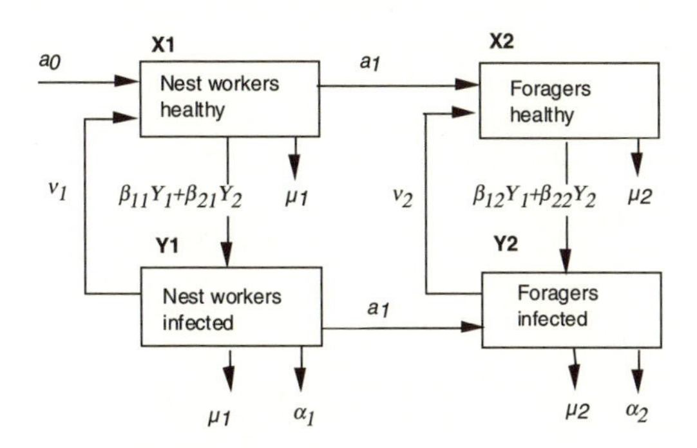

The parameters are as follows:

X_1 = number of healthy (uninfected) nest workers

X_2 = number of healthy (uninfected) foragers

Y_1 = number of infected nest workers

Y_2 = number of infected foragers

a_0, a_1 = rates of transition between age classes (recruitment)

μ_1, μ_2 = (background) mortality rates

α_1, α_2 = parasite-induced mortality rates

v_1, v_2 = recovery rates

β_{ij} = transmission rates

The transmission rates are defined in analogy with a WAIFW-matrix as follows:

β_{11} = transmission nest worker to nest worker

β_{12} = transmission nest worker to forager

β_{21} = transmission forager to nest worker

β_{22} = transmission forager to forager.

The dynamics of this system is started by introducing an infection that is picked up by the foragers outside the nest at an arbitrary time in the colony cycle. This marks the start of the epidemic process modeled here. Furthermore, one has to assume an age distribution at this point in time, where a fraction ξ of the workers are foragers. Then,

$$X_1(0) = (1-\xi)X_0;\ Y_1(0) = 0,$$
$$X_2(0) = \xi(1-f)X_0;\ Y_2(0) = \xi f X_0.$$

BOX 4.2 CONT.

The "birthrate" of healthy nest workers shall be proportional (with constant a_0) to the size of the colony given as $N = X_1 + X_2 + Y_1 + Y_2$. With this simplification, no further effect of the infection on colony productivity apart from increased worker mortality (rates α) is assumed. There is therefore no effect of caste proportions on productivity (as both classes, *X*, *Y*, contribute in the same way). Implicitly, it is also assumed that individual hosts cannot develop (i.e., change age class) and recover at the same time. With these simplifying assumptions, the following equations are a mathematical formulation of the flow of workers between compartments as depicted above, i.e., with inflow and outflow to and from the compartments:

$$\frac{dX_1}{dt} = a_0 N + v_1 Y_1 - X_1\left(\mu_1 + a_1 + \beta_{11} Y_1 + \beta_{21} Y_2\right),$$

$$\frac{dY_1}{dt} = X_1\left(\beta_{11} Y_1 + \beta_{21} Y_2\right) - Y_1\left(\alpha_1 + \mu_1 + v_1 + a_1\right),$$

$$\frac{dX_2}{dt} = a_1\left(X_1\right) + v_2 Y_2 - X_2\left(\mu_2 + \beta_{12} Y_1 + \beta_{22} Y_2\right),$$

$$\frac{dY_2}{dt} = X_2\left(\beta_{12} Y_1 + \beta_{22} Y_2\right) + a_1 Y_1 - Y_2\left(\alpha_2 + \mu_2 + v_2\right).$$

When parasites debilitate their hosts, the effective number of workers contributing to colony growth can be corrected as

$$X_1' = X_1 + d_1 Y_1;\ X_2' = X_2 + d_2 Y_2;\ hence,\ N' = X_1' + X_2'.$$

If, in addition, the caste ratio determines the effectiveness of the colony to raise new workers, then, without loss of generality, the best caste ratio is assumed to be equality (nest workers : foragers = 1 : 1); hence, a measure of effectiveness can be postulated with a deviation as

$$\varepsilon = \left(\frac{X_1'}{N'}\right)\left(\frac{X_2'}{N'}\right).$$

With these amendments, the first of the set of differential equations above is modified to become

$$\frac{dX_1}{dt} = a_0 \varepsilon N' + v_1 Y_1 - X_1(\mu_1 + a_1 + \beta_{11} Y_1 + \beta_{21} Y_2).$$

In the present scenario, division of labor is based on age-dependent polyethism. In addition to the effect of host birth and death rates that determine the size of each caste, differential transmission within and among castes is assumed to be a major factor affecting the dynamics of an infectious parasite. Evaluation of this set of equations will yield final colony size, $N(T)$, at the end of a colony life cycle, T, a fitness measure for colonies (see fig. 4.3).

With a system of division of labor, transmission within castes has more weight than between castes. This is different from the earlier scenarios where the parasite can pass between and within castes at the same rate. Furthermore, it may be assumed that parasites cause mortality and debilitation at the same time. The difference is not unimportant, since a debilitated but living worker will still be able to transmit a disease, whereas a dead worker removes a part of the infection.

In this context, it may be asked to what extent workers may commit "adaptive suicide" to remove an infection from the colony (Smith Trail 1980; McAllister et al. 1990) (cf. table 3.4). Adaptive suicide is unlikely to evolve when the social insect workers are intermediate hosts, e.g., as in the case of trematodes that utilize birds as final hosts (see chap. 2). Given the biology of the system, any colony in the local population would benefit from this suicidal act, and hence the selectively relevant benefit to one's own kin must be vastly diluted. In the case of intermediate hosts as considered by Smith Trail (1980), behavioral change is therefore unlikely to have evolved for the host's benefit. In other cases, where the worker is infected with a contagious disease, an immediate benefit seems more likely. For example, *Nosema*-infected honeybee workers that start to forage earlier might be more likely to remove the spores from the colony. Similar effects occur with sacbrood virus (chap. 3; Bailey and Ball 1991). These parasites are also more likely to be spread to other colonies in the population by acceleration of the aging process, which means that workers start leaving the colony and foraging earlier (Schmid-Hempel and Schmid-Hempel 1993). So, a general conclusion about the possible role of suicidal behavior is still outstanding.

The dynamical behavior of the model in box 4.2 is displayed in figure 4.3. It shows how the disease unfolds with and without division of labor. In the case considered here, division of labor in fact affects parasite dynamics twofold by the demography of age-related polyethism and by differential transmission rates. The two cases shown in the graphs produce very different dynamics. In the division of labor (fig. 4.3b), the infection is largely contained within the forager caste, as not all become infected when the recruitment of new foragers (i.e., the rate at which nest workers switch to foraging) is sufficiently higher than the rate of new infections. In a system without division of labor (fig. 4.3a), all foragers will eventually become infected (because they can also become infected by nest workers), but not all of the nest workers will be infected. In this case, the proportions of infected and noninfected caste members display a more complex pattern. In the two cases considered here, division of labor impedes parasite transmission and thus increases colony fitness (measured as colony size, N, in arbitrary units at the end of the cycle). Figure 4.3 also shows that the fraction of infected workers in any one caste varies along the colony cycle in a way that is determined by the dynamics of infection. Although more complex models can be built for social insects that have several castes, for the clearance

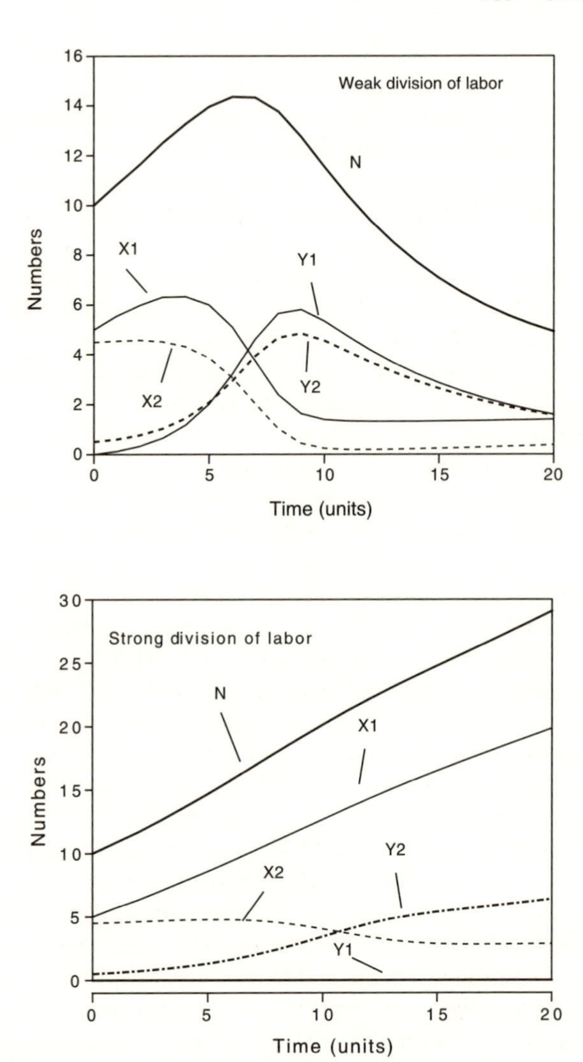

Figure 4.3 Numerical evaluation of the host-parasite dynamics described in box 4.2. Parameter values are: $X_0=10$, $\gamma=0.5$, $a_0=0.5$, $a_1=0.1$, $\mu_1=0.01$, $\mu_2=0.05$, $\alpha_1=\alpha_2=0.2$, $v_1=v_2=0$, $\beta_{11}=\beta_{22}=0.1$, f=0.1. (a) Weak division of labor, i.e., parasite transmission from foragers to nest workers and vice versa is intense; i.e. $\beta_{12}=\beta_{21}=0.05$. Final size of the colony at the end of the period (20 time units) is $N_{20}=4.93$ worker (units). (b) Strong division of labor; transmission between foragers and nest workers is negligible; i.e., $\beta_{12}=\beta_{21}=0$. $N_{20}=29.10$ workers (units). Symbols are X1, X2: number of uninfected nest workers and foragers, respectively; Y1, Y2: number of infected nest workers and foragers; N: total number of individuals (colony size). The length of the colony cycle after infection is set at $T=20$ units. See box 4.2 for details.

of parasites, or for different transmission matrices that result from serial and serial-parallel work schemes sketched in box 4.1, the pattern shown in figure 4.3 should capture the essence of the dynamics. Is protection against parasites therefore an incidental by-product of the division of labor, or is this system selected for by parasitism? Given the many determinants and consequences of the division of labor, the former is more likely. The second alternative will be considered below.

Worker Turnover

The average life span of workers varies considerably among species of social insects (table 1.3). This variable life span, together with variation in the rates of recruitment into different castes and with mortality rates that remove workers, determines the parasite dynamics within a colony. Note that in this section worker turnover, as for example set by mortality rates, is not assumed to be a consequence of life spans reduced by parasitism (e.g., Ponten and Ritter 1992). Rather, I will discuss this as a possible strategy adopted by the colony, i.e., to produce long-lived or short-lived workers and thus to accelerate or retard the schedule of age-related behaviors.

The basic model of box 4.2 can again be used to assess the effect of variation in these parameters. Figure 4.4a shows how final colony size (considered to be a fitness token, W) attained at the end of the cycle varies with worker birthrate. Birthrate is a reflection of habitat quality and/or the ergonomic strategy of the colony. As expected, as birthrate (a_0) increases, colony size increases too. But note that in these calculations, the mortality rates (cf. box 4.2) are set in proportion to birthrate so as to keep worker numbers in check. Hence, the increase shown in figure 4.4a is not a simple effect of higher birthrates, as the graph seems to suggest. Rather, it also results from the accelerated, externally driven but "strategically implemented" turnover rates of workers, as determined by birth and mortality rates, that outrun the epidemic. For example, when foragers die too quickly (high values of μ_2) and therefore cannot further spread the infection, fewer and fewer workers will become infected within the colony. In addition, as transition from one age-related caste to the next is accelerated (i.e., an increasing ratio a_1: a_0), final colony size changes. Under some combinations, too few workers remain in the nest and too many become exposed as foragers to parasites. Again, this is not just an effect of aging per se but an additional epidemiological effect. An important point, however, is that the interplay of these processes is not simple. In fact, figure 4.4a demonstrates that for intermediate transition rates (e.g., $a_1 : a_0 = 0.2, 0.5$), colony fitness is highest, because at these rates the epidemic is outpaced. When transition becomes too fast ($a_1 : a_0 = 1.0$), a negative effect sets in as workers pass to the forager stage too rapidly and acquire parasites too readily.

Figure 4.4b considers the combined effect of birth and background mortality

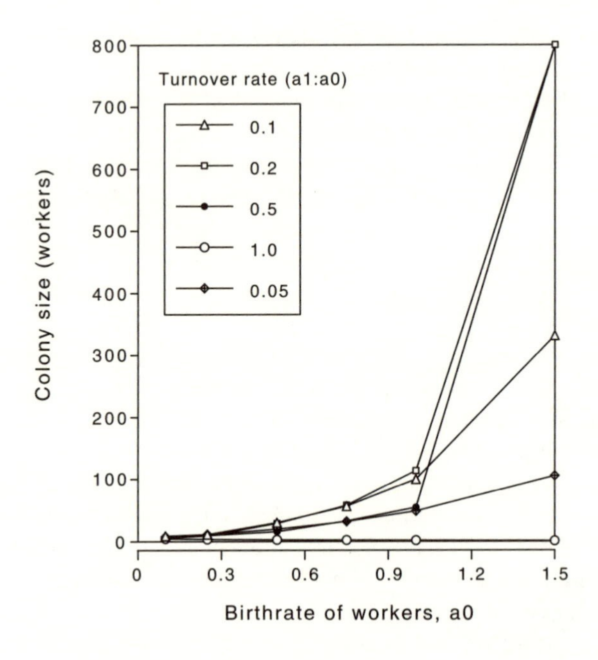

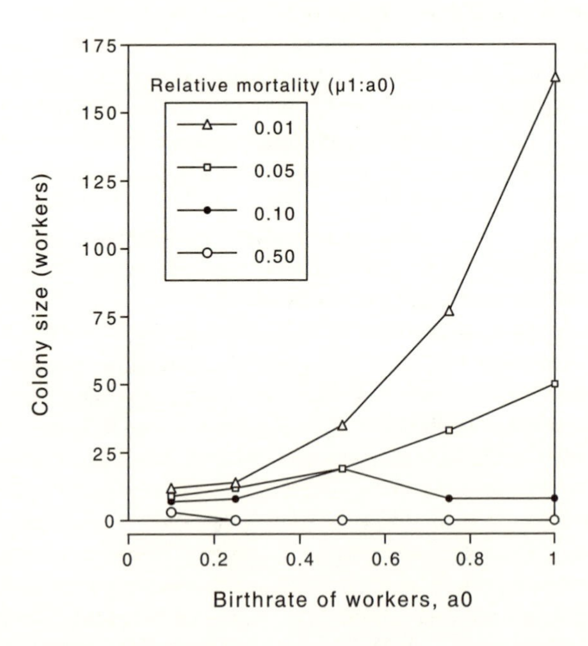

Figure 4.4 The effect of worker birthrate, a_0, (relative) worker mortality rate, μ_1, and transition rate (ratio $a_1 : a_0$) on colony size at the end of the cycle; this is proportional to colony fitness. The calculations follow the model shown in box 4.2. Mortality rates are chosen so that $\mu_2 = 5\mu_1$; hence, μ_1 is a general indicator of background mortality. (a) The effect of an increase in transition rate, i.e., the relative speed at which individuals pass from worker to the forager caste (given by the ratio $a_1 : a_0$), and of birthrate; $\mu_1 = 0.02\ a_0$. (b) The effect of mortality rate relative to birthrate (given by the ratio $\mu_1 : a_0$). Other values as in figure 4.3a (transition rate $a_1 = 0.1$).

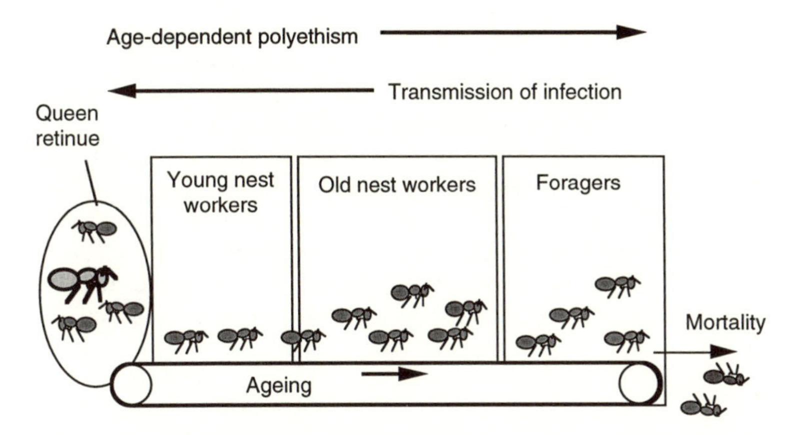

Figure 4.5 The conveyor belt model to visualize the importance of the internal population dynamics of colonies. The ontogenetic shifts in the activities of workers follows an age-dependent schedule. Young workers are first around the queen, then work in the nest and finally go out to forage. This is depicted as individuals being transported from inside to outside on a conveyor belt. Typically, the novel and potentially dangerous infections are acquired by the foragers from outside. The spread of the infection then follows the opposite, "inward" pathway. The major period of mortality is during foraging; therefore, the belt carries almost all individuals from start to end.

rates. As before, high birthrates lead to higher final colony sizes over a range of parameter values. In this case, the combined effects match the intuitive expectations. For example, colony success decreases with an increase in mortality rate relative to birthrate. However, this is again not a simple effect. As a high birthrate tends to outpace the epidemic (as in fig. 4.4a), this is no longer sufficient to compensate for a high background mortality. In fact, under certain circumstances a high birthrate that is unchecked by a corresponding increase in the mortality rate will eventually accelerate the epidemic because too many susceptible hosts become available. As the epidemic process explodes, this may even lead to very low colony success. The host-parasite dynamics displayed in figure 4.4 show that within-colony demographic strategies, i.e., fast or slow turnover of workers, profoundly affect the spread of an infectious parasite and hence determine the success the colony will ultimately achieve. The relationships are not always intuitively understood and an explicit model may be needed to gain insight.

Obviously, very many parameters affect parasite dynamics, even in a simple model like that in box 4.2. The situation can be visualized with the "conveyor belt" model shown in figure 4.5, which gives a good idea of the dynamics of the process. In fact, workers of many social insect societies have survivorship curves close to type III (see chap. 1 and fig. 6.3). Correspondingly, the conveyor belt takes individuals through their entire ontogeny and drops them at the forager end. From this metaphor, it can be appreciated that if, for example, the conveyor belt runs fast, i.e., workers age rapidly and start to forage early, this will confine the infection to the most "outward" compartments, because parasite transmission "inward" cannot keep pace with the belt's movement. The conveyor metaphor also illustrates that key individuals in the colony—the reproductives—are shielded from infections. Some observations to this effect could be mentioned: *Nosema*-infected honeybee workers can no longer produce royal jelly and start to forage earlier; they thus do not attend the queen (Wang and Moeller 1970). In addition, some simple predictions from these scenarios would follow. For example, high rates of aging should be associated with high parasite pressure. None of these expectations have been investigated empirically.

4.4 Parasites as Selective Factors for Social Organization

Polymorphism

Worker size does play a role for susceptibility to parasites. For example, large workers of the leaf-cutter ant *Acromyrmex octospinosus* are less susceptible than small ones to infection by nematodes (Kermarrec et al. 1990). Also, mites are known to infect not only younger nurse workers (e.g., in honeybees: Gary et al. 1989) but also smaller workers that act as nurses (as in the garden ant *Lasius flavus:* Franks et al. 1991). Such studies clearly need to be broadened to investigate to what extent different castes and morphs differ in their susceptibility to parasites.

With this in mind, the observation of a negative correlation between the degree of polymorphism, a morphological indicator of the sophistication of division of labor in a colony, and the degree of polygyny across ant species is interesting (Oster and Wilson 1978; Frumhoff and Ward 1992). The latter authors suggested an evolutionary sequence where polymorphism is first, and is less likely to be followed by polygyny than by monogyny. Keller (1995) interprets this association as consistent with selection for a reduction in parasite pressure in polymorphic species due to higher variability inside the colony, This would in turn lead to a higher threshold for the evolution of polygyny. Thus, Keller (1995) assigns polymorphism the major weight in generating variability for parasites. But despite the above cited examples, susceptibility should primarily be determined by genotype (see chaps. 5, 7). Moreover, in hymenopteran societies, genotype is not related to worker phenotype, since caste determination is almost universally regulated by phenotypic plasticity. Only to the extent that polymorphism generates a different dynamics, in the sense of box 4.2 and as discussed in figure 4.4, could the refractoriness of the entire colony against parasitic infections be affected. The argument is therefore not based on variability per se but on the resulting epidemiological processes.

In their discussion of the adaptive value of castes in social insects, Oster and Wilson (1978) concluded that, given its many advantages, polymorphism is less common than one might expect. Consequently, they looked for factors that reduce the value of polymorphism. For example, individual behavioral flexibility can make morphological castes superfluous. So far it has been assumed that polymorphism is advantageous in reducing parasite loads or effects (as in Keller 1995). However, could parasites also select against polymorphism? An example is provided by leaf-cutter ants that are attacked by phorid flies. The flies prefer to oviposit into larger workers (the majors) of the colony (Hölldobler and Wilson 1990; appendix 2.11). In the fire ant *Solenopsis,* the threat posed by the flies not only affects the workers directly but also reduces the competitive dominance of the host species over its competitors, since the large workers that normally defend a feeding site against other colonies or species tend to hide (Feener 1981, 1987, 1988). Preference for larger workers has also been found in conopid flies parasitizing workers of bumblebees (Schmid-Hempel 1994b; Schmid-Hempel and Schmid-Hempel 1996; Müller et al. 1996). Viewed from the colony's point of view, such caste-specific parasitization is an extra cost added to the production and maintenance of large worker morphs. Consequently, Feener (1987, 1988) suggested that such parasite-induced extra costs may select against polymorphism. For ants, this hypothesis can be checked. With the cases listed in appendix 2, and when a crude distinction into monomorphic and polymorphic caste systems is made, the standardized parasite richness is indeed higher for polymorphic than for monomorphic ants (table 4.2). Unfortunately, the phylogenetic effects cannot be sorted out in this data set, since too few contrasts are available. Nevertheless, in the closely related taxa in table 4.2, the differences go more or less in the same direction. On the other hand, polymorphic ants have, on average, larger colonies than monomorphic species. When both factors are considered simultaneously, the effect of polymorphism disappears (albeit with a rather low observed power of analysis), but colony size cannot explain the differences either. More data are thus urgently needed. For the time being, a rather tentative conclusion is that polymorphism may have an additional cost due to increased parasite loads. Polymorphic species may "collect" more parasites, perhaps because more niches are available for them. Parasites may thus help to explain why polymorphism is not as widespread as one would expect for purely ergonomic reasons.

Normally, insects show little external effects of parasitism by nematodes. In ants, the workers parasitized by nematodes often differ from normal ones only by a slightly distended gaster or a somewhat different color. But sometimes infection by nematodes manifests itself in more dramatic ways, as described in chapter 2. In particular, primary and secondary sexual characters can be modified by mermithids, resulting in intercastes (e.g., Wheeler 1910; Wheeler 1928). Such individuals will show female, worker, or soldier features to varying degrees. In addition, nematodes also parasitize the drones and queens (as in the

Table 4.2

Parasite Richness (Number of Parasite Species per Host Species; Appendix 2, Except Mites) in Mono- and Polymorphic Ants (Excluding Socially Parasitic Species)

Monomorphic Species[a]	*Richness*	*Standardized*[b,c]	*Polymorphic Species*[a]	*Richness*	*Standardized*[b,c]
Formica neogagates	1	−0.5603	*Atta texana*	5	0.4748
Iridomyrmex humilis[d]	5	−0.0083	*Cataglyphis cursor*	2	−0.4861
Lasius alienus	6	0.5000	*Cephalotes atratus*	3	0.2741
Lasius niger	12	0.6872	*Oecophylla longinoda*	1	−0.6964
Leptothorax acervorum	4	−0.0355	*Pheidole dentata*	7	0.6510
Leptothorax longispinosus	2	−0.3933	*Solenopsis geminata*	16	1.0332
Leptothorax recedens	1	−0.4754	*Solenopsis invicta*	21	0.8863
Myrmecia pilosula	1	−0.4762			
Myrmica rubra	2	−0.6843			
Plagiolepis pygmaea	2	−0.3473			
Tapinoma erraticum	1	−0.6744			
Tetramorium caespitum	9	0.6091			
Wassmannia auropunctata	1	−0.0691			
Average load ($N = 13$)		-0.1483 ± 0.133	($N=7$)		0.3053 ± 0.251

NOTES: Standardized parasite richness from fig. 2.1. Data on polymorphism from various sources.

[a]In this sample, polymorphic ants (mean log colony size $= 4.482 \pm 0.48$ S. E., $N=6$ species) have larger colonies than monomorphic species (mean $= 2.795 \pm 0.98$, $N=10$) ($t=3.11$, $df=14$, $P=0.008$).

[b]Monomorphic vs. polymorphic standardized richness: $t=1.76$, $df=18$, $P=0.048$ one-tailed. Too few contrasts available for comparative analysis.

[c]Taking into account colony size as a covariate ($F_{1,15}=1.09$, $P=0.32$; $r^2=0.134$), the effect of mono- vs. polymorphism is $F_{1,15}=0.52$, $P=0.485$, observed power of test $=0.12$.

[d] = *Linepithema humile*.

honeybee). External effects on morphology are also known to be associated with infestation by strepsipteran parasites (chap. 2), but so far these have only been described in solitary bees (Smith and Hamm 1914). Parasitism that affects worker morphology and even caste morphs should obviously leave a distinctive mark on caste structure and interfere with the colony's ergonomic design.

Polyethism

If an appropriate organization of the colony reduces the probabilities of parasite transmission, fewer parasites should be able to establish themselves. Sophisticated social organization should thus help to reduce parasite impact. For example, if workers can infect queens (e.g., Wang and Moeller 1970), then access to her should be restricted, particularly when the colony depends on a single queen, as under stable, long-lasting monogyny. In this context, it can be noted that in all species of social insects that display an age-based division of labor, it is the younger individuals that attend the queen while the older ones forage (e.g., Wilson 1971; Hölldobler and Wilson 1990). The advantage of this arrangement is considered to be that it places the highest mortality risks, imposed by hazards outside the colony, on the older and therefore less valuable classes of workers (e.g., Wilson 1968; Jeanne 1986). But the probability that a worker is infected by a disease must also increase with age. Therefore, older workers must pose a greater risk to the queen (or to the brood, for that matter). It is not known and it has never been explicitly tested whether strict age-related division of labor better protects the queen from parasites (fig. 4.5). In addition, see chapters 2 and 3 for several examples of the age-dependent susceptibility of workers. With parasites, therefore, queen attendance should be more specialized, i.e., be restricted to a more clearly defined worker caste (age or morphs), in strictly monogynous species as compared to polygynous or sequentially monogynous species (Schmid-Hempel and Schmid-Hempel 1993). In addition to queen access being reserved for young workers (via the age-related schedule), access could also be restricted to classes that are unlikely to be exposed to parasite-infested environments. An example may be the case of the leaf-cutter ants *Atta* spp., where the small number of workers that spend their entire life inside the nest stay primarily around the queen. Younger individuals are not necessarily more susceptible to disease (cf. chaps. 3 and 7), but, from an epidemiological stance, it is not only (physiological) susceptibility that determines the pattern of parasite prevalences inside the colony but also the force of infection, i.e., the chance of any particular class of hosts to become infected, and thus on the dynamics of the infection as discussed above (e.g., box 4.2).

Parasites also directly affect the division of labor in subtle but important ways. I have already mentioned the infection by the microsporidian *Nosema apis* that leads to an acceleration of the age-related division of labor in honeybees such that infected individuals start to forage earlier (Wang and Moeller

1970). The entire pattern of division of labor is changed at the same time, too (see also Waddington and Rothenbuhler 1976). Although not known from a social insect, *Nosema* has been reported to release juvenile hormone analogues into the host hemocoel, changing the development of the host (Fischer 1964). Similarly, workers infected by acute bee paralysis virus stop tending the brood earlier and their life spans are reduced by 25% (Ponten and Ritter 1992). In some ways, the effect of *Crithidia* on bumblebee workers is also an example of how parasitism interferes with the division of labor. Infection is associated with reduced conflict over reproduction such that infected workers are less likely to lay their eggs (Shykoff and Schmid-Hempel 1992). Similarly, the speed at which a parasite can infect, be established in, and transmitted to other nestmates is relevant for age-related polyethism (Schmid-Hempel and Schmid-Hempel 1993). Infections of the reproductive organs that lead to queen melanosis in honeybees (chap. 2) may in principle also lead to the production of worker-laid eggs and queen supersedure, with consequences for the social organization of the colony. Obviously, infections spread by parasites that can sterilize, such as the cytoplasmic incompatibility caused by *Wolbachia,* would also dramatically alter the social structure of a colony.

4.5 Reproductive Conflict under Parasitism

Active research over the past decade has demonstrated that social insect colonies are not the harmonious entity they were once considered. In fact, a great deal of competition and rivalry is always present. This leads to different degrees of reproductive skew, i.e., to biases in the reproductive output that benefits certain parties within the colony rather than others (Bourke 1988; Ratnieks 1988; Keller and Reeve 1994b). One of the most striking aspects of such conflict is the bias in sex ratio of offspring. This arises in social Hymenoptera because of the asymmetry in the genetic relationship between mother and her offspring relative to that between workers and sibs (see also chap. 5). For example, in monandrous, monogynous colonies, the queen prefers an equal investment in sons and daughters (due to equal relatedness to both), whereas the workers prefer a female-biased sex ratio (due to closer relatedness among sisters). Boomsma and Grafen (1991) extended this argument of Trivers and Hare (1976) into the "split sex-ratio theory" for social insects. Empirical research suggests that the workers seem to have won this evolutionary conflict in many cases and that such genetic asymmetries play an important role (Nonacs 1986; Sundström 1994; Sundström et al. 1996).

Sex ratio biases obtain an additional meaning in insect species where bacteria and "selfish" elements of DNA have been identified as agents of sex ratio distortion. For example, in the parasitic wasp *Nasonia vitripennis* bacterial infections are associated with a loss of sons among offspring, i.e., the distortion of the offspring sex ratio toward female production (e.g., Huger et al. 1985). *Spiro-*

plasma, known to be involved in sex ratio distortion in other insects (e.g., Williamson and Poulson 1979), is also found in the honeybee (Clark 1977; appendix 2.2). It is not known whether this is also common among social insects, but T. Wenseleers (pers. comm.) has recently documented the occurrence of *Walbachia* and similar agents in many species of ants. In principle, other kinds of parasites (including multicellular ones) could also distort the sex ratio produced by a colony. This would be advantageous if the overproduced sex is important for further transmission of the parasite. It can most likely be observed in annual host species, where parasites without durable stages must infect a particular sex, usually the daughters, to survive to the next season. Typical parasites that depend on such pathways should be internal parasites, i.e., protozoans (e.g., *Crithidia bombi* in bumblebees: Shykoff and Schmid-Hempel 1992) or viruses (e.g., paralysis virus), and mites (e.g., tracheal mites). Bourke and Franks (1995) have furthermore hypothesized that symbionts (in addition to parasites), for example, symbiotic fungi in leaf-cutter ants, could also act as sex-ratio distorters, since they are typically carried by the dispersing young queens to the new colonies. For the time being, there seems to be no good evidence for any of these cases. Nevertheless, given the fact that sex ratios in social insects are highly variable, among species and within a population, that they are under selection (by benefits for different parties), and that sex ratios may become affected through a variety of routes, these hypotheses certainly merit further attention.

Colonies of the wasp *Polistes exclamans,* which consist of several females capable of reproducing, illustrate another conflict potential. Normally, the older females will dominate and lay eggs. All others are restrained to foraging for the colony and to playing the role of a hopeful reproductive that may rise to dominance when the older individual dies (Strassmann and Meyer 1983). One obvious aspect of reproductive conflict is therefore who should forage for the colony rather than stay inside the nest and lay eggs. Not only will foragers lose opportunities to reproduce but, especially in social insects, foraging itself is also a high-risk activity (e.g., Porter and Jorgensen 1981). Obviously, foragers are also at risk of becoming parasitized, for example, by parasitoids (e.g., in *Polistes:* Strassmann 1981b, 1985), by mites (Schwarz and Huck 1997), or by infectious diseases (Durrer and Schmid-Hempel 1994). A typical solution of this conflict in advanced societies is through the age-related division of labor: the older individuals forage, the younger ones stay inside. In many other species, body-size-related dominance ensures that large workers often stay at home while the smaller, less dominant ones forage. In bees, for example, dominance is not only correlated with body size and the display of intimidating behaviors against other nestmates (Michener 1974), but also with the physiological development of the ovaries. Dominant workers have well-developed ovaries, and, because they stay in the nest more often, they manage to lay their own eggs, sometimes even in the presence of the queen (VanDoorn 1987; VanDoorn and Heringa 1986). Parasites may affect this pattern, as the example of the infection by the trypanosome *Crithidia bombi* in *Bombus terrestris* shows. Here infection compromises the dominance status of individuals and at the same time makes the colony more "social" (Shykoff and Schmid-Hempel 1992). Such effects may be more common than hitherto realized. For instance, Salzemann and Plateaux (1987) report that if a worker of the ant *Leptothorax nylanderi* is infected by a cestode, egg-laying rates of the other workers are also depressed. If dominance status is associated with the avoidance of infections, then dominance hierarchies should be particularly pronounced in disease-prone populations or species and whenever dominant workers can bias reproductive success to their advantage. This could involve nepotism (Page et al. 1989) or the laying of eggs in queen-right colonies (Bourke 1988). Dominance is moreover expected to correlate with the avoidance of foodsharing with low-ranking individuals if this carries the risk of contracting an infection, i.e., under high parasite pressure. Dominance should thus be doubly favored in the face of parasitism. Unfortunately, the available evidence does not allow one to check these expectations.

Some pathogens are transmitted vertically through the ovaries of infected females (e.g., the microsporidian *Thelohania* infecting females of the fire ant *Solenopsis:* Knell et al. 1977). Dominant females that manage to lay eggs facilitate this pathway of transmission for the parasite. Therefore, in addition to the usual arguments for the evolution of virulence (see chap. 7), vertically transmitted parasites should also be benign in allowing an infected female to become a dominant worker. However, in the few studies on the subject, dominance is not always correlated with low levels of infection. For instance, Hausfater and Watson (1976) found that dominant male baboons are more heavily infected; similarly, Halvorsen (1986) found that high-ranking reindeer had more parasites. Presumably this is because dominant animals have better access to food and simply eat more, which in turn leads to more infections. Clearly then, the role of parasitism in affecting dominance relationships with social insect colonies merits further attention.

4.6 Colony Size and Nesting Habits

Colony Size

Colony size is not a static quantity. In fact, colonies grow and develop with the addition of new workers. In addition, the proportions of different castes, individual behaviors, colony defense levels, and many other traits covary with colony size (e.g., Wolf and Schmid-Hempel 1990; Pacala et al. 1996). As the models discussed in the previous sections have shown, colony size may also change with parasite dynamics and is affected by them. It is also obvious that parasite effects will tend to reduce colony size; e.g., colonies of fire ants infected by the microsporidian *Thelohania* are smaller and have less sexual brood (Briano et al. 1995b). But in this section I will treat colony size as a fixed quantity, referring to the "typical" species-specific size of an established colony. This bold approach, while neglecting the detailed dynamics of the trait, has the advantage of providing insights into general processes that drive the evolution of colony size, a trait that is not well understood at all. Colony size is arguably the most important predictor for survival and reproductive success (e.g., see also Michener 1964; Cole 1984; Franks 1985; Seeley and Visscher 1985; Franks et al. 1990; Schmid-Hempel et al. 1993) (see fig. 4.6).

The typical colony size ranges widely (see table 1.4). Only a few individuals are found in the primitive society of the ant *Myrmecia,* whereas several millions of workers form a colony in the advanced army ants (*Eciton, Dorylus*) or leaf-cutter ants (*Atta*). There is a general trend toward larger colonies as one is progressing along the phylogenetic sequence. But colony size is not strictly associated with the evolutionary status of a lineage; rather, the ecological contingencies prevail. For example, size affects the ability of a colony to maintain a desired state (e.g., Seeley 1985). Kaspari and Vargo (1995) found that colony sizes of ants increase with latitude. They interpreted this as an instance of "Bergmann's rule" selected by fasting endurance. Similarly, bumblebees at northern latitudes have larger first broods (Richards 1973) which is known to increase colony survival under adverse conditions (Laverty and Plowright 1985). A similar pattern is also found for social wasps (fig. 4.7).

So, how does colony size affect parasitism? For example, the developmental rate of pathogens may play a role. Development is typically affected by temperature and by its temporal variation. Larger colonies will usually be better at maintaining the thermal environment of the nest (Seeley 1985). Hence, it is quite conceivable that species with typically large colonies are more prone to infections because they provide better thermal environments for the parasite. Larger colony size at higher latitudes (fig. 4.7) could thus favor the maintenance of parasites within the nest, while rates of transmission between colonies (due to low densities) as well as the probability that the parasite can overwinter (long winters) must necessarily be lower. For perennial species such as ants, this might translate into higher parasite loads with higher latitudes, contrary to intuition.

In a more quantitative way, and ignoring the details, suppose that the chance of a worker becoming infected by a parasite follows a Poisson process. The n foragers of a colony contract a parasite outside the nest with constant probability, λ, per unit time. With this assumption, the probability, P, that a worker has become infected until time t is proportional to

$$P \propto 1 - e^{\lambda n t}. \qquad (4.2)$$

Eq. (4.2) also describes the probability that an infection has occurred until time t as a function of colony size, because the number of foragers typically correlates positively with colony size. This function can be compared to the increase

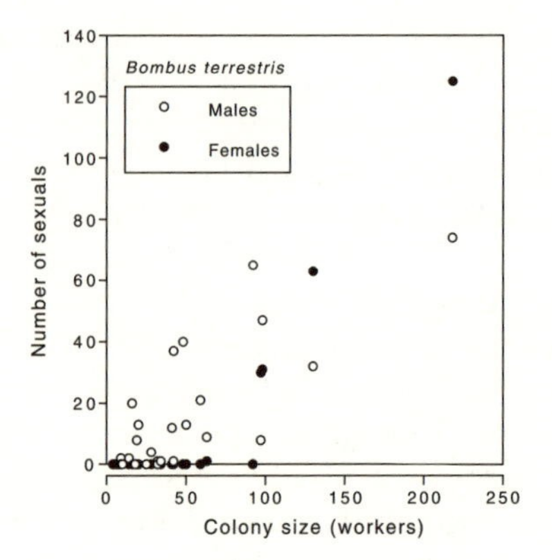

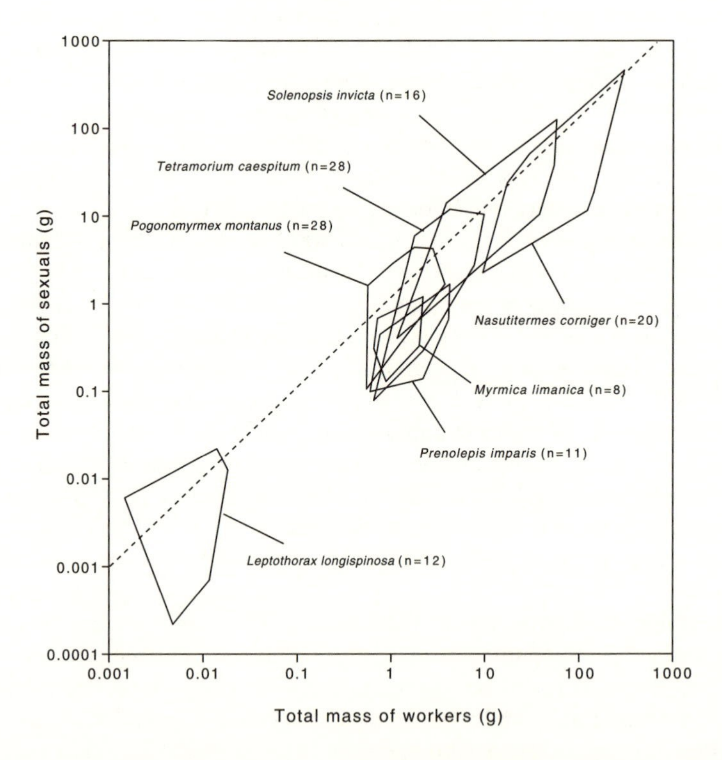

Figure 4.6 (a) Reproductive success of colonies of the bumble bee *Bombus terrestris* in relation to colony size (numbers of workers at time of reproduction). Males and females are drones and young gynes produced by the colonies in the same population. (b) Reproductive success (biomass of produced sexuals) for six species of ants and one termite species (*Nasutitermes*) in relation to colony size at reproduction (biomass of workers). The polygons connect the outermost points of the reported relationships for the respective species; *n* is the number of colonies measured. (Reproduced from Tschinkel 1993 by permission of Ecological Society of America.)

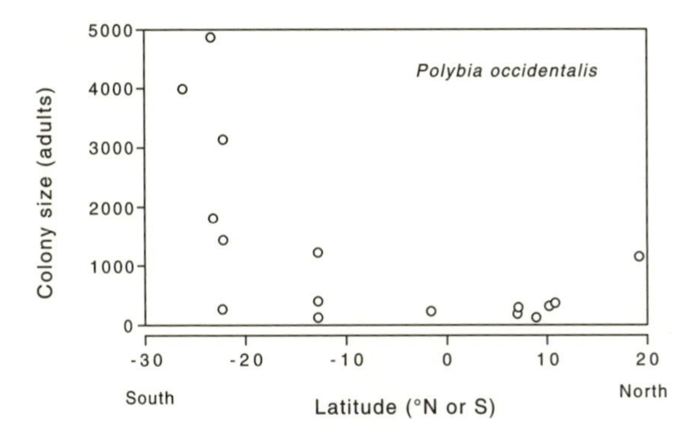

Figure 4.7 Colony size (number of adults) in the social wasp *Polybia occidentalis* in relation to latitude. (Reproduced from Jeanne 1991. Used by permission of Cornell University Press.)

in ergonomic efficiency, sensu Oster and Wilson (1978). If infection risk is considered a cost to the fitness of the colony and its members, then obviously a colony size intermediate between the two requirements is best (fig. 4.8). Jeffree (1955) noted that during the hibernation period of honeybees, *Nosema* had only moderate effects on colonies of moderate size, but affected large or small colonies more heavily. In this case, thermoregulation vs. the spread of the disease within the colony may be the reason.

Consider again the model formulated in box 4.2. An increase in colony size will inevitably lead to a more rapid spread of the parasite and a different dynamics over the colony cycle. For example, if under the parameter conditions described in figure 4.3 the initial size of the colony is doubled, the resulting increase in the final size (i.e., at the end of the cycle when reproduction takes place) is only marginally higher, because relatively more hosts have become infected with larger colony sizes. This leads to an accelerated spread of the parasite, which cancels out a part of the gain from initial colony size. Therefore, if producing more workers up to any one time is costly, which surely it must be, this extra investment may return very little under parasitism. Figure 4.9 quantifies the situation for the model in box 4.2. The size of the colony is varied by varying the value of X_0, which corresponds to colony size at first infection. There is also an additional effect of caste structure or division of labor. For example, in colonies with strict compartments, i.e., with division of labor or pronounced castes, we assume again that the infection cannot spread easily from foragers to nest workers. In this case, an increase in fitness with initial size results (fig. 4.9, "with"). The increase slows down after a certain size is reached (at ca. 5 units of initial colony size in fig. 4.9). This is because an ever higher rate of recruitment of individuals into the forager caste (as a result of larger initial colony size) will also ensure that infections are more frequently acquired.

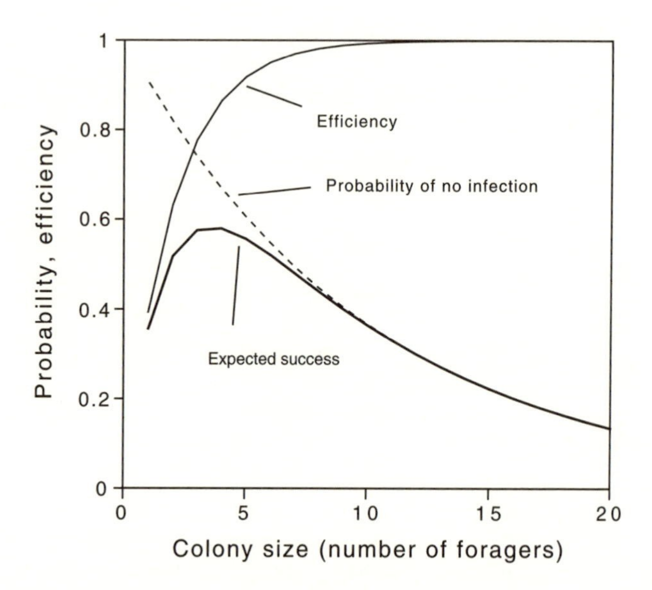

Figure 4.8 The effect of colony size (foragers, according to box 4.2) on the expected reproductive success under parasitism. The probability, P, to escape an infection depends on colony size according to eq. (4.2) and declines exponentially ($\lambda n = 0.1$). On the other hand, the efficiency at which the colony operates increases with colony size (here: *Efficiency* $= 1 - e^{-0.5N}$). The net effect, i.e., the expected success of the colony, is given by the probability of escaping an infection and being effective. This is the case at an intermediate colony size.

Hence, an ever larger toll is eventually taken by the parasite. Nevertheless, with a high degree of colony structure, it should inevitably pay to grow as fast as possible to maximize colony size at infection (fig. 4.9 "with"). In a system with weak compartmentalization, e.g., with no strict division of labor, the infection will be passed on at high rates to all colony members. When the colony is large at the time when infection occurs, infection will spread rapidly to the nest workers due to the large population sizes. This increases mortality and reduces the rate of recruitment of new foragers. As a result, the colony's growth rate declines, and final success may actually be smaller (fig. 4.9, "without"). In this case, it does not always pay for the colony to grow as fast as possible. In fact, the curve suggests that there may be two useful strategies: slow growth at the lower end of the axis in figure 4.9, or very fast growth at the higher end, where final size increases again (very far to the right in fig. 4.9, not shown). Thus, the interplay between colony organization, internal transmission dynamics, and colony size could select for colony growth strategies that may be counterintuitive at first sight. This should obviously be pertinent when parasite pressure is substantial and remains to be tested by collecting appropriate data. Recall that parasite

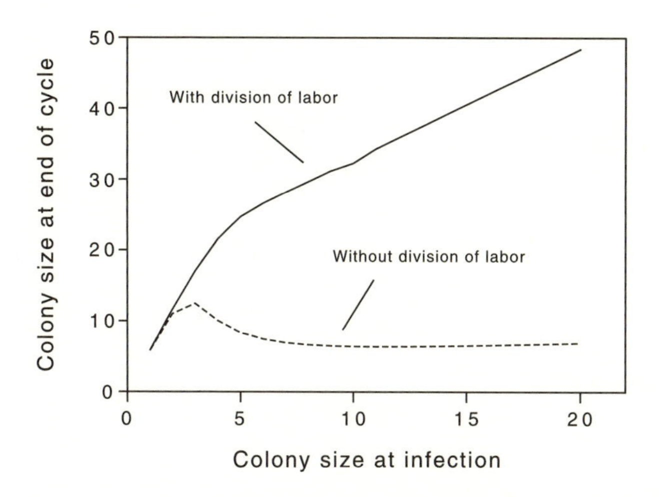

Figure 4.9 Colony size at the end of the cycle (a correlate of colony fitness) according to the model in box 4.2, as a function of colony size at the time when the infection is acquired. This point can also be thought of as the mean waiting time to infection for the colony after its foundation. Two scenarios (with and without division of labor) are calculated. Parameter values as in figure 4.3. With division of labor: many or distinct compartments in the colony. Without division of labor: few or weak compartments.

richness is probably not independent of caste structure (table 4.2), a fact that is not incorporated in the scenario of figure 4.9.

Studies in a range of socially living animals other than insects have investigated the relationship between group size and parasitism (Keymer and Read 1991). A positive relationship has been found in a range of taxa, for example, in colonially nesting swallows infected with ectoparasites (Brown and Brown 1986; Møller 1987; Hoogland and Sherman 1976), primates infected with malaria (Davies et al. 1991), quails infected with helminths (Moore et al. 1988), rodents with ticks (Hoogland 1979), and social spiders and their parasitoids (Hieber and Uetz 1990) (see also references in Loehle 1995). Côté and Poulin (1995) used meta-analysis to evaluate the existing evidence on social vertebrates. They found consistent positive correlations between social group size of the host (mostly mammals and birds) and both prevalence and intensity of infection by contagious parasites, i.e., those transmitted via individual contact or feces. On the other hand, intensity but not prevalence of infections decreased with group size in "mobile" parasites, i.e., those not requiring close contact for transmission. There was no effect of host mobility on either of these measures. This is valuable evidence and suggests that group size is indeed related to parasitism in the way discussed here.

Unfortunately, little is known for social insects and the evidence is somewhat conflicting, too. For example, in our studies of bumblebees and their parasites, a

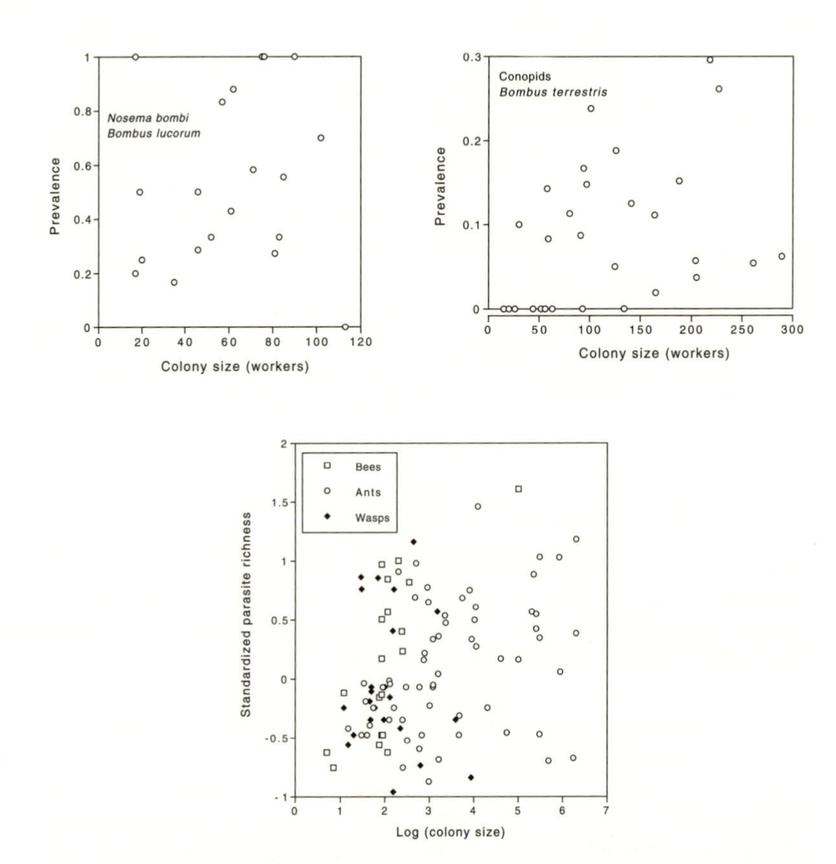

Figure 4.10 The relationship between the prevalence of infection (i.e., percentage of workers infected) and colony size (number of workers) in two different species of *Bombus*. (a) Prevalence of the microsporidian *Nosema* (Spearman's $r=0.437$, $N=29$, $P<0.05$). (b) Prevalence of parasitic flies (conopids) (Spearman's $r=0.466$, $N=29$, $P<0.05$). (Data from Müller and Schmid-Hempel 1993a.) (c) Standardized parasite richness (from fig. 2.1) in relation to colony size for ants, bees, and wasps. Colony sizes taken from table 1.4 and additional sources. The correlation is for bees: Spearman's $r_s=0.708$, $P=0.001$, $N=19$; for ants: $r_s=0.350$, $P=0.005$, $N=63$; for wasps: $r_s=0.113$, $P=0.6$, $N=23$; for all groups pooled: $r_s=0.289$, $P=0.003$, $N=107$.

weak but significant relationship between colony size of *Bombus terrestris* and parasite prevalence for a contagious microparasite (*Nosema,* fig. 4.10a) was present. The same was found for parasitoids (conopid flies) in field colonies of the sibling species *B. lucorum* (fig. 4.10b). In contrast, MacFarlane and Pengelly (1974) found that the prevalence of the conopid *Physocephala* was lower in large colonies of bumblebees. In the wasp *Polistes,* Gamboa et al. 1978 found no correlation between foundress colony size and rates of attack by parasitoids. Strassmann (1981b) observed that larger colonies of *P. exclamans* were more heavily attacked by parasitic wasps and a pyralid moth; the same was true for *Microstigmus comes* attacked by the brood-parasitic braconid wasp, *Heterospilus microstigmi* (Matthews 1991). In his review on social parasitism, Wcislo (1987) found that small colonies are less likely to be parasitized (see also Müller and Schmid-Hempel 1993a, for *Psithyrus* attacking *Bombus,* with the same result). The records listed in appendix 2 show that host species with typically large colony sizes have a higher parasite richness (fig. 4.10c). The pattern is similar for ants, bees, and wasps. For sixty-three ant species all data were available to run a comparative analysis by independent contrasts (Purvis and Rambaut 1995, based on the crude phylogenies of Hölldobler and Wilson 1990 and Baroni-Urbani et al. 1992). With this procedure, the resulting twenty-five contrasts produced no significant relationship between colony size and standardized parasite richness ($r_s = -0.004$, N. S.), suggesting that there are lineage-specific associations between colony size and parasite richness. Note that there is no strict reason to expect a positive relationship across species. In fact, the actual argument refers to within-species variation. More empirical data are thus clearly needed to substantiate this relationship. Given the evidence discussed here (e.g., fig. 4.10), the working hypothesis seems at least not unreasonable that, within species, colony size should correlate positively with parasite load.

Several expectations are thus pertinent. For example, defense against parasitism or any process that helps to eliminate an unavoidable infection should become more important in species with large colonies. Such processes could include self- and allogrooming, which help to eliminate at least external parasites, spores, and other infectious stages that settle on the surface of an animal. We should thus expect to see grooming to increase in frequency with colony size. Schmid-Hempel (1990) found that the frequency of self-grooming decreases with the typical colony size of ants, contrary to the expectation. However, figure 4.1 shows that the frequency of allogrooming increases with colony size. This would support the expectation if allogrooming is mainly effective as a hygienic behavior. Recall that in the discussion of figure 4.1, allogrooming was considered to facilitate parasite transmission. Which of these interpretations is appropriate remains an empirical question to be solved.

Not all workers in a colony of social insects work at any particular point in time. In fact, a large proportion of individuals stays idle and becomes activated only when important new contingencies have to be met. The existence of such a "reserve" worker force has been documented in many species of social insects (Wilson 1971). Parasites could affect this pattern. For example, idle workers will not be as likely to come in contact with infected workers, spores, or any other source of infection. In addition, a generally lower level of metabolic activity during idleness may reduce the rate of growth or replication of parasites within the host individual for those that depend on these activities or on elevated body temperatures. Whether such relationships exist is not known, and high metabolic rates, as far as they generate heat, would sometimes lower parasite development rather than promote it, as will be discussed later. But the division into active and inactive workers, rather than a graduated activity for all, could also have an additional, more exotic consequence. If multiple infection is frequent, the active part of the colony would collect most of the infections and perhaps "shield" others from the negative effects of parasitism. Obviously, this is a highly speculative perspective but due to the lack of forceful concepts of how such a pattern of inactivity can evolve, the effects of parasites can at least not be neglected.

The combination of colony size-dependent resiliency and of size-dependent reproductive success can lead to remarkable patterns (see box 4.3). Particularly when colony size-dependent reproductive success is strongly skewed toward large sizes (fig. 4.6), parasites do not necessarily select for the high levels of host defense. Hence, one would expect that under these circumstances parasites should be prevalent and/or virulent, or that individual host defense should be weak, especially in unpredictable environments (box 4.3). The pattern shown in figure 4.10 may also be considered in this light, although it obviously does not prove the case.

Nesting Habits

Social insects display a wide variety of different nesting habits, from open bivouacs, typical for nomadic army ants, to long-lasting ground nests in termites and in many ants (chap. 1, fig. 1.7). In addition, the organization of nests shows different degrees of sophistication. In ants, eggs and larvae are placed on the floor of the nest chamber. In advanced species, the nest consists of a number of chambers so that brood chambers are separated from other activities. In advanced wasps and bees, nests possess structures that support the brood in a more orderly way, with the hexagonal brood cells arranged on regular combs.

Some parasites have managed to directly utilize these systems. For example, the wax moth (*Galleria*) consumes the nest construction while simultaneously destroying the brood of social bees. Other commensals feed on debris or remains of food; examples include bee mites that reside in the nest and often feed on fungal hyphae that develop around the brood cells, but they also consume pollen and honey stores (Eickwort 1994). For the parasite, a well-structured nest may reduce transmission in ways that are similar to those resulting from a sophisticated division of labor and caste structure. Both aspects are in fact tightly correlated in that division of labor has a pronounced spatial component such that certain tasks are allocated in different regions of the nest (Seeley 1982).

Strassmann (1981b) has analyzed the effect of polydomy (several nests per colony) in the wasp *Polistes exclamans* in relation to brood parasitoids (e.g., the wasp *Elasmus polistis* Chalcididae and the moth *Chalcoela* Lepidoptera). But

BOX 4.3 RESILIENCE AND A SELECTION-WEAK SPACE FOR PARASITES

Systems such as colonies of social insects can absorb stress due to their organizational resilience. This fact opens the possibility that parasites exploit this characteristic. The following considerations may help to clarify the reasons. The scenario is given in verbal arguments to illustrate the logic of the process. The case illustrated here is modeled on an annual social insect where reproductive success is strongly skewed according to colony size (fig. 4.6).

Suppose that colony size at reproduction varies in a population. In particular, large colonies are always rare (graph 1: frequency in population). Colony size has two other correlates: (1) Resilience against external perturbation (including parasites) increases with colony size (graph 3: resilience); large colonies are expected to be relatively better buffered because of redundant organizational structure (chap. 4). (2) Reproductive success also increases with size (graph 4: offspring males, m, and females, f); this results from an increase in the biomass that can be invested into sexuals (fig. 4.7).

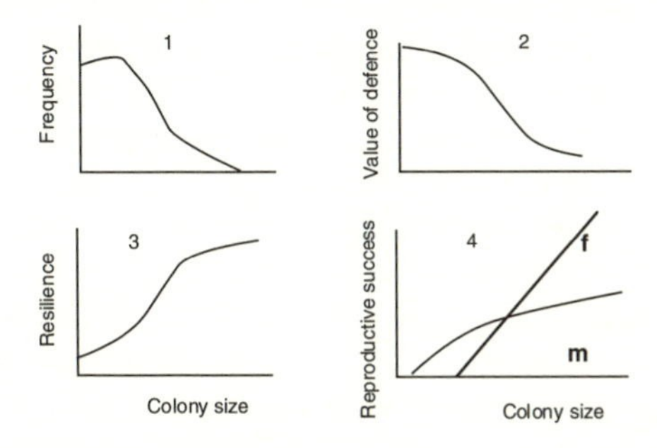

This scenario has several consequences, in particular if the relationships are as nonlinear as shown in the graph. The large colonies, if parasitized, do not lose much of their reproductive success because their resilience against perturbations is vastly greater than that of small colonies. Hence, although they are rarer and even when parasitized, large colonies will still contribute most of the offspring in the population, and incidentally also allow high reproduction by the parasites. Small colonies, if parasitized, may collapse and do not reproduce at all, and will hardly allow the parasite to reproduce, either. Therefore, in the population at large, selection for individual defense of workers (i.e., physiological immunity, behavior, etc.) against their parasites is low, since the process is dominated by the large colonies. Individual defense would of course be very valuable for small colonies. In large colonies, in contrast, individual defense is not as valuable (graph 2: value of defense; see also box 4.1).

The combination of resilience and the distribution of reproductive success in relation to colony size may therefore not select for levels of individual defense

BOX 4.3 CONT.

against parasites as high as one would intuitively expect. A (not very stringent) assumption is that the size-dependent distribution of resilience and reproductive success is primarily maintained by factors other than parasitism. Otherwise, the hosts would be more likely to adapt directly to the effect of parasites. Examples of such selective factors include environmental uncertainty or a strong colony size-dependent bias in the costs of producing offspring. Given the conditions depicted here, parasites would have a domain in the parameter space where selection against them is weak ("the selection-weak space").

no effect of protection through the construction of an extra, satellite nest was found, since these were just as heavily infected. Other nesting characteristics, such as temperature regime, recycling of nest material, or the reuse of nesting sites should also affect parasite persistence and transmission. For example, warm temperatures and high humidity often favor infections (e.g., by many fungi, chap. 2, 3). These conditions should be fairly typical in nests of social insects. In a number of cases, though, higher temperature actually reduces the success of parasites. For instance, the cabbage worm recovers from fungal infection when transferred to warm temperatures (Tanada 1967). Also the rate of development and multiplication of certain viruses (Watanabe and Tanada 1972), microsporidia (Weiser 1961; Wilson 1979), fungi (Gardner 1985), and perhaps nematodes (Kaya et al. 1982) is reduced at elevated temperatures. Microsporidia are particularly temperature-sensitive (Canning 1982). Appropriate behavior can also contribute to successful defense against parasites. Boorstein and Ewald (1987) found that if the cricket *Melanoplus* moves itself into a warm area, infections by *Nosema* are less successful.

Nest temperature is thus a potentially important factor contributing to host resistance against their parasites (Kaya et al. 1982). In fact, *Nosema* infection in honeybees depends on temperature (Karmo and Morgenthaler 1939). Because *Nosema* infections generally decline during the summer, the high summer temperatures in the nest may inhibit the multiplication of *Nosema apis* (Schulz-Langner 1958). However, the temperature needed to impede multiplication of *Nosema* seems to be rather high and has to exceed 35°C (Lotmar 1943). This temperature is rarely exceeded even in the center of the nest (Seeley 1985). Perhaps, this temperature regime has to be established in a way similar to "fever" as described for *Melanoplus* (Boorstein and Ewald 1987). In other cases, such as chalk brood of larval honeybees, caused by the fungus *Ascosphaera apis,* the parasite develops better in a cool brood (30°C) than in a brood kept at a temperature of 35°C (Bailey 1967a). The developmental rates and virulence of various other parasites of honeybees are also known to be affected by temperature (Bailey and Ball 1991). Acute bee paralysis virus multiplies faster at 35°C than at 30°C but will kill the host more quickly at the lower temperature (Bailey and Ball 1991). On the other hand, bumblebees that seek out cold areas reduce the chances of survival and development of parasitoids whose larva develops inside the abdomen of the bees. The workers thus increase their life span (Müller and Schmid-Hempel 1993b).

From the parasite's point of view, the thermal environment of the colony is a factor to which it has to adapt. Given the temperature sensitivity of developmental processes in organisms in general, we can readily imagine that different strains of a microparasite will differ in their rate of multiplication at a given temperature, and that some of this variation may be based on genotypic variation. Hence, the microparasites can be expected to adapt within certain temperature ranges. On the other hand, parasites that have adapted to a "normal" temperature regime in the nest face the thermal problems of dispersal and transmission to new hosts, or the problem of overwintering in a seasonal environment. This is more often the case in annual social insects and in temperate habitats. In such cases, the parasite will naturally be exposed to extreme temperatures. Indeed, if honeybee combs are heated to high temperatures or chilled to freezing, eggs of wax moths, for instance, are effectively killed. Such a regime would occur when the nest is abandoned and fully exposed to the weather conditions. Not only temperature but also other environmental hazards may pose a problem for the parasite. Spores of *Nosema* are known to be sensitive to UV light and to lose infectivity if exposed to sunlight for too long (Krieg et al. 1981; Canning 1982; Brooks 1988). The highly advanced social insects, on the other hand, where the colony persists for a long time and the environmental conditions in the nest are well regulated, should offer a rather stable and finely tuned habitat for a parasite. Such hosts include cavity-nesting bees in the genera *Apis, Trigona, Melipona,* leaf-cutting ants, and many termite species with well-regulated mounds. We should expect to find rather specialized parasites in these cases, which in turn are expected to be sensitive to deviations from their optimal temperature and humidity regime. This would also give the host a handle for combatting infections. Hence, such host species are expected to be more likely to utilize, for example, thermal strategies (i.e., cooling or heating) as a countermeasure against infections.

Many ants regularly move their nests to new sites several times a year, or at least several times during the lifetime of a colony (Herbers 1985). In some cases, increased rates of nest relocation have been found to be associated with infection by parasites. In fire ants, colonies move more frequently in areas of high occurrences of nematodes (Oi and Pereira 1993) and microsporidia (Patterson and Briano 1990). In the latter case, colonies in plots heavily infected by *Thelohania* (where 35% of colonies were infected) moved more often during the study period than those in less infected plots (3% infected) (85% vs. 75%

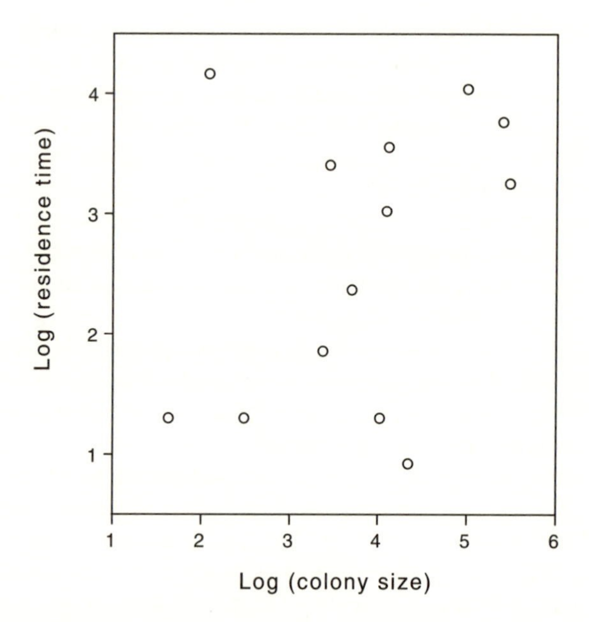

Figure 4.11 Mean residence time (in days) of nests in relation to colony size (number of workers). The statistics are as follows: Spearman's $r=0.260$, $N=22$, N. S. (without point at upper left: $r=0.535$, $N=21$, $P<0.05$). (Data from Hölldobler and Wilson 1990, Schmid-Hempel 1990, Golley and Gentry 1964.)

of colonies moved). They were also lost more often, i.e., they either perished or moved outside the study area (45% vs. 25% loss). The situation is ambiguous, though, as Briano et al. (1995c) in a later study found no effect of the microsporidia on colony movement in the same system. Roubik (1989) reports that gregariously nesting, soil-dwelling bees abandon their nests more often when fungal infections are present. Nest relocation should in particular reduce parasite pressure in cases where infection results from pools of durable stages in the environment, e.g., from cysts or spores in the soil. Nematodes are particularly prevalent in soil-nesting social insects, but opportunistic infections by bacteria or fungi may also add to the problem (chap. 2).

Herbers (1985) found that in some ant species nests last only for a few weeks at a time. In *Myrmica punctiventris, Aphaenogaster rudis,* or *Lasius alienius,* any one nest was abandoned in 80–100% of the cases during the study period. Hölldobler and Wilson (1990, their table 3.4) give mean residence times per nest in the range of 8 to 2800 days. A naive expectation is that large colonies should move more often because they attract and acquire more parasites than small colonies (fig. 4.10). However, as Figure 4.11 shows, if anything, the opposite is true. In other words, species that normally attain large colony sizes reside longer in a given nest than do those with small colonies. Experimental evidence for the effect of nest relocation in response to parasitism would be helpful and should be fairly straightforward to obtain. In the honeybee, for example, it has been shown that colonies often prefer to renest in previously occupied sites (Visscher et al. 1985) that potentially carry the risk of being contaminated. In fact, hives containing spores of American foulbrood are nevertheless accepted (Ratnieks and Nowakowski 1989). This suggests that nest relocations may not necessarily be a strategy to escape parasites. Nevertheless, the parasite hypothesis deserves further study, as moving a nest is a major feat for any social insect colony and therefore it seems important to understand its causes and its adaptive value.

This chapter should have made it abundantly clear that colony organization of social insects influences parasitism in many ways and, vice versa, that parasitism could be a selective force for the evolution of social organization. However, many of these expectations cannot be confirmed with the available evidence, so most questions remain unanswered and present a challenge for future studies.

4.7 Summary

The organization of social insect colonies is based on interactions among individuals, age-related changes, morphological differences, and the individual timetables of the workers' life and death. All of these elements relate to parasitism in various ways. For example, trophallaxis and provisioning of larvae is exploited by some parasites, and hence the exact pathway of these interactions also determines how parasites can be spread to nestmates. Specialized structures in the alimentary tract, such as the proventriculus, that act as filtration devices can provide some protection by keeping back appropriately sized particles. Grooming and related hygienic behavior is beneficial for the host but could also lead to the transmission of pathogens. Social organization, in particular the age-related division of labor and the allocation of work in parallel and serial chains, also profoundly affects the epidemiological dynamics of the transmission of contagious diseases within the colony. In general, social organizations that partition the colony into more compartments will impede parasite transmission more effectively. In addition, worker turnover determines the rates at which parasites are acquired from outside, passed on to nest workers, and removed through worker mortality. The dynamics of parasitism under division of labor is conveniently summarized in the conveyor belt metaphor. Parasitism thus suggests alternative ways to look at the age-related division of labor almost universally found in social insects, for the meaning of redundant task allocations, and for differences in life span and worker turnover within colonies. In addition, polymorphism could be associated with higher parasite loads. Parasites also impose extra costs to large worker morphs, and they can interfere with reproductive conflict within social insect colonies, either via direct effects on the reproductive capacity of individuals and/or via the dominance status of individuals. A number of expectations follow from these considerations.

Colony size, usually expressed as worker numbers, is a poorly understood trait in social insects, although it affects almost any aspect of their biology. Large colony size can facilitate parasite transmission and maintenance, perhaps also by providing a more stable thermal environment. Due to the dynamic nature of infections, however, the relationships are complex. Nevertheless, the evidence suggests that large colony sizes are typically associated with higher parasite loads. Colony size should thus be associated with a number of characteristics relating to parasitism, such as grooming behavior, activity levels, and so forth. Furthermore, size-related resiliency and skew in the reproductive success of colonies could lead to counterintuitive selection regimes and, for example, reduce defense against parasites. Nesting habits and nest architecture also affect the spread of parasites. Although the evidence is very ambiguous, traits such as polydomy and frequent nest relocations may have evolved in response to parasitism. This chapter can therefore mainly suggest new ways of looking at familiar traits. There is a need for more studies of how parasitism relates to colony organization.

5 Breeding Strategies and Parasites

In almost all the cases studied so far, the ecological and microevolutionary relationship between host and parasite depends on their genotypes (e.g., Anderson 1991; Thompson and Burdon 1992). This is because both host resistance and parasite infectivity vary genetically (chap. 7). Typically, insect societies are formed of groups of related individuals so that the populations of social insects are genetically highly structured and characterized by being divided into genetically homogeneous groups (the colonies). Insofar as genetics and genetic diversity affect host-parasite relationships, this indicates that the association of worker relatedness and selection by diseases is particularly interesting. Furthermore, the distribution of genetic variability within and among colonies is influenced by the breeding and mating strategies adopted by males and females. In this chapter I will discuss some examples of how the genetics of the mating and breeding system affects the biology of host and parasite and their interactions. Genetic relationships among workers of social insects have been studied in some detail, usually to analyze how social interactions among individuals depend on their genetic relatedness (Hamilton 1964). Not much is known in relation to parasitism. A major aim of this chapter is therefore to point out the problems and the kinds of studies still needed. In addition, some of the elements presented here are relevant to epidemiological and coevolutionary topics discussed in later chapters.

5.1 Breeding Systems and Colony Genetics

The breeding and mating system that prevails in a population determines the way in which genetic variance is distributed within and among groups of progeny. In the case of social insects, these groups are the colonies. The breeding system of a social insect can thus be summarized by the distribution of the number of mates of females, the relative contributions of males toward paternity in progeny, and the distribution of the number of females reproducing within and among colonies, together with their relative contributions to progeny.

Since the pioneering work of Hamilton (1964), the main interest in the breeding biology of social insects has come from considering the necessary conditions for the spread of alleles coding for altruistic, social behavior ("kin selection theory") and hence to understand the evolution of sociality. Such work has led to a substantial body of theoretical and empirical insights for the structure of colonies themselves, as well as for the effects of cooperation and conflict within insect societies. The current view is that members of social insect colonies both cooperate and are in conflict where necessary. Conflicts are solved or at least settled in different ways. For example, selfish interests of the workers may lead to "policing," whereby eggs of other workers are removed. This selfish behavior thus prevents worker reproduction and incidentally promotes reproductive harmony in favor of the queen (Ratnieks 1988). Similarly, cooperation among several foundress females may be established through a balance of stay and peace incentives from the alpha female (Keller and Reeve 1994b). Where such situations have been analyzed more thoroughly, it demonstrates the existence of conflicts of interest even in seemingly harmonious societies. For example, due to the haplo-diploid sex determination system, workers and queen disagree over the inclusive fitness value of male and female offspring. On the other hand, empirical research implies that the workers rather than the queen control sex allocation to offspring. Hence, this conflict and its solution can lead to substantial female biases away from an equal ratio that would be preferred by the queen (Nonacs 1986; Boomsma 1991; Sundström 1994). This is also an instance of parent-offspring conflict over resource allocation to further progeny. Conflicts among the workers themselves are associated with the breeding system. Potential conflicts can arise when several patrilines (where a queen has mated with several males) or matrilines of workers (several functional queens are present) are present in a colony.

For haplo-diploid social Hymenoptera, it is straightforward to calculate from a pedigree the genetic relationships between individuals within the colony under different mating systems(fig. 5.1). Box 5.1 gives an overview of different measures used in the literature. For example, the symmetrical coefficient of consanguinity between queen and daughter (worker) is $f_{Queen\text{-}Worker} = 1/4$. If there is no inbreeding, both r and g are twice this value, hence $= 1/2$. Similarly, the consanguinity between workers is $f_{Worker\text{-}Worker} = 3/8$, hence the coefficient of relatedness $r = g = 3/4$. If a queen mates with several males ("polyandry," fig. 5.1b), colonies contain both full sisters (from the same father) with relatedness as before, and half-sisters (from different fathers) with $f_{Half\text{-}sisters} = 1/8$, and $r = g = 1/4$. The average relatedness among workers of a colony decreases as paternity frequency increases. Finally, the case where more than one functional queen is present ("polygyny," fig. 5.1c) leads to a situation where several matrifilial families live together in the colony. The average relatedness among workers can be very low in such cases. Empirical data on the average genetic relatedness among workers in field colonies have accumulated in recent years (table 5.1; Crozier and Pamilo 1996, their table 4.7). Low values are primarily caused by breeding and mating systems that involve polyandry and polygyny (see examples in tables 5.2, 5.3). In many cases, therefore, average relatedness is much lower than the 3/4 expected for full sisters in social Hymenoptera (i.e., the case

BOX 5.1 MEASURES OF RELATEDNESS

The degree to which two individuals or groups are related to each other has many practical applications. In the social insect literature, the focus is often on how an actor's genes, including the copies in related individuals, are transmitted as a result of the actor's decisions, e.g., helping or not helping somebody else. In contrast to such sociobiological questions, the genetic relationships relevant to host-parasite interactions may not be the same. Genetic relatedness can be a complicated issue, since it does not only depend on recent pedigree, but also on population structure and evolutionary forces such as drift and selection. A typical assumption is that populations are large and selection is weak, in which case the effects of these additional factors become small and are usually ignored.

Measures of relatedness between two individuals or groups can either be symmetrical (i.e., based on correlation analysis), where the relatedness of A to B is the same as of B to A, or asymmetrical (based on regression analysis), where the relatedness of A to B is not necessarily the same as of B to A. The latter is particularly important when considering altruistic behaviors, where an act of help is directed from a donor to a recipient. The two estimates may differ because important genetic asymmetries are introduced by the haplo-diploid mechanism of sex determination in the social hymenoptera. In addition, the estimates can be based on known pedigrees and hence reflect absolute relatedness, or on genetic similarity against the background of the population, as derived, for example, from known genotypes in the form of allozyme data, and hence reflect relative relatedness. For further discussions, see Falconer (1989), Moritz and Southwick (1992), Bourke and Franks (1995), or Crozier and Pamilo (1996).

Symmetrical

(1) COEFFICIENT OF CONSANGUINITY (of kinship, of coancestry), f_{AB}, describes the probability that two loci taken at random, one from each individual, carry alleles identical by descent. If p_{ij} is the probability for the ith allele of individual A to be identical with the jth allele in individual B, and $n=$ the number of alleles in the population at the locus of interest,

$$f_{AB}=\frac{\Sigma p_{ij}}{2n}$$

then f_{AB} is identical to the inbreeding coefficient, F, of (diploid) offspring produced by a mating between A and B. In social hymenoptera, it is convenient to define the inbreeding coefficient of the haploid drones as $F_{drones}=1$.

The inbreeding coefficient, F_x, equals the probability that both alleles at a given locus of a diploid individual are identical by descent because of an ancestor common to both parents. It can be derived by considering the (diploid) pedigree connecting the individual of interest with its ancestors to be

$$F_x=\Sigma(^1/_2)^n\,(1+F_A)\,,$$

BOX 5.1 CONT.

where n = the number of ancestors in a path in the pedigree that includes both parents and F_A is the inbreeding coefficient of the common ancestor of a path.

(2) WRIGHT'S COEFFICIENT OF RELATEDNESS, r (Wright 1922), measures the average, symmetrical genetic relationship of two individuals A,B as

$$r_{AB} = \frac{2f_{AB}}{\sqrt{(1+F_A)(1+F_B)}}.$$

r is the correlation between the breeding values (i.e., additive genetic variance) of two individuals and would thus be found if all phenotypic variance were additive genetic.

Asymmetrical

(3) PEDIGREE COEFFICIENT OF RELATIONSHIP, g_{AB}, measures the proportion of A's genes identical by descent to those of B:

$$g_{AB} = \frac{2f_{AB}}{(1+F_A)} = \frac{p_{AB}}{p_{BB}},$$

where p_{AB} = probability that two gametes produced by A and B respectively share the same gene by descent, and p_{BB} = probability that two gametes produced by B alone share the same gene by descent. Hence, g_{AB} expresses the relatedness of A to B, i.e., the genetic "value" that A has to B. This measure usually requires that the pedigree is known.

(4) HAMILTON'S REGRESSION COEFFICIENT, b_{AB} (Hamilton 1972; Crozier and Pamilo 1980; Pamilo and Crozier 1982; Pamilo 1984), is asymmetrical and measures the probability that A's genes are identical by descent to those of B. The advantage of this coefficient is that it can be estimated with data, for example, allozyme variation in natural populations, without exact knowledge of the pedigree. It is useful to apply the genetic variances (s^2) and covariances (Cov), e.g., for the genetic markers, to derive

$$b_{AB} = \frac{\mathrm{Cov}_{AB}}{s_B^2}.$$

b_{AB} is the slope of the regression of B's individual genotype on A's individual genotype. It therefore measures the relative relatedness by estimating genetic similarity against the population background. Otherwise, b_{AB} (regression relatedness) is the same as g_{AB} (pedigree relatedness) above.

(5) THE QUELLER-GOODNIGHT ESTIMATOR, r (Queller and Goodnight 1989, who derived an estimator similar to Pamilo 1984's regression method), is based on allele frequencies observed at different levels of grouping, with i = index for groups (colonies), j = for individuals, l = for loci, k = for alleles at a locus

$$r_{AB} = \frac{\sum_i \sum_j \sum_l \sum_k p_{i(\notin j)} - p_{(\notin j)}}{\sum_i \sum_j \sum_l \sum_k p_{ij} - p_{(\notin j)}},$$

BOX 5.1 CONT.

where $p_{i(\notin j)}$ = allele frequency in group i excluding individual j, $p_{(\notin j)}$ = allele frequency of the population excluding group i, and p_{ij} = allele frequency within individual j in group i.

(6) LIFE-FOR-LIFE RELATEDNESS (weighted relatedness), W, where the "value" of individual A to individual B is given not only by genetic relatedness, b, but also by the reproductive values, V. If A and B differ in ploidy, then W is no longer identical to b due to differences in V. The V's reflect the asymptotic proportion of transmitted genes contained in individuals of different ploidy. In this case the weighted relatedness is

$$W_{AB} = b_{AB}\frac{V_A}{V_B}.$$

In particular, such a situation arises for the males, m, (haploid) and females, f, (diploid) of social hymenoptera. Then,

$$V_m = 1$$
$$Vf = (1+p)V_m,$$

where p = proportion of queen-produced males in the population. If all males are produced by the queen, then $p = 1$, and $V_f = 2\ V_m$. This simplification is used for the fitness calculations in chapter 5.

in fig. 5.1a). Such considerations are central to the discussions of kin-selected behavior but are also relevant for the discussion of host-parasite relationships.

Low average relatedness, such as caused by polyandry and polygyny, does not as readily favor the evolution of social behavior (Hamilton 1964). A number of hypotheses for the adaptive value of these mating systems have therefore been suggested (see Crozier and Page 1985; Bourke and Franks 1995; Crozier and Pamilo 1996). For example, Crozier and Page (1985) suggested that polyandry may be beneficial if the resulting genetic variation among workers in the colony allows the expression of a better-adapted colony phenotype, for example, a more elaborate form of division of labor or tolerance to a wider range of environmental conditions. The studies of Robinson and Page (1988; see also Robinson 1992; Calderone and Page 1991) indeed indicate that different worker patrilines within a colony differ in their response to task-specific stimuli. Hence, a larger genetic variation would potentially allow for a more adaptable or more generalized colony phenotype. Multiple mating may therefore expand the range of tolerable environmental conditions for the colony. Page et al. (1995) tested this idea and compared the performance of honeybee colonies headed by singly or multiply mated queens with respect to a number of criteria

Table 5.1

Examples of Estimates of Worker-Worker Relatedness in Colonies of Social Insects

Species	*Estimate*	*Method*	*References*
BEES			
Apis mellifera	≈ 0.25	P, M	Laidlaw and Page 1984
Bombus melanopygus	0.75	P, M	Owen and Plowright 1982
Exoneura bicolor	0.6	b	Schwarz 1987
Halictus ligatus	0.42	b	Richards et al. 1995
Lasioglossum zephyrum	0.64–1.0	b	Crozier et al. 1987
Melipona subnitida	0.75	P, b	Contel and Kerr 1976
ANTS			
Aphaenogaster rudis	0.75	P	Crozier 1977
Camponotus ligniperda	0.08	D	Gertsch et al. 1995
Conomyrma bicolor	0.33	b	Berkelhamer 1984
Formica exsecta	0.60	b	Pamilo 1991b
Formica sanguinea	0.1–0.4	b	Pamilo and Varvio-Aho 1979; Pamilo 1981
Formica truncorum	0.11–0.60	b	Sundström 1993
Formica polyctena	0.19, 0.44	b	Pamilo 1982a
Myrmecia pilosula	0.17	b	Craig and Crozier 1979
Myrmica rubra	≈ 0–0.54	b	Pearson 1983
Solenopsis invicta	0.75	P	Ross and Fletcher 1985
WASPS			
Ageleia multiplicata	0.69	b	Ross and Carpenter 1991
Mischocyttarus basimacula	0.44, 0.49	b	Strassmann et al. 1989; Queller et al. 1992
Mischocyttarus immarginatus	0.77, 0.58	b	Strassmann et al. 1989; Queller et al. 1992
Polistes annularis	0.30	b	Strassmann et al. 1989
Polistes bellicosus	0.34	b	Strassmann et al. 1989
Polybia occidentalis	0.27	b	Queller et al. 1988, 1993
Polybia sericea	0.28	b	Queller et al. 1988
Polistes versicolor	0.371	b	Strassmann et al. 1989
Polistes dominulus	0.652	b	Strassmann et al. 1989

Table 5.1, continued

Examples of Estimates of Worker-Worker Relatedness in Colonies of Social Insects

Species	*Estimate*	*Method*	*References*
WASPS, continued			
Vespula maculifrons	0.32	b	Ross 1986
Vespula squamosa	0.40	b	Ross 1986
TERMITES			
Reticulitermes flavipes	0.57	b	Reilly 1987

NOTES and SOURCES: Methods are regression method (b) with allozymes; after Pamilo and Crozier 1982, pedigree analysis (P), DNA micro- or mini-satellites (D, including regression method), morphological markers (M). Based in parts on Crozier and Pamilo 1996. The table displays the range of methods, estimates and taxa.

(e.g., worker numbers, produced mass, drone comb area, etc.). Since the multiply mated colonies fared better in some traits, Page et al. (1995) concluded that polyandry provides a selective advantage for the colony, since it is less likely to fail because of inappropriate colony responses to changing environments (see also Ratnieks 1990). On the other hand, quite drastic experimental manipulations of colony demography have often had negligible effects on colony performance (cf. table 4.1; Schmid-Hempel 1991b for a review). Hence, it is difficult to see how the small effects on colony efficiency typically associated with the presence of one as compared to several patrilines could actually select for the maintenance of genetic diversity. Multiple mating is also expected to decrease the variance in performance among colonies in the population, since compared to single mating it redistributes the available genetic variability within rather than between colonies. Indeed, Page et al. (1995) found that the variable colonies had more average phenotypes, as expected from the hypothesis of variance reduction.

For the same reason, polyandry is thought to reduce variance in fitness among colonies due to genetic load associated with (less viable) diploid males. This results from the sex determination system in Hymenoptera, where individuals that are homozygous at the sex locus are diploid male rather than female (workers or queens) (e.g., Cook and Crozier 1995). Diploid males are known from several social insect taxa (e.g., Crozier and Pamilo 1996, their table 1.4). For euglossine bees, for example, a group with primitively social species, Roubik et al. (1996) found very high frequencies in many populations. This poses a problem for social evolution, since a colony will fail if too many drones rather than workers emerge. In Roubik et al.'s (1996) study, the more social genera (*Eulaema, Euglossa*) were found to be genetically more variable than purely

Table 5.2

Examples of Polyandry in Social Insects

Species	*Number of Matings*	*Method*[a]	*Effective Matings*[b]	*References*
BEES				
Apis andreniformis	<4	S		Moritz and Southwick 1992
Apis cerana	14–30	S		Page 1986
Apis dorsata	>1	S		Moritz and Southwick 1992
Apis florea	<4	S		Moritz and Southwick 1992
Apis mellifera	7–20	S, O, A, D, M	6.6–17.9	Seeley 1985; Page 1986; Estoup et al. 1994
Bombus hypnorum	1–3	S, O, D		Röseler 1973; Estoup et al. 1995
Bombus huntii	1–3	O		Hobbs 1967
Bombus lapidarius	1	S		Sakagami 1976
Bombus terrestris	1	A, S, O, D		Röseler 1973; VanHonk and Hogeweg 1981; Estoup et al. 1995
Lasioglossum malachurum	>1	O		Packer and Knerer 1985
Lasioglossum rohweri	1–3	O		Packer and Knerer 1985
ANTS				
Atta sexdens	3–8	S		Corso and Serzedello 1981
Formica aquilonia	1–6	O	1.48	Pamilo 1993
Formica bradleyi	1–3	O		Halverson et al. 1976
Formica exsecta	1–2	A	1.16	Pamilo 1991b
Formica sanguinea	>1	A	1.31	Pamilo 1981, 1982b
Lasius flavus	>1	O	≤1.20	Hölldobler and Wilson 1990

Table 5.2, continued

Examples of Polyandry in Social Insects

Species	*Number of Matings*	*Method*[a]	*Effective Matings*[b]	*References*
ANTS, continued				
Lasius niger	>1	O	1.46	Van der Have et al. 1988
Myrmica rubra	1–7	O	1.21	Woyciechowski 1990
Pogonomyrmex badius	2–4	O		Hölldobler and Wilson 1990
Pogonomyrmex maricopa	2–3	O		Hölldobler 1976a
Polyergus lucidus	6	O		Hölldobler and Wilson 1990
WASPS				
Polistes exclamans	2–3	A		Page 1986, Starr 1984
Polistes metricus	>1	A		Page 1986
Polistes variatus	>1	A		Page 1986
Ropalidia marginata	>1	A		Page 1986
Vespa crabro	>1	O		Page 1986
Vespula germanica	1	O		Starr 1984
Vespula maculifrons	7.1–9.5[c]	O, A		Ross 1986, Starr 1984
Vespula squamosa	3.3–5.5[c]	A		Ross 1986

NOTES: The table displays the range of values, methods and taxa. The effective number of matings, as compiled by Boomsma and Ratnieks 1996 (for more references), should be most relevant in the context of parasitism.

[a]S: dissection; O: observed matings; A: allozyme study; D: DNA mini- or microsatellites; M: visible genetic marker.

[b]For ants, according to Boomsma and Ratnieks 1996.

[c]Harmonic mean.

nonsocial or parasitic ones (*Eufriesea, Exaerete*). On the other hand, where diploid males are recognized and removed by the workers at low cost, such as in the honeybee, polyandry is less likely to evolve as an adaptation to diploid male production (Ratnieks 1990). In addition, the frequency of diploid males also depends on the number of sex alleles maintained in the population, which should be free to evolve in any direction. However, low numbers and thus a high incidence of diploid males should be more common in rare or isolated species and correlate with polyandry. Given the problem of diploid males, Ratnieks (1990) found that variance reduction by multiple mating increases colony performance.

Table 5.3

Examples of Polygyny in Social Insects (Range of Estimates and Taxa)

Species	*Number of Queens**	*References*
BEES		
Bombus atratus	1–100	Michener 1974
Lasioglossum (Dialictus) laevissimum	1–6	Packer 1993
Lasioglossum (Dialictus) lineatulum	2.1	Packer 1993
ANTS		
Formica aquilonia	>500	Rosengren et al. 1993
Formica truncorum	1 . . . >40	Rosengren et al. 1993
Myrmica rubra	14.8	Elmes and Keller 1993
Myrmica sabuleti	1.5	Elmes and Keller 1993
WASPS		
Icariola montana	49–286	Itô 1993b
Metapolybia azteca	14–17	Hughes et al. 1993
Polybia occidentalis	8–35	Hughes et al. 1993
Polybia paulista	104	Hughes et al. 1993
Protopolybia minutissima	57	Hughes et al. 1993
Ropalidia fasciata	2	Itô 1993b
Ropalidia maculiventris	45	Itô 1993b
TERMITES		
Macrotermes michaelseni	>1–7	Thorne 1985; Roisin 1993
Nasutitermes corniger	>1–33	Thorne 1985; Roisin 1987
Nasutitermes costalis	>1–97	Roisin and Pasteels 1986
Nasutitermes polygynus	>1–105	Roisin and Pasteels 1986
Embiratermes chagresi	Many	Roisin 1993

*Queen number refers to mature colony, not to foundress associations.

Polyandry could also evolve under high potentials for queen-worker conflict in the colony to ensure more similar genetic interests between the queen and her workers. This is particularly relevant with respect to sex allocation for offspring. Observed sex ratios are generally biased toward the best value from the workers' perspective (Trivers and Hare 1976). Furthermore, the observed distribution of sex ratios among colonies is often bimodal (Nonacs 1986), and the

a) single queen (monogyny), one male (monandry)

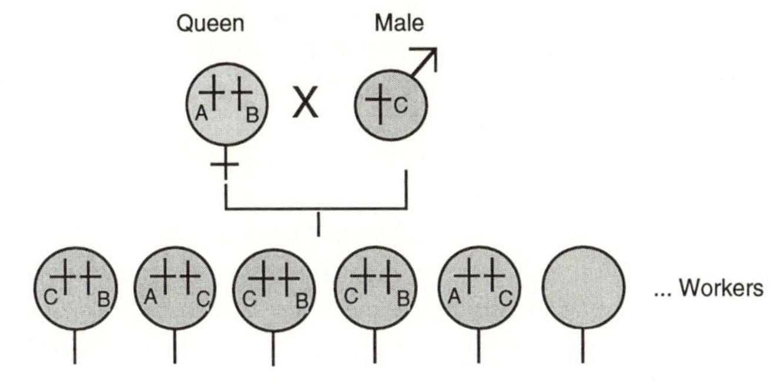

b) single queen (monogyny), several males (polyandry)

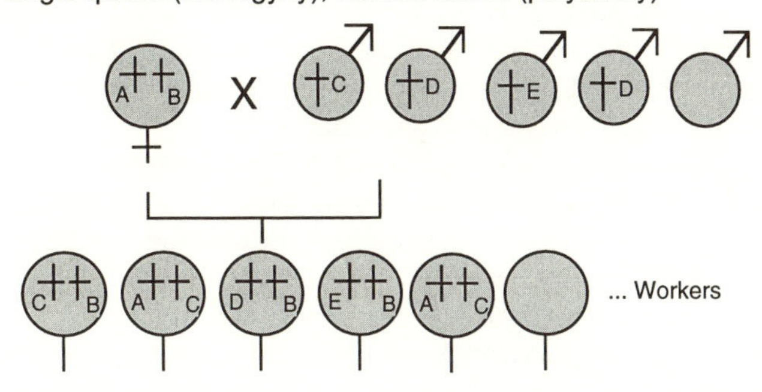

c) several queens (polygyny), several males (polyandry)

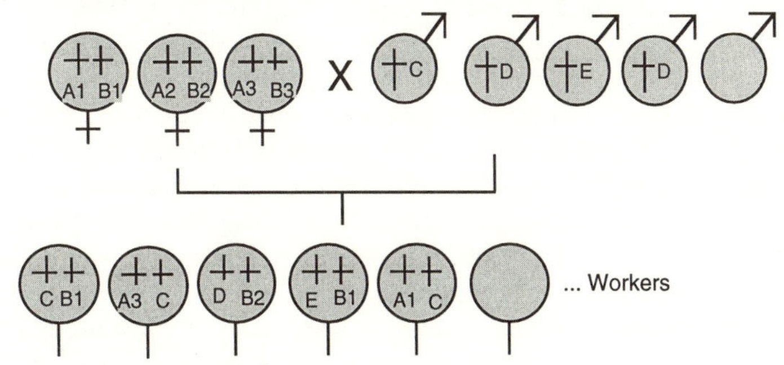

Figure 5.1 Mating schedules in social Hymenoptera and consequences for variability of workers. (a) Single queen mated to one male. (b) Single queen, multiply mated, i.e., polyandry. (c) Many functional queens (polygyny), mated to multiple males each (polygynandry). Letters symbolize different alleles at a given locus.

allocation in different colonies is in line with the expectations from split-sex ratio theory (Boomsma 1991; Boomsma and Grafen 1991; Sundström 1994; Sundström et al. 1996). With respect to polyandry, this hypothesis requires that workers facultatively bias the sex ratio in response to the mating frequency of the queen and that direct fitness benefits result for the queen due to excessive matings. With a theoretical analysis, Ratnieks and Boomsma (1995) indeed found that multiple mating by queens is selectively favored by worker allocation behavior, in agreement with earlier suggestions (Moritz 1985; Pamilo 1991a; Queller 1993b), even if some simplistic assumptions are relaxed.

At least among species of ants, the occurrence of polyandry correlates positively with typical colony size (Cole 1983). Boomsma and Ratnieks (1996) confirmed this empirically for monogynous but not polygynous ants. It has therefore been suggested that polyandry has evolved as a consequence of sperm limitation in males (Hölldobler and Wilson 1990) (fig. 5.2a). However, this hypothesis raises the question of why males are not capable of producing or transferring more sperm in the first place. Or why sometimes the males, in polyandrous aggregates, transfer far more sperm than the queen can accommodate, as, for example, in honeybees (Koeniger 1991). Evidently, queens in these species do not mate to get more sperm but to get more diverse sperm. Keller and Reeve (1995) hypothesized that sexual selection for more and more viable sperm is the driving force for polyandry.

Boomsma and Ratnieks (1996) reanalyzed the studies of paternity, mainly in ants. They identified several empirical problems, for example, that observed copulations tend to overestimate the genetically effective paternity frequency. In addition, the mean effective paternity frequency for nineteen species of ants for which accurate data are available is only 1.16 (range 1–1.48). The same trend was also evident from further, less detailed studies on additional ant species. Hence, it seems that, on average, the degree of effective polyandry is typically rather low. In addition, there was no relationship between mating frequency and number of queens per colony, contrary to Keller and Reeve's (1994a) claim. The authors conclude that multiple mating is simply not yet explained but that factors such as conflict over sex allocation or the avoidance of diploid males are likely to be important. Following Ratnieks and Boomsma (1995), it is quite possible that extreme mating frequencies, such as found in honeybees, are a derived character throughout the social Hymenoptera. In this context, one should at least mention that virtually all discussions to date have considered the social Hymenoptera, while termites have been ignored. Because termites are diploid, sex is determined in different ways, and the genetic load hypothesis, at least, could be ruled out on a priori grounds.

Polygyny in social insects has also been explained in similar ways through benefits for the colony due to genetic variability (Crozier and Consul 1976). Further hypotheses concentrate on the evolution of polygyny under high risk of dispersal and thus low probability of independent colony founding (Pamilo

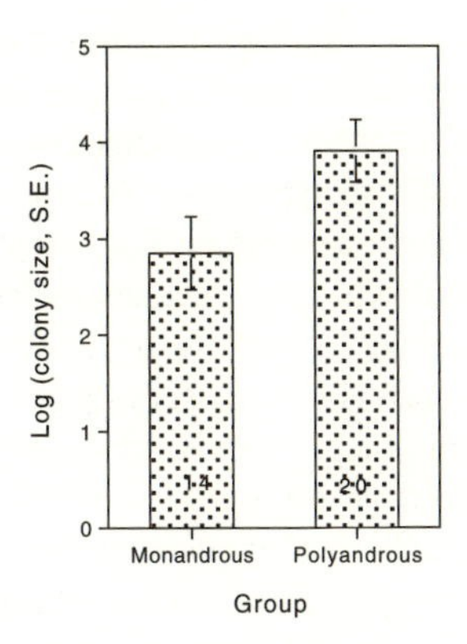

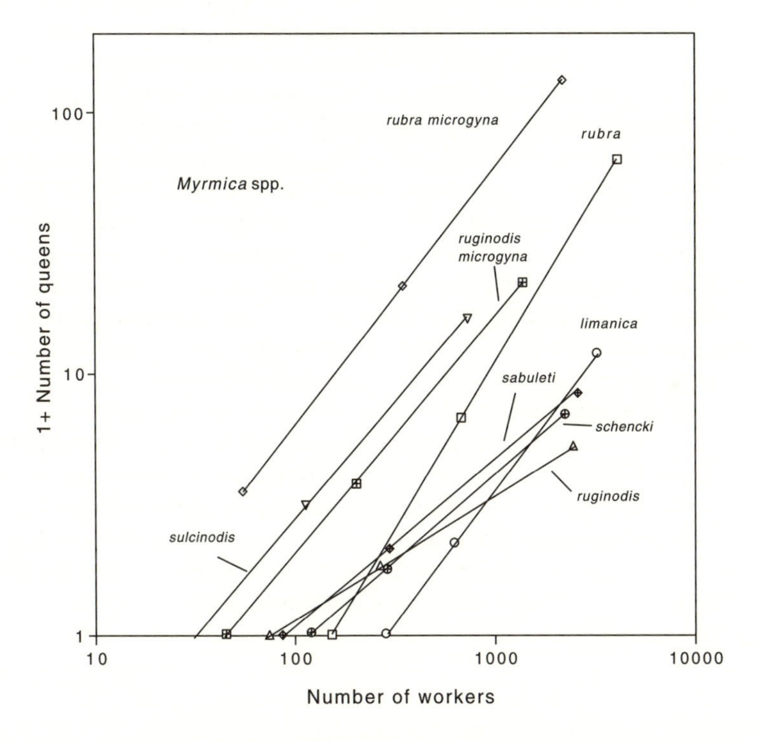

Figure 5.2 (a) Colony size in singly (monandrous) and multiply mating (polyandrous) species of ants. Multiple mating is also more frequent in species typically attaining large colony sizes. (Data from Dobrzanska 1959, Cole 1983, Hölldobler and Wilson 1990, Kaspari and Vargo 1995, Crozier and Pamilo 1996, and P. Frumhoff, pers. comm.) The statistics are as follows: Mann-Whitney $z=2.729$, $N=34$, $P=0.006$. A comparative analysis by independent contrasts (Purvis and Rambaut 1995), based on the phylogenies of Hölldobler and Wilson (1990) and Baroni-Urbani et al. (1992), yields the same result (5 contrasts with the positive mean toward polyandry, $P=0.032$). (b) Queen number in relation to colony size (number of workers) in ants of the genus *Myrmica*. (Reproduced from Elmes and Keller 1993 by permission of Oxford University Press.)

1991a), or where queens are comparatively short-lived and can be replaced by other related ones (Nonacs 1988). Queen number seems to be a conservative trait among the bees, less conservative in wasps, but rather labile in ants (Frumhoff and Ward 1992). Frumhoff and Ward (1992) note that in ants, polygyny is more likely to occur in ephemeral habitats. This fits with the other evidence that queen number is associated with ecological characteristics (Herbers 1993), although the exact relationships are not fully understood. Elmes and Keller (1993) found a positive correlation between queen number and colony size (worker number) in six out of eight species of the ant *Myrmica* (fig. 5.2b). The authors considered this relationship to reflect either a simple effect of colony age, or that colony size is affected by queen fecundity. In the latter case, they concluded that fecundity of queens is reduced in large colonies, which in turn correlates with higher queen numbers. In any case, Elmes and Keller (1993) concluded that queen number is a labile trait, depending on ecological factors. Obviously, a simple observation of polygyny does not necessarily imply that all queens contribute equally to offspring. In fact, systematic biases have been documented in several species (e.g., Ross 1988 in *Solenopsis,* Stille et al. 1991 in *Leptothorax*). On the other hand, cofounding queens are sometimes related, especially in moderately polygynous colonies (e.g., Craig and Crozier 1979; Pamilo 1981, 1982a; Rosengren et al. 1993). This tends to reduce genetic variability in the colony. Nevertheless, polygyny allows for a wide range of genetic relatedness depending on number and relative fecundity of functional queens.

I would like to end this section by pointing out that the discussion of why females mate with several males has also been rampaging outside the social insects, notably in birds. The respective hypotheses have concentrated more on the possible benefits of polyandry for securing paternal care for brood, sexual selection, sperm competition, or reduction of female harassment (see Birkhead and Parker 1996 for a recent review). In addition, Zeh and Zeh (1996) have suggested that polyandry is a hedge against genetic incompatibilities. Some of these contending hypotheses should be worthy of testing in social insects too but so far have not been intensively discussed. Other hypotheses are improbable or impossible due to basic differences in biology; nevertheless they raise intriguing questions.

5.2 Genetic Variability vs. Parasites

Several authors have proposed that genetic variability among the members of a colony might be beneficial in the presence of pathogens and parasites (Tooby 1982; Hamilton 1987; Sherman et al. 1988). This particular hypothesis is an instance of the general problem of how genetic variability, including sexual reproduction, can be maintained in natural populations. Haldane (1949), and later Hamilton (1980), suggested that pathogens may act as selective agents with negative frequency dependence, so that rare genotypes are at an advantage. Hence, polyandry (or polygyny) could have evolved to minimize the effect of parasitic infection of a colony ("the variability-vs.-parasites" hypothesis). To some extent, this is a variant of the tangled-bank hypothesis, whereby variability is advantageous under intense competition among offspring (e.g., Bell 1982).

Polyandry and polygyny tend to diminish the average relatedness among workers. On the other hand, an increase in the (among colony) variance in the number of males and queens contributing to the colony actually increases rather than decreases the population-wide average worker relatedness in a colony (Queller 1993a). There are other processes that do lead to high degrees of worker relatedness, e.g., cyclical oligyny in some social wasps (Hughes et al. 1993), and close relatedness among colony queens. Such differences affect how the available genetic variance is distributed within the population, i.e., within vs. among colonies. Typically then, high degrees of within-colony relatedness are associated with high degrees of among-colony genetic variance. As host-parasite interactions have a genotypic component, the (ecological) host-parasite dynamic is thereby also affected. In fact, from the parasite's perspective, not only the genetic diversity of a single host colony but also the overall distribution of genetic variance in the population matters, because parasites are under selection in all of the infected colonies at the same time.

For example, May and Anderson (1984) have shown that an increase in the variation of the transmission rate increases the net reproductive rate, R_0 (see chap. 6), of the parasite, even if the population-wide average transmission rate stays the same. For the sake of argument, suppose that transmission among colonies is primarily dependent on the genotypes of "source" and "target" colony and thus directly affected by the genetic variance among colonies. Then, in populations where workers are on average closely related and variance among colonies is consequently large, parasites would have a higher expected transmission rate and thus be more likely to be sustained. A population consisting of half-colonies with singly mated queens and half-colonies with trebly mated queens is not equivalent to a population where all queens have mated twice, although the average mating frequency would be the same. The former would imply a higher degree of average within-colony relatedness and higher variance among colonies, with an increased effective reproductive rate for the parasite, all other things being equal. Among other things, this argument depends on the relative importance of within- and among-colony transmission for the persistence of the parasite (see also Lipsitch et al. 1996 and chap. 7). Nevertheless, the argument shows the importance of the pattern in which genetic variance is distributed in a population. Substantial variation among colonies in the prevalence of parasites has in fact been observed in well-documented cases such as the honeybee and its pathogens, e.g., *Nosema apis* (e.g., Doull and Cellier 1961), or APV (acute bee paralysis virus; Bailey and Gibbs 1964). This does not prove

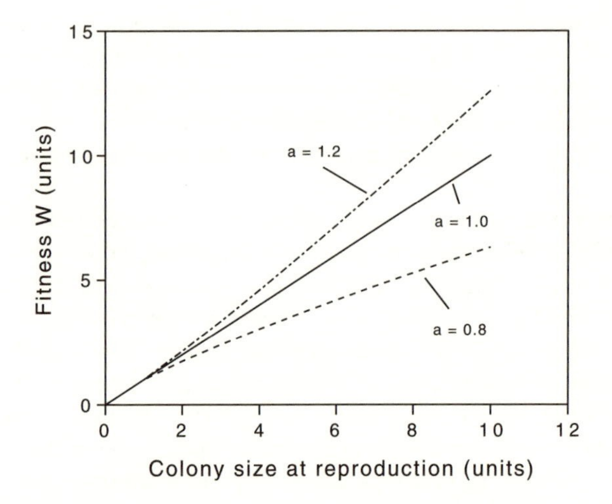

Figure 5.3 Hypothetical relationships between colony size (e.g., number of workers) at reproduction and colony reproductive success (*W*, fitness). The curves are simple functions of $W = (\text{size})^a$, where a takes on different values above and below unity; $a = 1$ corresponds to a linear relationship, values of $a < 1$ produce concave functions, $a > 1$ convex functions.

the case but it is an indicator that such heterogeneities may exist at the relevant level. The argument would also indicate that an increase in genetic variation is not inevitably disadvantageous for the parasite, as has been discussed in other contexts (e.g., Hamilton 1980; Ebert and Hamilton 1996). A crucial difference is in the assumption about how and where (e.g., within or between hosts) parasites are under selection.

Selection for Polyandry and Polygyny by Parasites

In their treatment of the adaptive value of polyandry, Sherman et al. (1988) assumed that parasitic infections ultimately reduce the number of workers in the colony. At the same time, the colony is supposed to survive only when its size is above a certain threshold. Assuming further that susceptibility of workers is genetically based (see also table 7.1), multiple mating reduces the variance in the proportion of surviving workers and hence should be selected for under certain kinds of truncation selection. The assumption of truncation selection seems reasonable for colonies of social insects that experience bottleneck periods early in colony development, during overwintering, or during drought periods. However, truncation has never really been shown empirically in social insects. Therefore, the scenario is extended here following the treatment of Schmid-Hempel (1994a). It focuses on colony-level selection, taking the success of the colony queen(s) as the unit of interest.

Consider a situation where k different pathogen strains can in principle infect colonies of hosts, and where in the host population there are k corresponding resistance alleles. In the polyandrous case, colonies are supposed to contain a single queen mated randomly with a total of m males. Such a mating system will produce a mixture of homozygous and heterozygous workers in the colony (as in fig. 5.1b). If the colony happens to become infected by an appropriate pathogen strain, i.e., a strain that can infect hosts with the corresponding alleles, infected homozygotes are assumed to have reduced viability and thus to contribute less to the growth, survival, and eventual reproduction of the colony. Let this be with factor $w = 1 - s$ ($w < 1$), where s is the selection exerted by the pathogens against their individual hosts. Heterozygotes, infected or not, and those homozygotes not infected by the appropriate strain are assumed to contribute to colony success with factor one. Imagine that these worker contributions eventually affect the state, X, of the colony (e.g., its size) at time of reproduction. State X is mapped into fitness through a monotonically increasing "fitness function" (fig. 5.3), given by the number of sexuals produced and their chances of mating. As shown in the appendix to this chapter, with these assumptions the expected state and variance of the colony at time of reproduction, $E(X)$, $Var(X)$, is

$$E(X) = 1 - \frac{sp}{2} \text{ and } Var(X) = \frac{1}{4}\frac{s^2}{m}pq, \qquad (5.1)$$

where $p = 1 - q$ the probability that a queen mates with a male having an allele identical to one of hers. In this simple model, the infectious parasites are assumed to be clonal. There is a lively debate as to whether this is true of natural populations of parasites of this kind (Dye et al. 1990; Keymer et al. 1990; Tibayrenc et al. 1990). If parasites were not clonal, the analysis would of course become more complicated. But the standard scenario depicted here nevertheless captures the essence of the problem.

The fitness function is assumed to be the result of processes that act mostly independently of the parasites' direct effects. For example, large colonies may be disproportionately successful because they are well defended against predators, better survive drought periods, or are ergonomically more efficient. For simple functions, fitness can be readily calculated: (1) Linear function, i.e., fitness proportional to colony state X: $W_{\text{linear}} = E(X) = 1 - sp/2$, and fitness will be independent of mating number. (2) Quadratic fitness function $W_{\text{quadratic}} = E(X^2) = (s^2/4m)pq + (1 - sp/2)^2$. This will select against multiple mating with factor $1/m$. (3) Truncation selection: colonies below a certain threshold size, $X < X_0$, are assumed to produce no reproductives, while those above all have fitness 1. This is the explicit case considered by Sherman et al. (1988); fitness will be proportional to the variance of the distribution as given by eq. (5.1). Hence, multiple mating will be favored if $X_0 < E(X)$, and its benefits increase with $1/m$, because this decreases variance and less is truncated. Conversely, multiple mating will be at a disadvantage if $X_0 > E(X)$, because this reduces the tail of the distribution that escapes truncation.

The benefit of multiple mating thus strongly depends on the actual fitness function (see also Pamilo et al. 1994 for a similar concept in the context of diploid male production). If only a few large colonies get the lion's share of the reproductive success in the population ($a > 1$), selection through pathogens should actually select for single mating, i.e., monandry. Polyandry is selected for when fitness is more evenly distributed among the colonies, notably when fitness is decelerating the function of colony size ($a < 1$), such that moderately sized colonies get a disproportionate share of the reproductive success. For instance, in social insects, limitation in the size of available nesting sites, internal colony conflicts, and the reproductivity effect disproportionately depresses the reproductive success of large colonies, which would produce such a concave function (fig. 5.3; Michener 1964). Figure 5.4 shows empirical examples of this function. The particular fitness functions just mentioned do indeed occur in natural systems. From figure 5.4a, we would predict that *Bombus terrestris* is a single mating species, while *Polistes annularis* (fig. 5.4b) should be multiply mating, everything else being equal. From all we know, *B. terrestris* is indeed a singly mating species (Estoup et al. 1995), while *P. annularis* has multiple queens, hence qualifies for a species with genetically variable colonies. Further data on the relationship between colony size and actual fitness in a population are highly desirable and will make it possible to test these scenarios.

In the general case, fitness W scales with colony state X like $W = X^a$, where a = the shape parameter (fig. 5.3). Furthermore, the number of infections per colony (by the same or different strains) is a random variable. Mating can be costly too, since females in search of mates become exposed to predators, hazards of the weather, or simply have to spend a lot of time and energy. In addition, infection by cytoplasmic elements, or the risk of contracting sexually transmitted diseases (Lockhart et al. 1996), would also increase the mating cost. At least for the model situation, the costs of multiple mating need not only refer to actual costs incurred when mating. In the honeybee, Hillesheim et al. (1989) and Oldroyd et al. (1992) claim an ergonomic cost of multiple mating in the honeybee, i.e., that the genetically variable colonies are less efficient due to bad representation of appropriate genotypes (see also Schmid-Hempel 1990 for a nepotism-based argument).

The appendix shows that for the general case, the expected fitness from mating strategy m is

$$W_m = e^{-c(m-1)} \bullet \left(e^{-\mu f} + (1 - e^{-\mu f}) \sum_{i=0}^{m} \left(1 - \frac{i}{2m} s\right)^a \binom{m}{i} p^i q^{m-i} \right), \qquad (5.2)$$

where μ = average number of infections per colony (a Poisson variable), c = reduction of fitness for each additional mating, and f = frequency of a particular strain among the parasites. This equation has been evaluated numerically

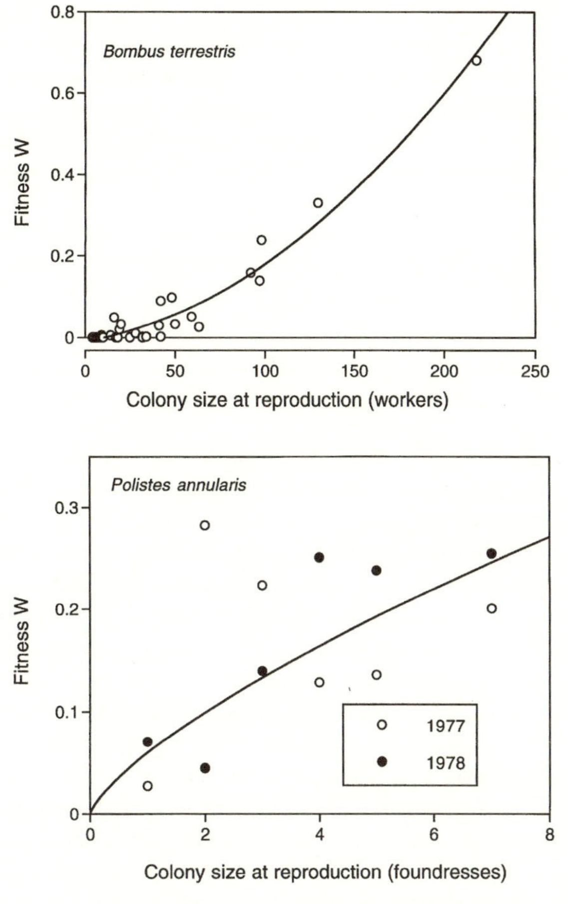

Figure 5.4 (a) Calculated fitness (W) in relation to colony size (N) in *B. terrestris*. Same data as in figure 4.7a, but drones and gynes are weighted according to the Shaw-Mohler equation (e.g., Charnov 1982): $W = m/M + f/F$, where m, f = number of males and females produced by a colony of size N; M, F = total number produced in the population, assuming all males are by the queen. (Modified, after Schmid-Hempel 1994a, courtesy of C. B. Müller.) (b) Fitness W as a function of colony size in the wasp *Polistes annularis* (defined as number of foundresses). W is calculated as $p{\cdot}f/S(p{\cdot}f)$, where p = survival probability of foundress association to reproduction, f = number of females produced by the colony if surviving, $S(p{\cdot}f)$ = sum for the population. Values are for the years 1977 and 1978. (Data from Queller and Strassmann 1988.)

for a range of values of the parameters s, c, p, k, μ, and m. Figure 5.5 shows characteristic results. Multiple mating is advantageous under conditions where the costs of mating are low relative to parasite pressure (as indicated by a low ratio c/μ). More interestingly, multiple mating is favored for values of a close to 0.5. The advantage of multiple mating is thus most evident when the shape of the fitness function implies diminishing returns from size (a concave function; fig. 5.3).

So far, the problem has been discussed under the assumption that every colony in the population is infected. It is easy to see that a relaxation if this requirement does not alter the conclusions, as

$$W = P_{\text{Not infected}} \times W_{\text{Not infected}} + P_{\text{Infected|Not-A}} \times W_{\text{Infected|Not-A}} + P_{\text{Infected|A}} \times W_{\text{Infected|A}} \,. \tag{5.3}$$

$W_{\text{Not infected}} = 1$ per definition. $P_{\text{Infected|Not-A}}$ and $W_{\text{Infected|Not-A}}$ both refer to colonies that are infected but contain no worker lines that are susceptible to the pathogens ("Not-A"), hence $W_{\text{Infected|Not-A}} = 1$. $P_{\text{Infected|A}}$ and $W_{\text{Infected|A}}$ refer to colonies that are infected by a strain to which at least one of the patriline worker groups is susceptible, i.e., that have the matching allele A ("A"). The calculations furthermore suppose that no more than one parasite strain to which a worker group is susceptible is present at any one time. This should approximate the situation where levels of infection are not too high to produce frequent superinfections, or when allelic diversity is high. The calculations in the appendix additionally demonstrate that the qualitative conclusions remain the same if the distribution of pathogens over the host colonies is no longer Poisson (e.g., follows a negative-binomial distribution), or if a different genetic system (e.g., dominant-recessive for resistance alleles) is assumed. In numerical simulations, further assumptions can be relaxed. For example, what happens if queens are allowed to mate with a random number of males, m, and colonies are subject to a Poisson-distributed number of infections of which any number of strains, 0 . . . n, could be virulent (i.e., when a strain matches any allele among the different worker lines)? For theses cases, numerical simulations based on populations of five hundred colonies were run (unpubl.). The results led to the same qualitative pattern, suggesting that the qualitative conclusions of the model described above are robust.

To date, there are few empirical tests of this scenario. Nevertheless, a number of results speak in favor of the hypothesis that within-colony genetic variability is related to selection by parasitism. For example, Oldroyd et al. (1992) report that in honeybee colonies where the queen has been inseminated by one or several drones, the presence of some drone lines had a significant positive effect on colony growth. Although nothing on parasite effects is known in this case, it demonstrates that there is genetic variance for colony performance in the sense discussed in this section. More to the point, drones of the honeybee transmit

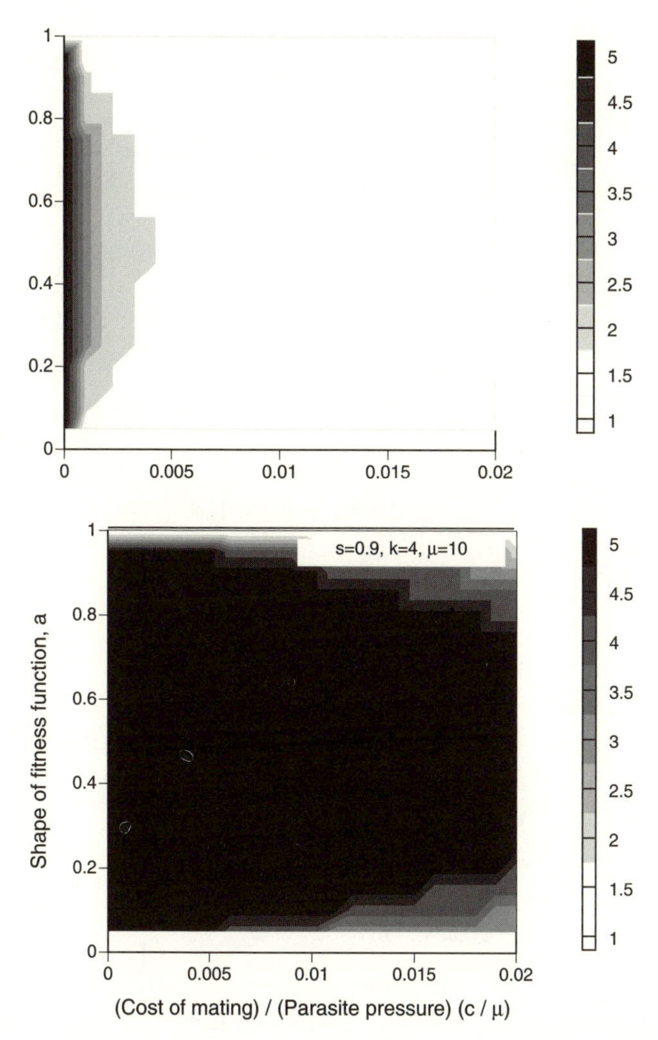

Figure 5.5 Predictions for multiple mating, according to eq. (5.2) in text. Parameter values are in (a) selection coefficient $s=0.1$, allele number $k=2$, and average number of strains infecting the colony $\mu=10$. In (b) $s=0.9$, $k=4$, $\mu=10$. The shadings indicate the optimal number of matings according to the scale on the right. Ordinate: parameter *a*, describing the shape of the fitness function (see fig. 5.3). Abscissa: fitness cost per mating relative to parasite pressure, i.e., ratio c/μ. (In parts reproduced from Schmid-Hempel 1994, by permission of the Royal Society.)

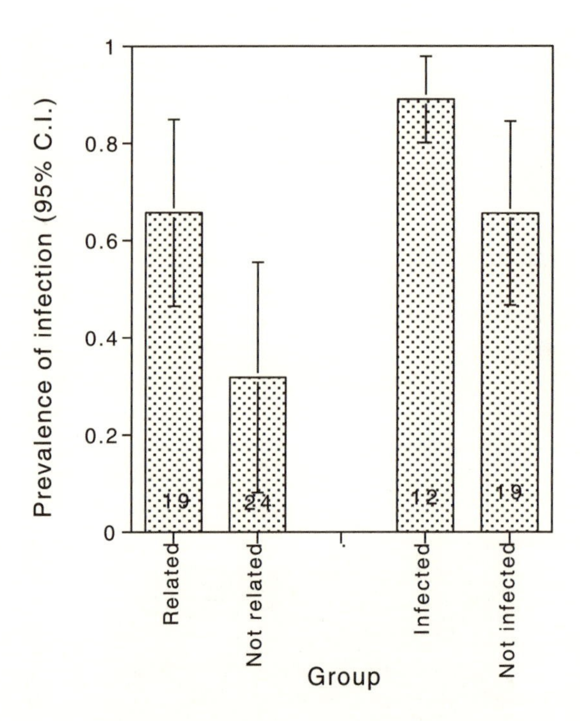

Figure 5.6 The trypanosome *Crithidia bombi* infecting the bumblebee *Bombus terrestris*. (*Left*) The effect of relatedness between infected and susceptible hosts on the probability of parasite transmission and infection. The parasite is spread more readily in related groups (t-test: $t=2.26$, $df=41$, $P<0.03$). (*Right*) The effect of natural susceptibility of queens in the field on the probability of parasite transmission and infection among her colony workers. The parasite is spread more readily among workers if they come from a queen that was herself naturally infected ($t=2.05$, $df=29$, $P<0.05$). Small figures are number of worker groups tested. (Data from Shykoff and Schmid-Hempel 1991b.)

characteristics of parasite resistance to offspring that they sire (see reviews in Cale and Rothenbuhler 1975 and Taber 1982). In bumblebees, Shykoff and Schmid-Hempel (1991a,b) demonstrated that the prevalence of infections by *Crithidia bombi* (Trypanosomatidae) among workers of a group that are related is higher than among unrelated workers (fig. 5.6), as expected from the hypothesis. Further observations demonstrated that this was not due to differential behavior towards kin or non-kin. In addition, the data suggested that there is a genotype-genotype interaction underlying this host-parasite system (Schmid-Hempel and Schmid-Hempel 1993; Wu 1994; Shykoff, unpubl.; see also fig. 5.6 for the queen effect). Perhaps the best evidence so far in favor of the variability vs. parasites hypothesis is provided by the study of Liersch and Schmid-Hempel (1998). In experimentally manipulated colonies of bumblebees studied under field conditions, the genetically variable colonies in fact acquired fewer parasites (fig. 5.7) and had higher success measured in terms of colony size.

On the other hand, a number of studies have remained inconclusive. For example, Briano et al. (1995b) found no difference in the degree of polygyny (i.e., number of queens, but also in the prevalence of parasites) between colonies of fire ants infected or not infected by the microsporidium *Thelohania*. Similarly, no difference in the performance of polyandrous colonies of honeybees infected by the microsporidian *Nosema apis* as compared to monandrous ones was found by Woyciechowski et al. (1994). Ratnieks (1989, cited in Boomsma and Ratnieks 1996) found no difference in the amount of disease in colonies of low or high genetic diversity. At present, therefore, the experimental evidence is ambiguous. But it should be noted that the bumblebee studies refer to natural populations, while at least the honeybee studies have used animals managed by breeders for a long time. It is probable that the different lines used in the honeybee studies do not differ sufficiently in the parasite-related aspects due to long-term breeding for increased general performance.

With the compiled information in appendix 2, a number of comparative analyses can also be made. Standardized parasite richness, i.e., the number of parasite species per host species corrected for sampling effort (see fig. 2.1), is conveniently used as a measure of parasite pressure. Obviously, this is not the same as parasite effect, but it is the best estimate for parasite pressure that we currently have on this broad scale. It is furthermore not unreasonable to assume that if more and more parasite species are present in a given host, it also becomes more likely that particularly virulent parasites will be among them. As more studies accumulate on parasites as well as on effective paternity frequencies, this comparative approach will be more powerful. At present, only ants are usually well enough investigated to make such comparisons possible. With the comparative evidence, it is found that parasite richness in monandrous ants is presumably higher than in polyandrous ants (fig. 5.8). Moreover, Figure 5.2a shows that polyandrous species tend to have larger colonies, which is associated with more parasites (fig. 4.10). Due to a lack of data, colony size as a confounding factor cannot be removed from the analysis of figure 5.8. At the very least, then, the difference found for the twenty-six species in this graph makes the pattern even more telling. It would support the hypothesis if polyandry is indeed effective in reducing parasite load; it would of course oppose it if one assumed that high parasite loads would lead to polyandry. The difference in figure 5.8 is statistically not significant, which is not surprising given the many sources of variation that enter the analysis. Nevertheless, the pattern is encouraging enough so that future studies are promising. For example, in some ants, effective paternity frequencies vary among populations. This could be exploited for testing the hypothesis by relating mating frequencies with parasite loads.

Polygyny is somewhat different from polyandry. Each additional queen in the association poses the risk of being a carrier of a novel infection, either by herself

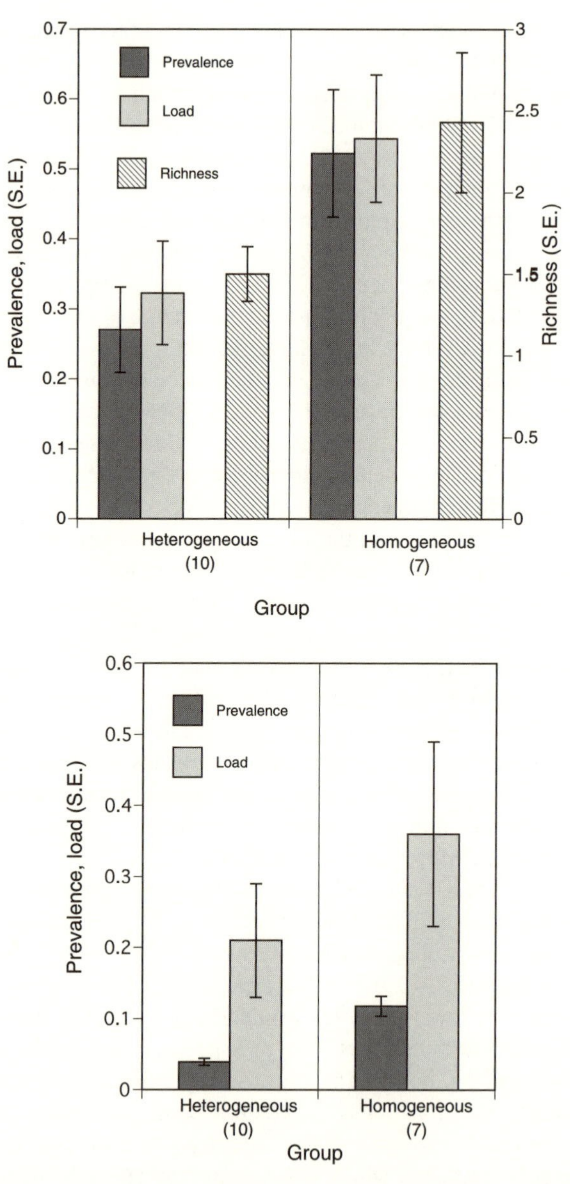

Figure 5.7 Parasites in experimentally manipulated colonies of *B. terrestris* in the field. Homogeneous colonies consisted of full sisters. Heterogeneous colonies contained workers from three to four different origins. Prevalence is the proportion of infected workers; parasite load is the average number of parasite species per individual worker per colony; richness is the number of different parasite species found per colony. (a) Statistics for all parasites combined: Prevalence (arcsin-transformed values, $t=2.37$, $P=0.032$, $df=15$), parasite load ($t=1.83$, $P=0.028$ one-tailed, $df=15$), parasite richness (log-transformed values, $t=2.24$, $P=0.041$, $df=15$). (b) For infectious parasites only (protozoa): all differences go in the same direction but are not significant ($P>0.05$ in all cases). Small numbers are sample sizes (colonies). (Data from Liersch and Schmid-Hempel 1998.)

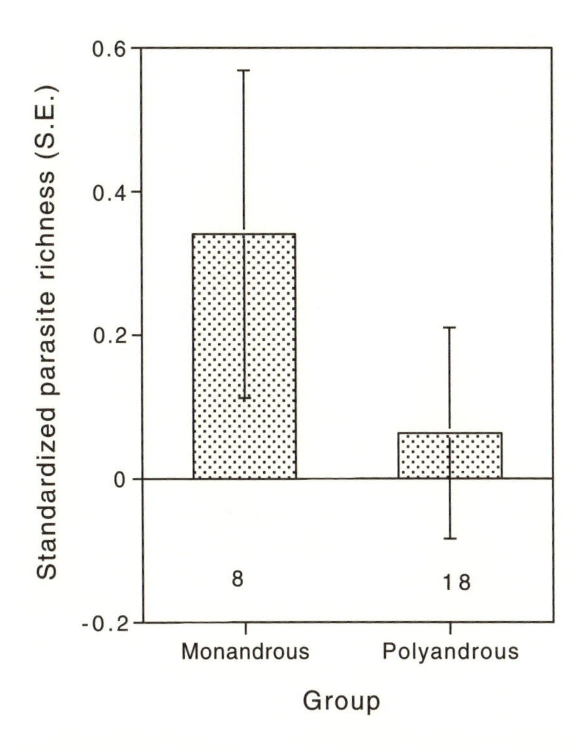

Figure 5.8 Standardized parasite richness (S. E., see fig. 2.1) for monandrous and polyandrous species of ants. The formal test is t=1.03, df=24, P=0.31. (Parasite data from appendixes; mating system from various sources—cf. table 5.2.) Small figures are number of species in group. The comparative analysis by independent contrast for the same question (see fig. 5.2 for more details) has only four contrasts available and is not significant.

or by her worker offspring, although the latter, with the same arguments as above, are assumed to "dilute" the risk at the same time. Polygyny is also different because with multiple queens, genetic variability within the colony increases, but it does not necessarily do so among a queen's own offspring (ignoring correlated levels of polyandry). Hence, polygyny must be advantageous for other reasons to begin with. The basic requirement is that each queen within a multiple female association would on average enjoy a greater success than as a singly nesting queen. This could occur as a result of nest site limitation or predation pressure on incipient colonies.

For the time being, the simplest formal treatment of polygyny is that the expected fitness per queen is an increasing function, at least up to some point, of the number of females (F) present in the colony. If queens are not related to one another and are each mated to a single, unrelated male, the probability that a given queen has mated with a male that carries an allele matching one of the queen's is again $p = 2/k$. Hence, among F monandrous queens in random association, the expected number of queens with a matching male is $E(F') = F \cdot p$, with variance $Var(F') = Fpq$, where $q = 1 - p$ as before. The expected frequency of homozygous workers in the colony is then $p/2$, that of heterozygotes $1 - p/2$. Hence, eq. (5.2) holds if m is replaced by F, and expected fitness corrected by $1/F$. It follows that concave fitness functions would again increase the benefits of polygyny, while convex fitness functions would favor an intermediate degree of polygyny. The number of queens (F) that maximizes per capita fitness can be calculated by combining the corresponding equations with assumptions about the increase in success with the size of the association. This would also allow to take into account the combined effects of polygyny and polyandry.

Keller and Reeve (1994a) proposed that polyandry and polygyny act compensatorily to increase genetic diversity within a colony. Therefore, high degrees of polygyny should be negatively correlated with polyandry to ensure a certain degree of variability. Comparing a number of ant species, such a relationship was found. However, Boomsma and Ratnieks (1996) found no support for this correlation in a reanalysis of the data. Keller and Reeve (1994a) also did not discuss the possibility that polygyny is different in kind from polyandry as mentioned above. In fact, it is not clear why genetic variability is selected for by the compensatory action of polyandry (a mating strategy) and polygyny (a breeding strategy involving permanent associations).

Nevertheless, the combined genetic effect of polygyny and polyandry can be approximated by the average actual worker relatedness within a colony. Crozier and Pamilo (1996, their table 4.7) have summarized the knowledge. For ants and wasps, there are enough data so that these values can be plotted against the typical colony size. As figure 5.9 shows, for wasps a significant negative relationship emerges that fits the expectation from the increase in polyandry and polygyny with colony size. If only single genera are considered (e.g., *Polistes* in wasps), a similar relationship still holds. There are obviously many hypotheses that predict such a pattern (see, e.g., Crozier and Pamilo 1996). Yet, chapter 4 has substantiated the working hypothesis that large colony sizes are associated with higher parasite loads. So, for the wasps, the pattern of fig. 5.9 is in line with selection by parasites for increased within-colony genetic variance in large colonies. On the other hand, for the ants no such relationship with colony size is apparent from figure 5.9. This somewhat contradicts the pattern found in figure 5.2. Perhaps the compensatory effect of reduced polygyny, as suggested by Keller and Reeve (1994a), reduces the genetic effect of polyandry in large colonies of ants, although this compensatory behavior has not yet been confirmed (Boomsma and Ratnieks 1996).

However, enough data are available so that the standardized parasite richness from appendix 2 can be directly plotted against the average within-colony relatedness. A positive correlation emerges, even when colony size is parceled out

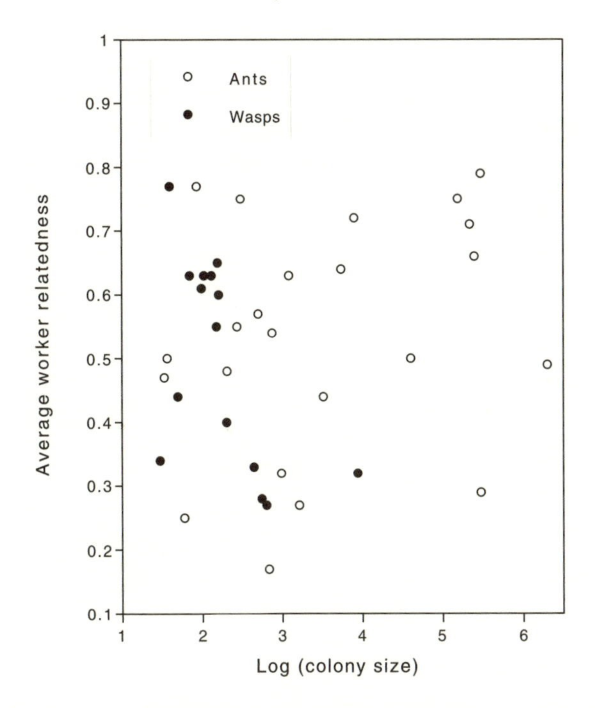

Figure 5.9 Relationship of average genetic relatedness among colony workers and typical colony size of the species. The formal relationship is, for all species pooled as follows: Spearman's $r_s=0.024$, $N=40$, $P=0.88$; for ants alone: $r_s=0.239$, $N=23$, $P=0.293$; for wasps: $r_s=-0.613$, $N=15$, $P=0.015$. (Data as in figure 5.10; relatedness from Crozier and Pamilo 1996.) A comparative analysis by independent contrasts for ants (see fig. 5.2) showed no correlation (contrasts of colony size vs. relatedness: $r=-0.095$, $N=13$, $P=0.76$). No such analysis is possible for wasps.

(fig. 5.10a), and which is maintained even when lineage-specific effects are controlled for by comparative analysis with independent contrasts (fig. 5.10b) (Harvey and Pagel 1991; Purvis and Rambaut 1995). This suggests that species with colonies consisting of closely related workers indeed have higher parasite loads. If variability reduces parasite load among species, this is as expected under the variability vs. parasites hypothesis, although supportive, more meaningful tests must address variation within species (sensu fig. 5.7).

The combined evidence of experimental and comparative studies is suggestive of a picture where lower parasite loads are associated with species that have low colony sizes, that are polyandrous, and/or have genetically variable colonies. This pattern supports the idea that parasitism is associated with the breeding system of social insects, which is an important point, since there are few other glimpses into what drives the evolution and maintenance of different

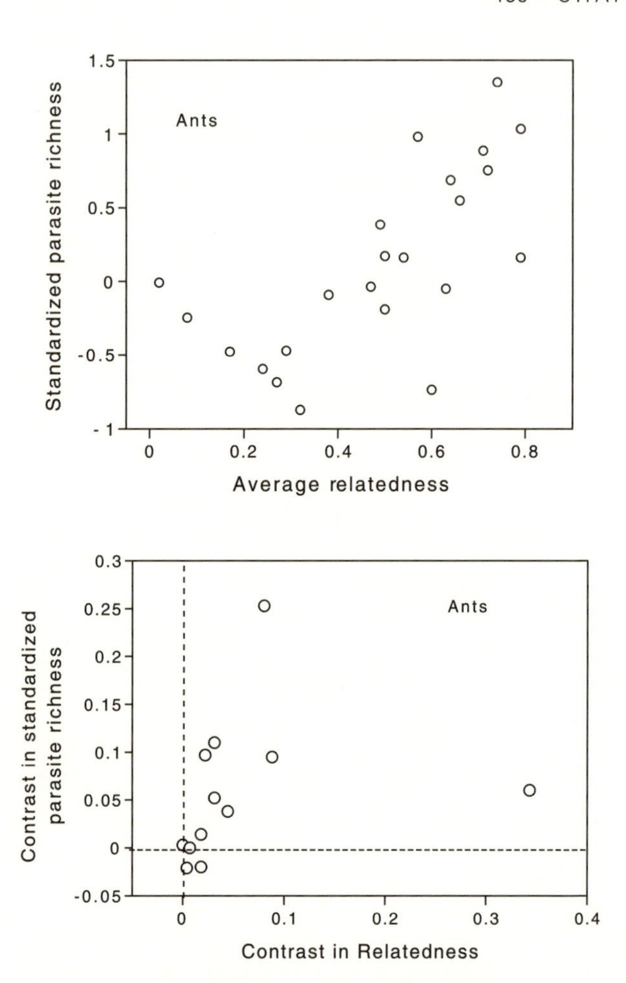

Figure 5.10 Standardized parasite richness (from fig. 2.1) in relation to average within-colony relatedness in colonies of ants. Relatedness values as in fig. 5.9. (a) All species considered independent. The formal statistics are as follows: Spearman's $r_s=0.704$, $N=23$, $P<0.001$. In a multiple regression with relatedness and colony size (both transformed to normalize variances) as covariates, only relatedness is significant. (b) Comparative analysis by independent contrasts (see fig. 5.2); the positive association is significant (Spearman's $r_s=0.673$, $N=12$ contrasts, $P=0.017$; without point at right: $r_s=0.764$, $N=11$, $P=0.006$).

breeding systems in social insects. Interestingly, the degree of exposure to severe pathogens also correlates positively with the degree of polygyny across a number of human societies (Low 1988).

Given that polygyny evolves in social insects, parasites could counteract the kin-selected advantages of associating with related females for the same reasons as discussed before, i.e., increased transmission as a result of genetic homogeneity of the colony. Relatedness under polygyny does indeed vary considerably. On one hand, the multiple queens in the ant *Solenopsis invicta* seem not related at all (Ross 1993). In some epiponine wasps, such as in *Polybia, Protopolybia,* and *Parachartergus,* in contrast, a high degree of relatedness is maintained by cyclical oligyny (Hughes et al. 1993). In many species of *Ropalidia, Parapolybia, Polistes, Mischocyttarus,* and *Belonogaster,* serial polygyny (short-term monogyny) leads to relatively low degrees of relatedness (Gadagkar et al. 1993). Such different polygyny schedules could be under selection by parasites for the same reasons as before. This hypothesis could be tested by comparing parasite loads under these different systems. For example, cyclical oligyny as in Epiponines should typically be associated with low parasite loads, whereas serial polygyny as in *Ropalidia* should have evolved under high levels of parasite pressure. Serial polygyny is considered to be a primitive trait for Polistine wasps that is lost in temperate habitats (Carpenter 1991). Hence, useful variation may be available but the phylogenetic effects obviously have to be accounted for. The available parasite records in appendix 2, however, do not allow one to check this idea, since very little is known for parasites of epiponine wasps. Nevertheless, social wasps may be good study subjects for most of the questions posed here, since they have a wide range of nesting habits, polygyny schedules, and life histories. In addition, they lend themselves readily for experimental tests that involve the manipulation of polygyny schedules.

Boomsma and Ratnieks (1996) favor alternative explanations for the maintenance of polyandry mainly because of the rather low average frequency of effective paternity. The discrepancy between observed mating frequency and effective paternity results from unequal contributions by the male or unequal use of sperm by the queen. To analyze this, Taber (1955) inseminated queens with wild-type drones and with drones carrying a visible genetic marker. The results suggested precedence of the first male, though this could not be confirmed in later studies (e.g., Laidlaw and Page 1984), and reanalyses showed errors in Taber's analysis (Crozier and Brückner 1981; Page and Metcalf 1982). Nevertheless, males do contribute in drastically different proportions, as several studies of the honeybee have now shown (Page and Metcalf 1982; Woyke 1983; Laidlaw and Page 1984; Estoup et al. 1994). Unequal contributions were also known from social wasps, e.g., Wilkes (1966) for *Dahlbominus* and Metcalf (1980) for *Polistes variatus.* This inevitably lowers the effective paternity frequencies as compared to mating frequencies.

In addition, many queens in social insects are very long-lived and therefore have to store sperm for a long time (fig. 1.3). Over the lifetime of the queen, sperm could thus be used in a nonrandom manner. In the honeybee, a detailed study with hypervariable microsatellites (Estoup et al. 1994) showed such temporal fluctuation to occur, although no systematic bias could be found. This pattern could nevertheless have some consequences. For example, in honeybees Laidlaw and Page (1984) observed a relationship between the fluctuation in male contribution to the worker force and the onset of brood disease (American foulbrood). Their colony contained six patrilines. As the disease set in, one of the patrilines dramatically increased in frequency while the others diminished. Only after antibiotic treatment did frequencies of patriline contributions return to their previous values. This is very suggestive for the effect of male contributions to a colony, susceptibility to disease, as assumed in the above models. In fact, it has been known for a long time that susceptibility to American foulbrood varies among inbred lines of honeybees (e.g., Rothenbuhler and Thompson 1956).

Socially parasitic species produce no colonies of their own but utilize the services provided by their hosts. If polyandry has evolved to generate a variable colony as a defense against parasitism, it is to be expected that social parasites are typically singly mated whereas their close, nonparasitic relatives may be multiply mated. So far, too few data are available to test this idea. However, the standardized parasite richness of socially parasitic species (mean $= -0.242 \pm 0.11$ S.E., $N = 14$ species in the sample of appendix 2) is generally lower than that of nonparasitic species (mean $= 0.001 \pm 0.02$, $N = 502$ species, fig. 2.1) ($t = 2.05$, $P = 0.041$, $df = 514$). Although this result is very preliminary and should be more thoroughly tested when more data become available, it at least suggests that socially parasitic species are less prone to parasites.

5.3 Patterns of Variability

Because of the importance of genetic variability for host-parasite interactions, it is worthwhile to have a brief look at natural populations. All hymenopteran species show lower degrees of genetic variability as compared to other insect orders when measured, for example, with isozyme analysis. Berkelhamer (1983) suggested that genetic variability is particularly low in the social Hymenoptera. Unfortunately, her study was flawed in several ways. But as Graur (1985) and Owen (1985) later showed, eusocial Hymenoptera indeed have lower levels of genetic diversity as compared to other insects, and as compared to solitary Hymenoptera. Graur (1985; see also Shoemaker et al. 1992) found an average expected heterozygosity $H_{exp} = 0.045 \pm 0.005$ S.E. for fourteen species of solitary Hymenoptera, but $H_{exp} = 0.026 \pm 0.005$ ($N = 16$) for eusocial species. The authors concluded that not haplo-diploidy per se but the small effective population size of social insects contributes to this difference. Alternatively, the production of males by workers may enhance this effect. For instance, Owen (1985 and references therein) demonstrated that if a substantial proportion of the males is derived from workers rather than from the queens, this would—except in the case of codominant loci—also lead to a decrease in the average genetic variability at loci maintained by balancing selection. To what extent this affects the interaction with parasites must await further studies.

When discussing host-parasite interactions, it is also useful to briefly consider the genetically effective population size of the host. This is because small effective sizes would, for example, reduce the effect of selection by parasites relative to random drift. Many of the observed patterns would thus be misleadingly assigned to parasites. In turn, parasites would not encounter a tightly coevolving host. Thus, the assumptions of negative frequency-dependent selection would not be met. Recall that social Hymenoptera, but not the termites, are characterized by the haplo-diploid genetic system. This property means that the genetically effective population sizes, N_{eff}, are generally lower than those in a diploid population of the same number of individuals. With a standard approach (e.g., Hartl and Clark 1989, p. 74; see also Crozier 1979) it can readily be appreciated that the effective population for populations of monandrous, monogynous colonies is typically only 75% of the value of a diploid-diploid population. Also, deviations from a random distribution of colony fecundity, such as must often be the case in social insects (fig. 5.11), should further reduce effective population size.

Effective population sizes of social insects have rarely been estimated or measured. Kerr (1974) estimated $N_{eff}=4542$ for the honeybee (with $N_{eff}=428$ and $N_{eff}=257$ in more isolated areas), and for meliponines (stingless bees), *Scaptotrigona postica* ($N_{eff}=575$) and for *Melipona rufiventris* ($N_{eff}=1064$). Estoup et al. (1996) assumed an effective size of $N_{eff}=986$ for honeybees in Europe (corresponding to 460 colonies with multiply mated queens), based on dispersal of sexuals and the associated neighborhood size. Widmer (1996) estimated population size based on a model of mutation-drift balance for the maintenance of allele numbers (i.e., Ewans's sampling formula, Hartl and Clark 1989) and found a N_{eff} around 800 for the bumblebee *B. pascuorum.* According to Wright (1943), for population sizes around a few thousand, the effects on drift would become virtually negligible. Hence, it is potentially significant that these estimates are in the same order of magnitude. These estimated sizes are in fact not very much smaller than those estimated for various vertebrates, snails, or even *Drosophila* (see Begon et al. 1980 and the discussion therein). The empirical evidence is thus somewhat at variance with the usual assumption of small effective populations in social insects. Bearing in mind that estimation of effective sizes of natural populations is notoriously difficult, this discrepancy may not be surprising. Because of the importance of host population variability and

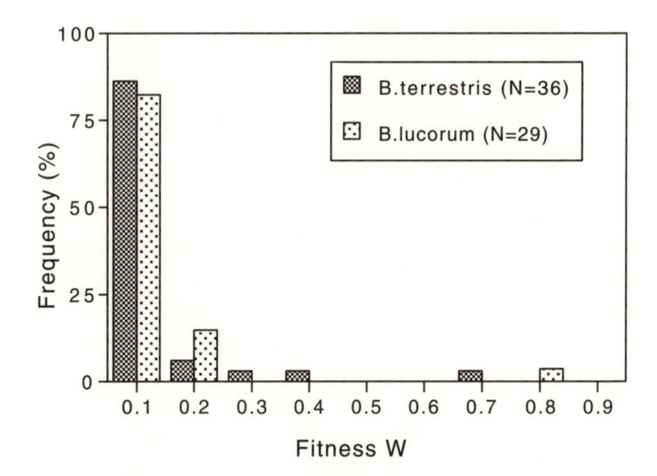

Figure 5.11 Fitness distribution in field populations of colonies from two bumblebee species, *Bombus* spp. Fitness is calculated with the number of males and females released by each colony and according to the Shaw-Mohler equation (as in fig. 5.7). (Data from Müller 1993.) Both distributions deviate significantly from Poisson. *B. terrestris*: $N=$ 36 colonies, mean$=0.056 \pm 0.130$ (S.D.), Var $(x)=0.017$, C.V.$=2.335$. *B. lucorum*: N$=29$ colonies, mean$=0.071 \pm 0.133$, Var $(x)=0.018$, C. V.$=1.860$.

effective size for the interactions with parasites, more empirical data are desperately needed. For the time being, however, we can assume that for the characteristics of parasites and hosts, selection rather than drift is important, and that therefore the standard assumption of coevolution is justified.

Chromosome Number and Parasites

Some studies have attempted to test the tangled bank hypothesis by analyzing the relationship between the average number of excess chiasmata (i.e., the number of chromosomal crossovers during meiosis in excess of one) and characteristics of parasite pressure. More excess chiasmata imply that offspring may have higher genotypic variability and hence this should correlate positively with parasite pressure. So far, the results have been ambiguous (Burt and Bell 1987; Koella 1993). Nevertheless, the scenario is pertinent to social insects, because offspring (i.e., the workers) interact within the selection arena of the colony and so eventually determine the success of the parental strategy (e.g., whether to produce variable or nonvariable offspring). Moreover, the selection regime in this arena is determined by the presence of parasites. If so, we would expect that excess chiasmata frequency positively correlates with indicators of parasite pressure also for social insects.

We cannot test this idea directly since data on chiasmata frequencies and parasites in the same species are rare. However, it has often been discussed in which way chromosome number relates to eusociality. The two aspects—chromosome number and genetic variability in the colony—are in fact not independent of each other. The number of chromosomes in a genome affects the distribution of genotypic variability among individuals. For a given map length of the chromosome, an increase in chromosome number leads to a decrease in the variance of the coefficient of consanguinity (kinship; box 5.1) between full or half-sisters (Templeton 1979). In other words, with an increase in chromosome number, the actual relatedness among any two randomly chosen workers of a colony becomes less variable around the mean. This would benefit kin-selected social behavior and should thus be more likely in highly social species. On the other hand, less variability due to higher chromosome numbers would also benefit the spread of a disease (sensu figs. 5.6, 5.10).

In a comparative study, Sherman (1979) found that the average chromosome number is indeed somewhat higher in eusocial taxa (diploid number 2n for Formicidae: $x = 31.2 \pm 12.9$ S.D., $N = 178$ species; Vespidae: $x = 30.2 \pm 11.9$, $N = 9$; Apidae: $x = 29.8 \pm 7.5$, $N = 53$; termites: $x = 42.9 \pm 4.4$, $N = 32$) than in other, nonsocial groups. The study suffers from a lack of statistical independence among the taxa studied. For example, families had a strong effect. Within taxa, a general tendency toward increased chromosome numbers in the phylogenetically advanced species seems to occur (see Sherman 1979). This has also been confirmed by a detailed study of the primitive Australian ants of the genus *Myrmecia* (Imai et al. 1994). In this species complex, chromosome evolution as a whole goes toward higher numbers by centric fission. On the other hand, simple geographic patterns are often found, e.g., New World species of the wasp *Polistes* have almost double the number of chromosomes of East Asian species (Reeve 1991).

With the working hypothesis suggested in chapter 4, let us assume that species with large colonies are generally more prone to parasitism (fig. 4.10). Hence, taking the argument above, colony size should correlate negatively with chromosome number in order to increase variability in the genetic relationship between pairs of workers. Figure 5.12a shows that when karyotypes are compared across a number of ant species, this trend is found, although there is a lot of scatter and the correlation is weak. Of course, the argument is only appropriate if it reflects processes that are affected by local ecological forces and are not simple lineage-specific constraints. Thus, the analysis by independent contrasts is needed. Now, the relationship is reversed, with large colonies being associated with higher numbers of chromosomes (fig. 5.12b). This is not expected under the parasitism hypothesis. But large colony size is confounded with generally increasing degrees of polyandry; hence, for the time being these conclusions must remain tentative. It is clear, though, with respect to parasitism, that the subject of chromosome evolution merits further elucidation.

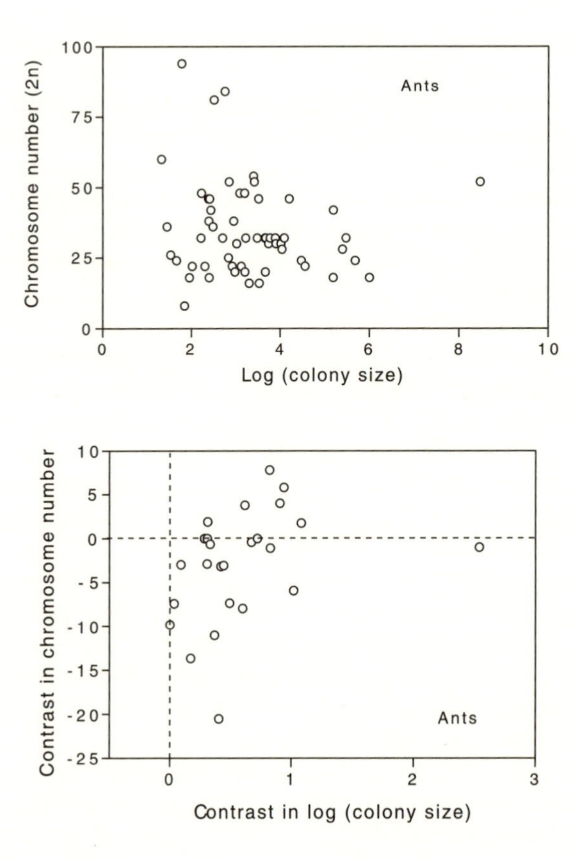

Figure 5.12 (a) The relationship between chromosome number (2n) and typical colony size (logarithmic scale) of different ant species. Pearson's correlation coefficient is $r=-0.132$, $N=58$ species, $P=0.322$. (b) The comparative analysis with independent contrast (see fig. 5.2) of the same data yields a positive slope ($r_s=0.409$, $N=26$ contrasts, $P=0.038$). (Data for colony size of ants taken from Dobrzanska 1959, Schmid-Hempel 1983, Hölldobler and Wilson 1990, Yamauchi et al. 1991; those for chromosome numbers from Kumbkarni 1965, Crozier 1968, 1970, Crozier and Pamilo 1996, Imai and Kubota 1972, Imai et al. 1977, Hauschteck-Jungen and Jungen 1983, Imai et al. 1984, 1988, 1994, Taber et al. 1988, Palomeque et al. 1993.)

5.4 Inbreeding and Mate Choice

Inbreeding does increase the average relatedness in a group of individuals and thus increases population "viscosity" (Hamilton 1964). In social insects, where kin selection plays an important role, a naive expectation would be that inbreeding could to some extent be advantageous. However, several theoretical studies have suggested that the consequences of inbreeding and population viscosity are neither simple nor necessarily favorable for kin selection (e.g., Queller 1992; Michod 1993). Nevertheless, because of the importance of genotype for host-parasite interactions, the issue merits some attention.

As Crozier and Pamilo (1996) discuss, inbreeding can come about in social insects in many possible ways. These include mating behavior, colony founding by sibs, breeding by related individuals in semisocial colonies, and so forth. Despite this range of possibilities, the actual degree of inbreeding in social Hymenoptera is extremely low or absent (see Crozier and Pamilo 1996, their table 4.9), with the possible exception of socially parasitic ants. This is primarily driven by mating biology. In the flying species—bees and wasps—both sexes are obviously well equipped to disperse (e.g., Mikkola 1978). But also in ants, sexuals have strong dispersal capacities and are released into large mating aggregations (Hölldobler and Bartz 1985; Hölldobler and Wilson 1990; Bourke and Franks 1995). On the other hand, termites, although they have wide-ranging, dispersing sexuals, may be more liable to inbreeding (Reilly 1987).

In addition to dispersal behavior and large mating aggregations, active avoidance to mate with close relatives further reduces the degree of inbreeding. Inbreeding avoidance should have a number of advantages. For example, it reduces the chances for the production of diploid males or typically increases general viability and fecundity. Some studies have shown that mating with close relatives is indeed avoided, for example, in the sweat bee *Lasioglossum zephyrum* (Smith 1983), in bumblebees (Foster 1992; although with an unconvincing experimental paradigm), in *Polistes* wasps (Ryan and Gamboa 1986), and in ants (*Iridomyrmex humilis:* Keller and Passera 1993). The fact that these few examples are observed in all major taxa suggests that avoidance may be widespread. Given the discussions in the preceding chapters, inbreeding avoidance should be particularly advantageous under parasitism. This point has not often been discussed in an explicit way. Rosengaus and Traniello (1993) hypothesized that immunity against local pathogens may select for outbreeding in the termite *Zootermopsis angusticollis,* but the evidence for the involvement of the parasite (a fungus) in their study is very weak, since no experimental infections and only post hoc examination of dead termites were made, with no direct proof for the action of the fungus or the alleged bacteria.

On the other hand, mate choice based on parasite-related characteristics has been intensively discussed as a major component of sexual selection in many taxa other than social insects (e.g., Thornhill and Alcock 1983; Andersson

1994; Stubblefield and Seger 1994). There is now good evidence that females are typically choosy, and that in many cases they choose males that are likely to ensure increased genotypic resistance for their offspring (e.g., in birds or fish: Andersson 1994). For example, the genes of the MHC-complex in vertebrates code for proteins involved in cell-cell recognition (Hedrick 1994), which is important in the defense against parasites. In seminatural populations, mice indeed choose to mate according to MHC-type (Yamazaki et al. 1976; Potts et al. 1991). As these examples from vertebrates show, mate choice is potentially more than just inbreeding avoidance.

In the social Hymenoptera, males only contribute sperm. In addition, females typically encounter many potential mates in a single mating period (e.g., Hölldobler and Bartz 1985; see also Bourke and Franks 1995 for a detailed discussion). The female depends on this once acquired sperm complement for her entire life. As already mentioned, several of the phenotypic characteristics of her subsequently emerging colony are known to be affected by genotype, e.g., division of labor and resource collection (Robinson and Page 1989a) or parasite resistance (Kulincevic 1986), and hence by the male she has mated with (e.g., Oldroyd et al. 1992). It is a peculiarity of social insects that the female depends intimately on her family (the colony) for her success and thus on how the respective characteristics are affected by the genetic contribution of her mates. The situation in termites, where the male also contributes to parental care, is an interesting contrast but cannot be discussed further for lack of data.

For a priori reasons, females of social insects should therefore be selective in their mate choice. For example, they may choose mates that ensure their sons are well equipped for intrasexual conflict (see Andersson 1994 for a review of the different models of sexual selection). However, it appears that in many species, males lack particular adaptations for intrasexual fighting (e.g., Bourke and Franks 1995). More typically perhaps, intrasexual conflict among males is "passive," as illustrated by the honeybee. Mating occurs in aggregations where drones must chase the queen; hence, selection should be on flight performance or on effective mating signs that prevent further matings rather than on direct fighting ability. Alternatively, females could choose males that ensure their sons are attractive, according to the Fisherian model of runaway sexual selection. This should be particularly favored in "lekking" species where females visit drone congregations and mate with selected males, as, for example, in some bumblebees (Haas 1949; Alford 1975). Keller and Reeve (1995), discussing polyandry, imply that sperm of males is under sexual selection for numbers and viability. Finally, females could choose mates according to parasite resistance for her offspring—the "good-genes" model of Hamilton (Andersson 1994). This seems particularly relevant given that parasites can infect a colony, depending on worker genotypes, and thus reduce the queen's fitness if she has mated with the "wrong" male.

Whether or not mate choice in social insects thus extends beyond avoidance of mating with close relatives is unfortunately not really known. For ants, Bourke and Franks (1995) have concluded that female choice in this sense is absent. Arguably, this is because there are virtually no studies (Bourke and Franks 1995), in particular with respect to parasite resistance. Given the discussions in this and the previous chapters, there may be many surprises in the store once such studies are seriously undertaken.

5.5 Summary

In social insects, the distribution of genetic variability among workers within a colony, and among colonies in the population, is important for the evolution and maintenance of of sociality as well as for the interaction with their parasites. In particular, polyandry and polygyny increase genetic variability within the colony. Since average worker relatedness is thus decreased, such breeding strategies are not easy to understand from first principles of kin selection theory. A number of hypotheses have been formulated for the adaptive value of polyandry. These include the expression of a better general scheme of division of labor or work attendance, reduction in genetic load due to diploid males, sperm limitation or worker-queen conflict over sex allocation to offspring. Similarly, polygyny has been assumed to be adaptive under high dispersal risks or nest-site limitations.

Infectious diseases are a good example for studying the role of breeding strategies with respect to parasitism. Theoretical analyses suggest that genetic variability within colonies can increase colony fitness under parasitism. The benefits depend on factors such as the diversity of alleles coding for parasite infectiousness and host susceptibility, on mating costs, or on the fitness function that relates colony state at reproduction with achieved fitness. In particular, polyandry should be selected for when fitness is a decelerating function of colony state. The assumption of genotypic variation in both host and parasite is well justified, but to date few tests of these concepts have been done. Experimental studies in bumblebees show that parasites are more easily spread in groups of related workers, and that experimental, heterogeneous colonies acquire fewer parasites over their life cycle than homogeneous colonies. On the other hand, data from honeybees have remained inconclusive. Comparative data, mostly on ants, supports the notion that polyandry and/or polygyny is related to parasitism. In wasps, species with large colonies (which are typically associated with more parasites) are more genetically variable. In ants, species with polyandrous colonies have fewer parasites than those with monandrous colonies. Similarly, species that have overall genetically more heterogeneous colonies have fewer parasites, as expected from the hypothesis that breeding strategies have evolved in response to parasitism. On the other hand, effective mating frequencies in ants, as measured by genetic markers, indicate generally low levels of genetically effective paternity.

A underlying assumptions is that selection by parasites is an important evolutionary force compared to genetic drift. Although populations of social insects seem to be less heterozygous than solitary species and may show a very skewed distribution of reproductive success among colonies, their effective population sizes are large enough to make selection important. A number of additional considerations can relate chromosome number to colony size and parasitism. In addition, inbreeding avoidance and perhaps mate choice by females should play a role in the avoidance of parasitism. The material discussed in this chapter suggests a relationship of genotypes and of genetic variability with parasitism in social insects. This should be clarified in future studies.

Appendix to Chapter 5

This treatment follows Schmid-Hempel (1994a). The worker's contribution is to increase the size (more generally: the state) of the colony, X, proportional to the effort a worker is able to give. For convenience, the colony size reached if all workers are uninfected is scaled to unity and to $w = 1 - s$ respectively, if all workers are infected and susceptible, s being a measure of the average fractional weakening caused by the pathogens to a susceptible host. The weakening caused to individual workers is assumed to be additive in reducing the total size (state) of the colony. Each colony is exposed to a random sample of the k pathogen strains in a population, but at most one of the strains is assumed to infect a given, matching, homozygote worker (this latter assumption can be relaxed in numerical simulations). It is assumed that the host population contains the same number, k, of alleles, matching the parasite strains so that workers homozygous with allele i will succumb to parasite strain of type i.

Each queen, through her polyandrous mating behavior, can be considered to be sampling the genes of the population at random. Under the assumed additivity of reductions the expected achievement of queens building their colonies, $E(X)$, will obviously be independent of the number of males (m) they sample but will depend on (1) the chance, p, that a sampled male has an allele identical to one of the queen's (because the number of alleles, k, in the population is large, the frequency of homozygous queens is assumed to be negligible), and on (2) the independent chance that the queen's gamete that has been contributed to the worker carries the same allele as the male ($= 1/2$), thus making the worker homozygous at the critical locus. The expected achievement is therefore $E(X) = 1 - sp/2$.

The variance of achievements, $Var(X)$, however, reduces with m. If colony sizes are large and queens mate with single males, the variance of colony size is that of a binary variate taking values 1 and $1-s/2$ with frequencies $q=1-p$ and p, thus $Var(X)=(s^2/4)pq$. For the case where the queen mates with m males, the variance of the mean of m of such varying sets is therefore $Var(X)=(s^2/4m)pq$. For the case of quadratic dependence of colony fitness on colony size we need

$$E(X^2)=\sum_i X_i^2 P(X=X_i)\,.$$

This expression can be found by reversing the "correction" that provides a variance from a simple sum of squares

$$E(X^2)=Var(X)+\left[E(X)\right]^2.$$

Hence,

$$E(X^2)=(s^2/4m)pq+(1-sp/2)^2\,.$$

In the general case, when colony fitness depends on colony size with a power function of exponent a, it is easier to consider the explicit formulation: the queen's achievement when she mates with m males whereof i have an allele identical to one of hers is

$$X_i=\frac{1}{2}\left(\frac{i}{m}w\right)+\frac{1}{2}\left(\frac{i}{m}1\right)+\left(1-\frac{i}{m}1\right)=1-\frac{i}{2m}s\,.$$

Under these conditions, the expected colony fitness, Wm, is

$$W_m=E(X)=\sum_{i=0}^{m} X_i\times P(X=X_i)=\sum_{i=0}^{m}\left(1-\frac{i}{2m}s\right)^a\binom{m}{i}p^i q^{m-i}\,. \qquad \text{(A1)}$$

Colonies are infected by an average of μ independently acquired, Poisson-distributed parasite strains. The probability of the colony not being infected is therefore $P_0=e^{-\mu}$. Suppose that among the strains the frequency of a particular one (or a set of several) that is dangerous for a colony is f. The number of infections by this strain (set) will also be Poisson distributed with mean μf. Hence, the probability of a colony receiving no infections from this strain (set) is $P_A=1-e^{-\mu f}$. Since infection is assumed to build up rapidly within a colony, all cases having other than zero dangerous infections can be considered equivalent, i.e., the colony becomes totally infected. Moreover, P_1 = probability that the colony is infected but the corresponding strain to which workers are susceptible is absent, hence $P_1=e^{-\mu f}-e^{-\mu}$. The expected fitness for the colony is therefore

$$W = P_0 \times 1 + P_1 \times 1 + P_A \times W_m = e^{-\mu f} + (1 - e^{-\mu f}) W_m .$$

Furthermore, suppose that mating has an independent cost such that a single mating is scaled to unity and each additional mating reduces fitness by a constant amount, *c*. In this case, the expected colony fitness for a queen accepting *m* matings is

$$W = e^{-c(m-1)} \times \left[e^{-\mu f} + (1 - e^{-\mu f})^{-\mu f}) \sum_{i=0}^{m} \left(1 - \frac{i}{2m} s\right)^a \binom{m}{i} p^i q^{m-i} \right] \qquad \text{(A2)}$$

Dominant-Recessive Genetic System

In seeking the mean colony size for the whole population, recall that in whatever colony they are produced, all worker contributions in the population are additive. Thus since susceptible *aa* workers have frequency q^2, to obtain the mean, the standard colony fitness (1) is decreased proportional to this: $E(X) = 1 - sq^2$. In the case of a linear fitness function, this is also the mean colony fitness. In the case of the quadratic fitness function, we first need to find the variance, and for this we need to know the frequencies of the different kinds of female broods that can emerge from matings with a single male. Given Hardy-Weinberg frequencies, *aa* broods have frequency q^3, broods with *A* phenotypes and *aa* in a 1 : 1 ratio have frequency $2pq^2$, and all other broods are entirely of *A* phenotype, frequency $1 - 2pq^3 - q^3$. The sizes of those types of colony are $1 - s$, $1 - s/2$, and 1, and on the basis of these, together with their frequencies, via some algebra, we find $E(X^2) = 1 + sq^2\left[s\left(\frac{1}{m}p + q\right) - 2\right]$. Hence, similar conclusions would also obtain for this genetic system.

Aggregated Distribution of the Parasites

Assuming that parasites are distributed according to the negative-binomial distribution, the probability of having *x* infections, given an average of μ infections per colony, is

$$P(X = x) = \left(\frac{\mu}{\mu + \gamma}\right)^x \left(\frac{(\gamma + x - 1)!}{x!(\gamma - 1)!}\right)\left(1 + \frac{\mu}{\gamma}\right)^{-\gamma}, \qquad \text{(A3)}$$

where, *g* = the aggregation parameter. Specifically,

$$P(X = 0) = \left(1 + \frac{\mu}{\gamma}\right)^{-\gamma}, \text{and } P(X > 0) = 1 - \left(1 + \frac{\mu}{\gamma}\right)^{-\gamma}.$$

Hence, the probability of a colony to be infected *and* to contain at least one strain of type “A” is

$$P_I=\sum_{x=1}^{\infty}\left(\frac{\mu}{\mu+\gamma}\right)^x\left(\frac{(\gamma+x-1)!}{x!(\gamma-1)!}\right)\left(1+\frac{\mu}{\gamma}\right)^{-\gamma}(1-g^x)$$

$$=\sum_{x=1}^{\infty}\left(\frac{\mu}{\mu+\gamma}\right)^x\left(\frac{(\gamma+x-1)!}{x!(\gamma-1)!}\right)\left(1+\frac{\mu}{\gamma}\right)^{-\gamma}$$

$$-\sum_{x=1}^{\infty}\left(\frac{\mu}{\mu+\gamma}\right)^x\left(\frac{(\gamma+x-1)!}{x!(\gamma-1)!}\right)\left(1+\frac{\mu}{\gamma}\right)^{-\gamma}g^x 1\,.$$

After some algebra, we find

$$P_I=1-\left(1+\frac{\mu}{\gamma}\right)^{-\gamma}\left(1+\frac{1}{\gamma(\gamma+1)}\left(\left(\frac{\gamma+\mu/k}{\mu+\gamma}\right)^{-\gamma}-1\right)\right).$$

Note that this is independent of *m*, as is assumed in the model. Given P_I it is straightforward to calculate expected fitness given a linear, or a quadratic fitness function.

Note that in the treatment given here, only one strain of pathogen (*a*) and its corresponding susceptible genotype in the host (*aa*) is considered. Multiple infection as well as multiple susceptibility is assumed to be negligible. This property is given in the model by the requirement that only one locus with *k* alleles exists. However, it should also cover cases with multiple loci where the joint probability of multiple infection vs. susceptibility is negligible by comparison. The present treatment thus serves as a first approximation to the problem of multiple mating.

6 Host-Parasite Dynamics

With few exceptions (Royce et al. 1991; see also Schmid-Hempel 1995), the quantitative epidemiology of social insect parasites has received little or no attention. On the other hand, the theory of host-parasite dynamics is quite well developed and has been used for many different purposes (Anderson and May 1991; Grenfell and Dobson 1995). The aim of this chapter is therefore to apply some of the standard tools of epidemiology to the study of disease in social insects in order to identify some promising areas for future research. This seems particularly important, since social insect research has often concentrated on single colonies rather than on populations of colonies. On the other hand, host-parasite interactions cannot be understood without consideration of the overall population dynamic. The model case is a parasite that can be transmitted in proportion to infected and susceptible individuals, e.g., an infectious disease.

For this purpose, a macroscopic view of a population of social insects is adopted. Colonies take the place of "individuals" within populations and are regarded as units. The main benefit of this approach is that it becomes easier to link social insect biology with the theory of host-parasite population dynamics. Needless to say, the macroscopic properties of populations vary considerably among species. For example, many bees and wasps form annual, semelparous colonies (the short-lived, annual type). For this situation, models with discrete, nonoverlapping generations of hosts (the colonies) are adequate. Alternatively, ants and termites typically form perennial and iteroparous colonies (the long-lived, perennial type). Hence, models should be based on overlapping, continuous generations of hosts. Many variations on these two basic themes exist. For instance, some ants form supercolonies of many hundreds or thousands of nests, tightly knit together through the exchange of food and workers (e.g., in *Formica lugubris:* Chérix 1980). Some bumblebee species, such as *Bombus pratorum,* are partially bivoltine, while army ants reproduce by fission and thus the generations do not overlap in the same sense. Obviously, the hierarchical structure of populations, the pattern of discrete generations, within-colony dynamics, genetic heterogeneities, multiple infections, or the ephemeral nature of colonies themselves make the dynamics of host and parasites potentially very complex. Nevertheless, we can often readily recognize the level of individuals within colonies and of colonies within populations, as well as continuous- or discrete-generation populations. Simple models capturing these parameters therefore provide a good approach in gaining some understanding of the host-parasite dynamics in social insects. In this chapter, the ecological dynamics is of prime interest; any evolutionary change in the host and parasite that occurs in the same time frames is neglected.

6.1 The Basics of Host-Parasite Dynamics in Social Insects

Models of the dynamics of parasite infections, in particular of microparasites, are part of standard epidemiological theory (Anderson and May 1991). Its application to the analysis of disease in wildlife populations is reviewed in Barlow (1995). When using this body of theory, some preliminary remarks are in order. For example, models of this kind normally assume that parasite-induced mortality is negligible, host population size is constant, and that therefore the regulation of the host population is due to factors other than parasites. The dynamics of the parasite are then considered against this background. This assumption is made for mathematical convenience, as it greatly simplifies the solution of the corresponding equations. Obviously, for many of the interesting cases, this assumption is blatantly wrong and one usually has to turn to numerical simulations instead. In this framework, the net reproductive rate, R_0, is a useful quantity that describes the success of a parasite. It is analogous to Fisher's reproductive value (Fisher 1958). However, R_0 is only an accurate measure of parasite fitness at the start of an epidemic (i.e., when the parasite is still rare) but loses accuracy later on, although it remains a helpful measure for the parasite's spread. R_0 cannot cope well with vertical transmission (i.e., to offspring), since horizontal and vertical transmission scale differently for the spread of the parasite. Nevertheless, it serves as a useful approximation in most cases. A further important restriction in epidemiological models, with some exceptions (e.g., Levin and Pimentel 1981; Bremerman and Pickering 1983; Antia et al. 1994; Nowak and May 1994), is that only single infections of the host individual are considered. Parasitism by more than one strain, or infection by different parasite species (Hochberg et al. 1989) at the same time, is usually not allowed. In addition, the ecological properties of host and parasite are assumed to be constant over time.

How do these assumptions compare with social insect biology? Approximately constant numbers of colonies over the years are often found in many species that live in not too ephemeral habitats. For example, this is the case in the long-term population studies of ants in England (Pickles 1940). The number of workers, on the other hand, does fluctuate drastically with season and over the years (Wilson 1971). Also, in many cases the effects of pathogens are moderate rather than highly destructive (e.g., Bailey and Ball 1991 in honeybees; appendix 2). Hence, the standard model assumption that factors other than parasites, e.g., the availability of nest sites or food, are the prime factors that limit the number of colonies in a given area seems not too unreasonable for many social insect populations.

On the other hand, colonies of social insects are likely to be infected by more than one strain of a pathogen, because they form a large and persistent target. For example, several mites can infest a single brood cell at the same time, as discussed earlier (chap. 2). Multiple infections are in fact quite common in many host-parasite systems, e.g., in the detailed studies of malaria in humans (Day et al. 1992). Unfortunately, most studies of social insects do not report whether multiple infections occur, i.e., whether several variants or strains of a parasite per host or per colony are present. Clark (1977) identified ten different strains of *Spiroplasma melliferum* in the honeybee, but each case pertained to a different colony or location. A case of multiple infection by the trypanosome *Crithidia bombi* in its host *Bombus terrestris* has been found by Wu (1994), but the situation is probably not as simple (Schmid-Hempel and Reber, unpubl. data). Of course, even if each colony member were infected by one strain only, but different strains would reside in different workers, the colony as a whole could be multiply infected. With the macroscopic view, this would count as multiple infection of the host "individual," i.e., the colony. Multiple infections are important because they affects the dynamics and evolution of the system (e.g., Nowak and May 1994). Simultaneous infections by different parasite species rather than different strains/inocula of the same parasite are of course easier to recognize and in fact are the rule (e.g., chaps. 2, 3, 7, and appendix 2). The weakness of the standard models in dealing with multiple infections is therefore an obvious shortcoming that is pertinent in the analysis of social insects.

The ecology of many social insect species is furthermore characterized by low survivorship of founding queens and incipient colonies but long persistence after establishment, perhaps followed by a rapid decline of colonies toward the end (fig. 6.1). If the whole population is considered, the survivorship curves are close to the idealized Type I or Type II (fig. 6.2). If one ignores the incipient colonies for the epidemics, then Type II or Type III is more appropriate. This seems justified when the number and force of infection for incipient colonies are negligible. For many long-lived perennial social insects this should be the case, because incipient colonies are rare and not as exposed to infections due to their small-scale operations. In addition, we also have to assume that vertical transmission is not too frequent, and hence incipient colonies cannot introduce an infection in this way.

An important process in classical epidemiology is that hosts recover from an infection and reenter the class of susceptibles (i.e., hosts that potentially can be infected) after some time. Insects can also mount defenses against disease and recover (Gupta 1986). The hygienic behavior of honeybee workers, e.g., the uncapping of brood infected by *Bacillus larvae* (Rothenbuhler 1964b), or the removal removing corpses from the colony by specialized undertakers (Robinson and Page 1988), is an example of a behavioral mechanism that leads to resistance and perhaps to the recovery of the entire colony. In addition, microparasitic infections can be cleared by individual host bees (Imhoof and Schmid-Hempel, unpubl. data). For entire colonies as the unit of consideration, however, it is more likely that the recovery rate is generally low because workers are permanently exposed to a parasite established in the nest. Since an immunological memory is typically weak or absent, even when workers clear their infection, they will become reinfected so that the colony as a whole remains infected.

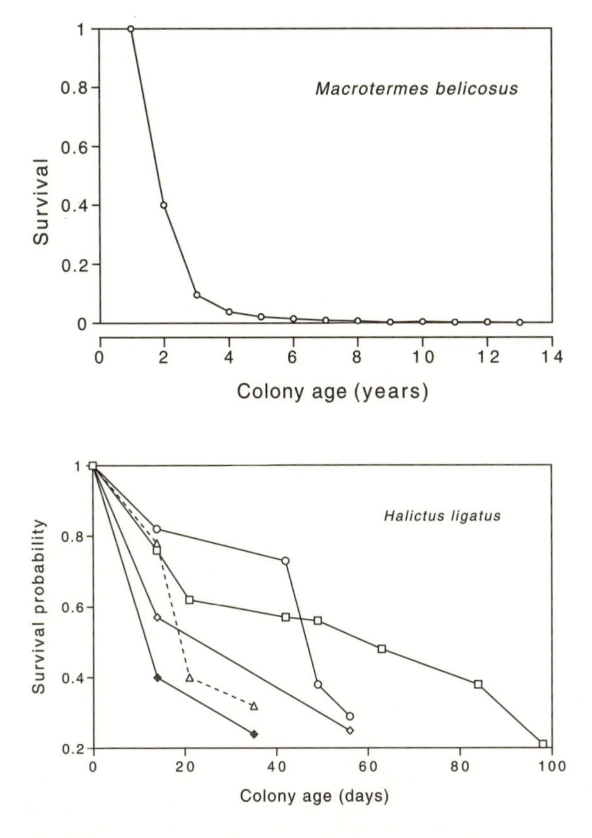

Figure 6.1 Survival of colonies. (a) The termite *Macrotermes bellicosus* (data from Collins 1981). Survivorship is close to an exponential decay. (b) Nest aggregations (colonies) of the primitively social bee *Halictus ligatus* (data from Litte 1977). Survivorship declines in the beginning but levels off later. Overall, survival rate is roughly constant per unit time. Each line corresponds to a different colony/aggregation.

6.2 The Dynamics of Microparasitic Infections

In the epidemiological sense, microparasites are parasites whose dynamics can be best described by considering different classes of host, i.e., susceptible (not infected), infected, recovered, or resistant hosts (due to immunological protection) (e.g., Anderson and May 1979). In such a system, an epidemic starts when a parasite enters a wholly susceptible population, i.e., where every individual can potentially be infected. The infection spreads among hosts (colonies in our case) and among workers within colonies, and builds up to a given infection level. At this point, the first wave of infection can grind to a halt, as few new susceptible hosts are available. The epidemic will continue when hosts recover or

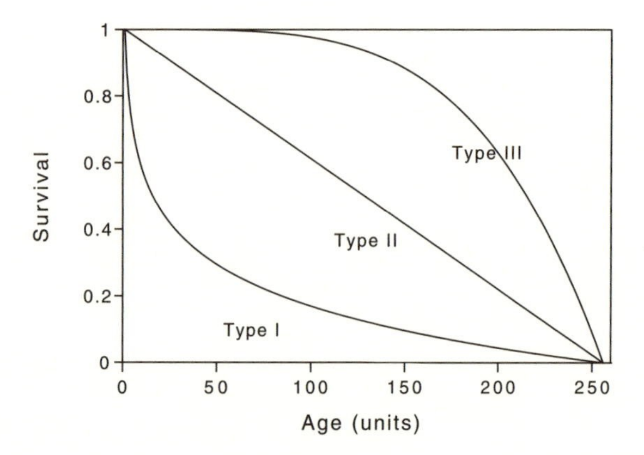

Figure 6.2 Characteristic survival curves recognized in the standard treatment of age dependence in natural populations. Type I is characterized by high rates of mortality in the younger age classes, Type II by a constant rate of mortality, and Type III by high mortality rates in old age.

new ones become available through recruitment into the population, i.e., through birth or immigration. In this case, the epidemic may pick up again, starting a second wave. This process can repeat itself several times, but it is likely that the fluctuations will get smaller, eventually transforming the epidemic cycles into an endemic state. The epidemic phase thus refers to the initial nonequilibrium situation when the infection builds up in a host population. In the endemic state, in contrast, the relationships approach an equilibrium and the disease is maintained at some level. Eventually though, persistent oscillations may emerge. Figure 6.3 is a graphic illustration of the typical course of an infection, based on the standard model in box 6.1.

Box 6.1 describes the standard epidemiological model with overlapping generations, stable population sizes, no immunity, and a fraction of vertical transmission to newly founded colonies. This is typical for many ants and termites. The most important result of box 6.1 (see there for definitions of symbols) is that the net rate of reproduction of the parasite, describing the number of secondarily infected colonies generated per infected colony, is

$$R_0=\frac{\beta N}{\alpha+\mu-bf+v}, \tag{6.1}$$

and the threshold population size of colonies (i.e., number of colonies in the population) needed to maintain the parasite is

$$N_T=\frac{\alpha+\mu-bf+v}{\beta}. \tag{6.2}$$

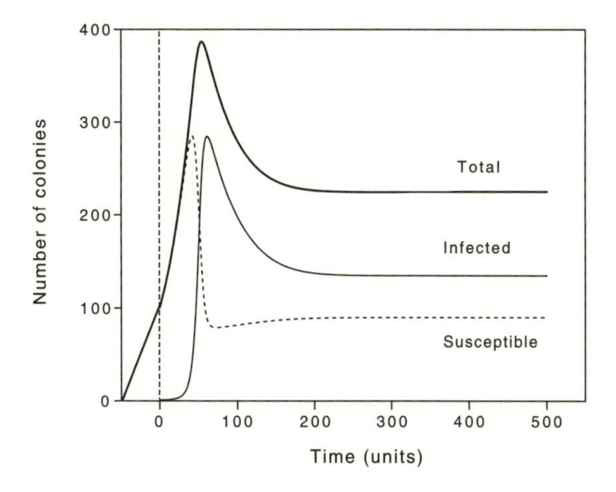

Figure 6.3 The course of a typical epidemic in a social insect population, shown as the number of susceptible, infected, and total number of colonies. The graph results from evaluation of the equations in box 6.1 with an infection entering the population at Time = 0. Parameter values are $X_0 = 100$ (population size at time of infection), $b = 0.04$, $\mu = 0.02$, $\beta = 0.001$, $f = v = 0$, $r = 0.02$, $\alpha = 0.05$.

It was assumed above that parasites have little effect on colony mortality rate (in contrast to mortality of individual workers) and that colonies as a whole have typically low recovery rates. Thus, α and ν are typically small. Furthermore, in long-lived, perennial colonies, the background mortality rate, μ, and birthrate of new colonies, b (which must, on average, balance mortality), are also small.

For the time being, it is assumed that the rate at which newly founded colonies are already "born" infected (the term bf) is small. The term bf is the chance that a daughter queen (or the royal pair in termites), having become infected in the mother colony, successfully starts a new colony. In social insects that reproduce by fission (e.g., honeybees, stingless bees, army ants), the term bf is the rate at which infected colonies are able to produce new, infected colonies. Hence, this term is small when the infection reduces the size and reproductive success of the colony and/or the chances for the sexuals or daughter colonies to survive a diapause and to successfully get established in the next season. The assumption is of course not true for all cases. In fact, vertical transmission could occur in a number of ways. For example, workers of stingless bees (Meliponini) typically carry food provisions, cerumen, mud, excreta, and other materials from the old nest to the new nest sites when the mother colony splits into daughter colonies (Michener 1974). Hence, any parasite stage that resides in this material can readily get transmitted vertically. On the other hand, New

BOX 6.1 STANDARD EPIDEMIOLOGY OF MICROPARASITES

In the simplest case, colonies are considered to be the individual hosts of standard theory. Colonies are either infected or not infected (susceptible). In a population with overlapping generations, vertical transmission and recovery (but no immunity), like in the ant-termite type, the standard notation is as follows (e.g., Anderson and May 1991):

- X number of uninfected (susceptible) colonies
- Y number of infected colonies
- b rate at which new colonies are founded ("birth rate"); assumed independent of f.
- μ background mortality rate of colonies due to other reasons than parasites
- α mortality rate of colonies due to infection by parasites (virulence for whole colony)
- β rate of horizontal transmission to other colonies
- f fraction of vertically transmitted infections among newly founded colonies
- v recovery rate, i.e. rate at which infected colonies loose the infection
- N population size, $=X+Y$
- N_T threshold population size for establishment of an infection

and the dynamics of the system are

$$\frac{dX}{dt}=b(X+Y)-\mu X-\beta XY-bfY+vY$$

$$\frac{dY}{dt}=\beta XY-(\alpha+\mu-bf+v)Y,$$

with $N=X+Y$: $\frac{dN}{dt}=rN-\alpha Y$, where as usual $r=b-\mu$. The term βY is often referred to as the "force of infection."

The threshold population size at which an infection can spread (i.e., $N\approx X$, $Y\approx dY$) is given as

$$N_T=\frac{\alpha+\mu-bf+v}{\beta}.$$

The net reproductive rate of the parasite can be approximated as $R_0=N/N_T$, and hence

$$R_0=\frac{\beta N}{\alpha+\mu-bf+v}.$$

BOX 6.1 CONT.

The initial rise in the fraction of infected hosts, $p(t)$ (see fig. 6.3), with negligible vertical transmission during this phase, can be approximated by a simple exponential growth (Anderson and May 1991):

$$p(t) = p_0 e^{\Lambda t}, \text{ where } \Lambda = \frac{R_0 - 1}{D},$$

where p_0 = fraction of infected hosts at the start of the epidemic, and D = duration of the infectiousness of the colony, i.e., the time over which the colony can infect others. This equation can be understood intuitively. The parameter Λ describes the rate of multiplication for the infection beyond the replacement of the current infection (i.e., $R_0 - 1$), which must be weighed against the time, D, during which the host remains infectious and thus cannot be infected by further parasites.

Eventually in the course of an epidemic, an equilibrium is reached where new individuals are infected at a rate in balance with the rate at which new, susceptible colonies are founded, or at which clearance of a previous infection combined with the lack of protecting immunity occurs. In the stable case, each infected colony will produce exactly one new infected colony, and hence $R_0 = 1$. If the population is homogeneously mixed such that infections can be passed freely among colonies having identical properties, and if we assume that a fraction $(1 - I)$ of colonies is still susceptible in the population at equilibrium, the following relationship holds:

$$R_0(1 - I) = 1 .$$

When R_0 is large, as in some of the cases discussed above, this indicates that a large fraction of the population needs to be infected to sustain the parasite.

World army ants and honeybees also reproduce by colony fission but no brood or stores are transferred to the daughter colonies (Seeley 1985; Hölldobler and Wilson 1990). Hence, only parasites of adults can transmit vertically to the new colony. Where only queens overwinter to produce the next generation (as in many bees and wasps), vertical transmission may be the only way that parasites can survive into the next year. The model of box 6.1 is then not appropriate for analyzing this scenario. In these cases, we may even actually expect that colony founding success is not too heavily affected by parasite infection, because of selection on the parasite to survive hibernation (chap. 7; see also Rennie 1992).

With these considerations in mind, the value of R_0 in eq. (6.1) can potentially be quite large, depending on the rate of transmission, β. Therefore, it can be inferred that parasite dynamics in a population of social insects appears to be driven to a large extent by the rate of horizontal transmission between colonies. With the same reasoning, NT is expected to be rather low (eq. 6.2), especially when β is large. It is therefore important that future studies attempt to get an estimate of the rate of horizontal transmission, β.

Horizontal Transmission

The transfer of directly transmitted parasites between host colonies requires some form of contact between members of different colonies. This is more likely to occur in species that share food resources and regularly exert physical aggression over their possession. It should not be common when foraging grounds are separated, as, for example, in territorial species and where the territorial dispute does not associate with raiding behavior (Hölldobler 1976b). In addition, where pathogens have specialized dispersal or resting stages (e.g., spores like *Nosema:* Cantwell 1970), transmission should be more effective. The same would be expected when parasites are able to manipulate the behavior of their host to their own advantage (Dobson 1988) (see table 3.4).

There are very few studies on how parasites are transmitted between colonies in a population. Experiments have shown that *Crithidia bombi,* which infects bumblebees, is transmitted via flowers visited by workers of different colonies and of different species (Durrer and Schmid-Hempel 1994). Further studies have also suggested that infection of *C. bombi* by horizontal transmission must be quite common in populations of bumblebees, as all colonies become newly infected sooner or later in their life cycle. This can be measured by placing sentinel colonies, i.e., noninfected colonies that are experimentally exposed to parasites, into the field at different times of the year. Such an experiment showed that the later the colonies are exposed in the field, the sooner they become infected (fig. 6.4). This indicates an increasing force of infection (approximated by $\lambda = \beta Y$; see box 6.2) as the season progresses. The pattern in figure 6.4 may also be due to an increase in the number of workers in the colony as it grows over the season and thus to an increasing chance to pick up the parasite (chap. 4).

The situation just described should be fairly common in social bees or wasps. They are not territorial and do not maintain exclusive access to food resources (although fighting over resources can occur). For bees, the food source (flowers) is so persistent that it can be used by many different colonies or even species at the same time. In fact, pathogens of the honeybee (e.g., spiroplasms: Mundt 1961; Raju et al. 1981) or bumblebees (ABP virus: Bailey and Ball 1991; trypanosomes: Durrer and Schmid-Hempel 1994) have been found in the pollen and nectar of flowers. As discussed in chapter 3, parasites may also live saprophagously on flowers from where they may be picked up by a passing forager and

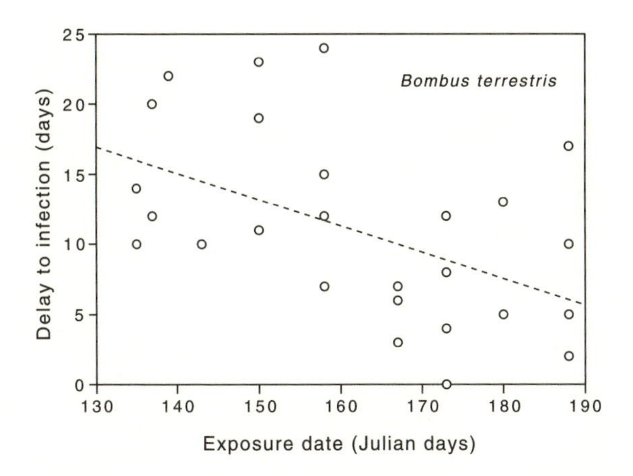

Figure 6.4 Time to first infection (delay) with *Crithidia bombi* in workers of sentinel colonies of *Bombus terrestris*. Each dot refers to a different colony. Noninfected colonies were placed in the field as "sentinels" and exposed to the natural force of infection at different times of the year (exposure date). The time until infection occurred was then observed. The regression line is: (Delay) = 41.34 − 0.19•(Exposure date) ($F_{1,29}=9.86$, $P=0.004$, $r^2=0.23$, $N=32$ colonies). (B. Imhoof and P. Schmid-Hempel, unpubl. data.)

thus enter the social insect hosts, e.g., as with the trypanosome *Leptomonas* (Tanada and Kaya 1993).

In a situation where the parasite can be picked up by routine daily activities such as foraging for nectar and pollen on flowers, the transmission rate, β, should be high. According to eq. (6.2) this implies that the threshold population size needed to sustain the parasite is small, and, as a corollary, the prevalence of the parasite should be high. For example, Durrer and Schmid-Hempel (1995, 1996; Durrer 1996) have estimated the average prevalence of infection by *Crithidia bombi* in co-occurring bumblebee species at different localities in northwestern Switzerland. For this case, it is known that transmission between colonies is possible when foragers of two colonies visit the same flower (see above, Durrer and Schmid-Hempel 1994), and that it is more likely if the waiting time between visits is short (Schmid-Hempel et al. 1998). In the field, parasite prevalence was indeed negatively correlated with the mean waiting time between visits to flowers (fig. 6.5), suggesting that frequent transmission opportunities lead to high parasite prevalences. Obviously, the pattern of figure 6.5 may also be explained by high host population densities ($N > N_T$, see eq. 6.2), which in turn are associated with frequent visits to flowers. However, this is not necessarily the case in the system described here. In fact, other studies have also shown that host density of social insects does not always correlate positively with parasite prevalence. For example, Briano et al. (1995d) found that the average preva-

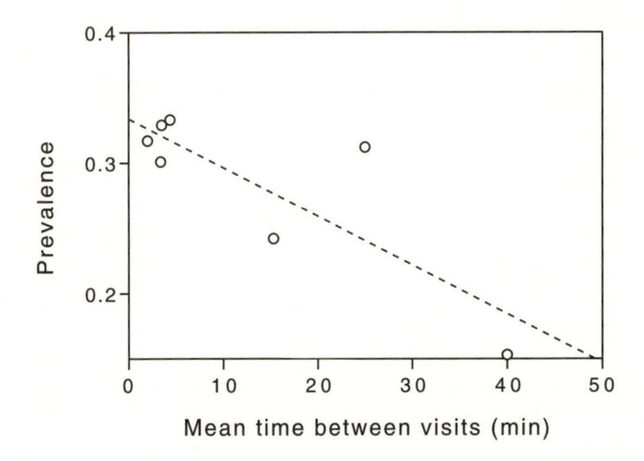

Figure 6.5 The relationship between the average prevalence of the parasite *Crithidia bombi* in workers of coexisting species of bumblebees and the mean time between visits to flowers in the same area. The mean time is calculated from the observed waiting time between subsequent visits to the same flower, corrected for the abundances of the observed plant species. Similarly, the average prevalence is corrected for the abundances of the different *Bombus* species in any given locality. Each dot refers to a different locality in northwestern Switzerland and to observations during the summer months. The regression line is Prevalence = 033 − 0.004•(waiting time) ($r = -0.814$, $N = 7$, $P = 0.023$). (After Durrer 1996.)

lence of infection by the microsporidian *Thelohania solenopsae* correlates negatively with the density of its host, the fire ant, in the native habitat of South America.

Box 6.1 describes how the initial rise of an epidemic depends on R_0 and D, the average duration for which a colony remains infectious. In these epidemiological models D is actually the time period during which the colony can no longer be infected because it is already parasitized. Note that this is a consequence of the assumption that only a single infection is present, and that therefore hosts can be classified into susceptible and infected. With this constraint, a long time period D inevitably slows the increase of an epidemic, as susceptible hosts become rarer as the epidemic unfolds. In social insects, little is known about the turnover of infections in workers or in entire colonies. Social insect colonies are different from individual hosts, though, because colonies produce new susceptible workers as they grow and develop. In addition, worker life spans vary inversely with colony size (fig. 1.3). Hence, in species with typically large colony sizes (e.g., in some ants and bees), D may actually be short (as workers die soon, remove the infection, and are replaced by new hosts) and the epidemic is expected to rise quickly. In the best-studied social insect, the honeybee, several serious epidemics have indeed been reported—for example, American foulbrood (Keck 1949; Eckert 1950), chalkbrood (Thomas and Luce 1972), tracheal mites (*Acarapis;* Weiss 1984), and the mite *Varroa* (Kraus and Page 1995). The biology of these parasites and the host (see chaps. 2, 3) is indeed likely to lead to low values of D and high values of R_0.

Given the standard situations just discussed, the conditions for the start of an epidemic process in social insects may sometimes be quite favorable. However, a major effect results from variation in the transmission parameter β. For example, species with small colony sizes should generally support a slow increase in the epidemic, since D should typically be large (e.g., because of long-lived workers that contain the infection for long periods). On the other hand, small colony size is typically correlated with high population densities (in terms of colony numbers) and thus high values of β, which would compensate for the effect of refractoriness. As box 6.1 shows, we should furthermore expect that for parasites with a high value of R_0, a large fraction of the colonies should be infected to maintain the parasite in the population. More studies are clearly needed to identify how appropriate these epidemiological considerations are in social insects and whether the epidemics follow these expected patterns.

Vectored Infections

Vectors can also ensure horizontal transmission. For example, mites are transferred between colonies of honeybees and are also vectors of some viral diseases (see table 3.2). Schwarz and Huck (1997) give evidence that the phoretic stages of the mite *Parasitellus fuscorum,* parasitic on bumblebees, use flowers to disembark and may subsequently transfer to the next worker that visits the flower. Although it is not known whether *P. fuscorum,* in particular, carries viruses, viral infections that are known to be vectored by mites in the honeybees (e.g., bee paralysis virus) are also present in bumblebees. To what extent social parasites, such as *Psithyrus* parasitizing bumblebees, act as vectors is unknown. Since these parasitic queens will eventually permanently associate with its social host colony, transmission of a host pathogen could occur only during the early stages when the inquiline "samples" host colonies until the parasitic queen finally settles. This "sampling" behavior, since it is in the interest of the parasite that uses the inquiline as a vector, should therefore be more carefully analyzed. If the vectored host pathogen is in control, we would for example expect that inquilines in heavily parasitized host populations would express more sampling and be less host-specific. This could be tested if more data became available.

Parasites can also be horizontally transmitted to other colonies by the drifting of workers. "Drifting" describes the fact that workers enter a foreign colony rather than returning to their own. This is quite common in honeybee apiaries where colonies are close to one another (Free 1958; Jay and Warr 1984).

Drifting probably explains why the more widely scattered feral colonies of honeybees are usually less infected with *Varroa* mites than those densely packed in apiaries, and why isolated colonies are less infected than clustered ones (Goncalves et al. 1982). This pattern is also found in other diseases, such as American foulbrood (e.g., Bailey 1958b; Ratnieks and Nowakowski 1989). Rates of drifting of workers in natural populations are unknown and may be low (Jay and Warr 1984; Bailey 1958b). Drifting is usually considered to be an "error," because the worker enters the "wrong" colony. However, for the parasite the chances of horizontal transmission are affected by the worker's behavior. In this light, drifting may not represent an "error" committed by the worker but rather the result of a behavior rigged by the parasite to its own advantage (e.g., Moore 1984). Interestingly, drifting is in fact more common in *Apis* colonies infested with *Varroa* mites (Sakofski 1990).

The rate of transfer of mites among colonies is correlated with the incidence of robbing in bees, i.e., where workers steal nectar from their neighbors (Sakofski 1990; Sakofski and Koeniger 1988; Sakofski et al. 1990). Robbing in turn could easily lead to the transmission of parasites. In ants, raiding behavior is the equivalent of robbing in bees, at least in terms of parasite transmission. Robbing and raiding behavior may thus be an important element of the epidemiology and persistence of parasites in social insect populations. Species using these behaviors could also have evolved special adaptations, e.g., filtering devices, choice of raided or robbed items, and so forth, to avoid becoming infected by their victims. On the other hand, robbing and raiding aids the parasites' transmission similar to drifting. Hence, if parasites could affect this, robbing in parasitized populations may be different, e.g., shorter but more frequent bouts, to increase transmission rates of the parasite. The following other possibilities could also be tested in future studies.

Defensive and nesting behavior is another factor that may contribute to variation in transmission of parasites among colonies. For example, some bee species within the same genus are highly aggressive toward predators and nest robbers (e.g., the Africanized honeybee, the African honeybee, *Apis scutellata,* and also *Bombus hypnorum*), while others are much less so (e.g., the European honeybee, *Bombus pascuorum*). In the honeybee, the differences are known to be related to differences in general metabolic activity. If intruders are potential carriers of parasites, fierce nest defense would probably be more likely to keep infections out. Specialized robber bees, such as *Lestrimellita, Tayra,* and *Mellivora,* would certainly qualify as possible carriers of parasites, and their success must depend on nest defense by the hosts. Future studies might reveal whether the expected relationship between habitual robbing, parasite transmission, and defense efforts exists. Finally, communal nesting behavior, such as is typically found in several species of social bees (e.g., in the Asian *Apis* and the meliponine genus *Plebeia*), should also favor horizontal transmission. The stingless bee

Scaura habitually nests in nests of termites (*Nasutitermes:* Michener et al. 1994) and thus could pick up parasites across the species boundary. Unfortunately, it is unknown how species that form such clusters and associations prevent an epidemic from breaking out, or how large the cost of parasitism is in the first place. Perhaps we will find that the general investment in defense, e.g., immune reactions, are more pronounced in such species.

The epidemiological theory for vectored parasites follows the standard scenario described in box 6.1 and typically results in (e.g., Anderson and May 1981)

$$R_0 = \frac{ra^2bc}{\mu\gamma}, \tag{6.3}$$

with r = the ratio of vector to host population size; a = the "attack" rate of the vector, i.e., rate at which the vector interacts with a host; c = the fraction of attacks that lead to parasite propagules within the vector, i.e., transmission from host to vector; b = the fraction of attacks that result in a new host being infected, i.e., transmission of vector to new host; μ = the background mortality rate of hosts; and γ = the average duration of infection in a host.

In this case, it is not immediately obvious how eq. (6.3) should be translated to social insects, because vectors do not attack colonies as a unit but individual members. Hence, eq. (6.3) could be interpreted at two levels. If colonies are used as the units, then only those events of vectoring that happen between colonies have to be taken into account. In cases where the vectors also transmit within the colony (e.g., *Varroa* mites in honeybees), the model of eq. (6.3) must be carefully interpreted, since the process involves a highly structured population. Typically, though, the ratio of vectors to host population size should be high (i.e., many vectors per colony). On the other hand, a may be rather small and γ may be large. Hence, it is unclear whether social insect populations are favorable ground for a vectored disease. If we look at individual workers as the host units (thus neglecting grouping and within-colony processes), r in particular may become quite small (social insect populations have a lot of workers). Again, it is difficult to assess whether this leads to high or low values of R_0. Interestingly, the dynamics of such models are not critically influenced by superinfection, i.e., infection of more than one strain per host. The evidence assembled in chapters 2 and 3 shows that vector-transmitted disease in social insects is probably not very common.

6.3 Colony Age

Populations of perennial social insect colonies are age structured. As discussed above, the survival curves for the epidemiologically relevant, established colonies may be reasonably approximated by a Type II or Type III curve (fig. 6.2). Box 6.2 summarizes the simplest epidemiological approach to the problem of age structure under these conditions. The formulation assumes that the force of infection, colony mortality rate, and recovery rates remain constant. Again, the population as such is regulated by factors other than parasites, which means that population size remains constant. The model then simplifies to a purely age-dependent dynamics.

Figure 6.6 shows the situation one can expect in an age-structured population. With survival of type II the proportion of colonies infected increases with age, but at a diminishing rate (fig. 6.6a). Not surprisingly, most of the colonies are infected at an earlier age if the force of infection is high. On the other hand, the absolute number of infected colonies that are found in a population peaks at some intermediate age (fig. 6.6b). This is because young colonies have not yet been infected while older colonies have already died for other reasons. Although the model is overtly simple, it does allow one to appreciate that collecting colonies in the field and checking them for parasites will inevitably lead to estimates and inferences that may be flawed if the dynamics of the process are not taken into account. For example, as colonies are sampled in proportion to their abundance, prevalences will be calculated according to the most abundant age class, i.e., the younger ones, thus indicating a lower figure than the "true" value. Also, differences among populations in background mortality may be interpreted as differences in force of infection, or differences in susceptibility to disease, although they merely reflect a different outcome of the age-dependent dynamic. This problem should be particularly pertinent in long-lived species such as ants and termites.

The treatment sketched in box 6.2 furthermore shows that with $R_0 = 1$ (steady state), the force of infection should be inversely proportional to the life expectancy of colonies. Under these considerations, we expect that the force of infection is rather low in species such as ants that typically form long-lived, ant-type colonies. The force of infection should be considerably higher in short-lived species such as annual wasp or bee colonies. When colonies follow a Type II curve, then the expected age at infection is late when colonies have low mortality rates (box 6.2, fig. 6.6). This suggests that the long-lived ant and termite colonies may generally show low levels of infection because of their population structure, age dynamics, and reduced horizontal transmission in the case of territorial species. Note that this would primarily be an epidemiological phenomenon rather than an indicator of good defense (cf. Hölldobler and Wilson 1990).

Equilibria may seldom be attained in natural systems, though. We should probably rather envisage a situation where new strains or new parasite species are constantly introduced into a susceptible population. At the beginning, newly introduced parasites realize their maximum net reproductive rate. Typically, the spread of the disease will stop when a large fraction of all hosts has

BOX 6.2 STANDARD AGE-STRUCTURED MODEL

The notations of box 6.1 are used here, but now $X(a, t)$ . . . denotes both age and time dependence of the variables; $\lambda(t)$ = force of infection, given as $\lambda = \beta \int Y(a,t)\, da$, integrated over all age classes. The force of infection defines the per capita rate of acquisition of an infection. Following Anderson and May 1991, a simplified model of the dynamics can be written as

$$\begin{aligned} \frac{X}{t}+\frac{X}{a} &= -\lambda(a,t)X(a,t)-\mu(a)X(a,t)+v(a)Y(a,t) \\ \frac{Y}{t}+\frac{Y}{a} &= \lambda(a,t)X(a,t)-\alpha(a)Y(a,t)-\mu(a)Y(a,t)-v(a)Y(a,t)\,, \end{aligned} \tag{1}$$

with the boundary conditions that new colonies are produced by those in age class a, with fecundity $m(a)$, and that some infected mother colonies produce a fraction f of vertically infected daughter colonies (with fecundity unaffected by infection). Hence, for the newborn colonies ($a=0$):

$$\begin{aligned} Y(0,t) &= f\int m(a)Y(a,t)da \\ X(0,t) &= \int m(a)X(a,t)da+(1-f)\int m(a)Y(a,t)da\,. \end{aligned} \tag{2}$$

Similarly, the boundary conditions for $t=0$ are the age structures of X and Y at the beginning of the process.

The dynamic model so defined is rather complicated and has nontrivial solutions. Some simplifications are therefore needed. In particular, it is assumed that the force of infection, λ, and recovery rate, v, are constant over time and age. Furthermore, the population is supposed to be regulated by factors other than parasites, hence, $\alpha \approx 0$. Furthermore, with constant population sizes, colony death rate is balanced by the rate of new colony foundations:

$$B = \int \mu(a)N(a)da\,. \tag{3}$$

In the case of constant mortality, $\mu(a)=\mu$ (Type II survival, fig. 6.2), we find $B=\mu N$, or more generally, $B=N/L$, where L = length of life of a colony. With these restrictions, eqs. (1) simplify to:

$$\begin{aligned} \frac{dX}{da} &= -\lambda X(a)-\mu X(a)+vY(a) \\ \frac{dY}{da} &= -\lambda X(a)-\mu Y(a)+vY(a)\,. \end{aligned} \tag{4}$$

With $N=X+Y$, it is readily calculated that

$$\frac{dN}{da} = -\mu N(a), \text{ and } N(a)=N(0)e^{-\mu a}\,. \tag{5}$$

BOX 6.2 CONT.

If recovery rate is very low (i.e., $v \approx 0$), eqs. (4) are solved to give

$$X(a) = N(0)e^{-a(\mu+\lambda)}$$
$$Y(a) = N(0)e^{-\mu a}\left(1 - e^{-\lambda a}\right). \tag{6}$$

The fraction of infected colonies is

$$y(a) = \frac{Y(a)}{N(a)} = 1 - e^{-\lambda a}. \tag{7}$$

If recovery rate is not zero, an intractable inhomogeneous differential equation results that requires numerical simulations.

When host and parasite have settled into an equilibrium state and the host population is stationary in size and parasite-induced mortality rate is low, then (Anderson and May 1991),

$$\text{for type II survival curves:} \quad R_0 \approx 1 + \lambda L \tag{8a}$$
$$\text{for type III survival curves:} \quad R_0 \approx \lambda L, \tag{8b}$$

where λ = force of infection and L = the life expectancy of the host, i.e., the colony. Hence, with $R_0 = 1$ (steady state) the force of infection should be inversely proportional to the life expectancy of colonies.

Finally, the standard theory can also give some indication for the average age at infection, A, in such a static population. In particular, with the same symbols as before,

$$\text{for type II survival:} \quad A \approx 1/(\lambda + b) \tag{9a}$$

$$\text{for type III survival:} \quad A \approx 1/\lambda. \tag{9b}$$

Therefore, the age at which a colony becomes infected is, not surprisingly, typically late when the force of infection, λ, is low, for instance, by virtue of low transmission rates or small population sizes.

become infected and the disease may even die out when no new susceptibles appear. Long-lasting dispersal stages of the parasite would then be necessary to generate a new epidemic when the disease spreads to new, susceptible colonies in subsequent generations. Some parasites of social insects indeed have durable stages, e.g., *Bacillus larvae* (although for within-colony transmission the relevant time may be much shorter, i.e. days; Wilson 1972). Alternatively, vertical transmission through the host's offspring could carry the disease to the next

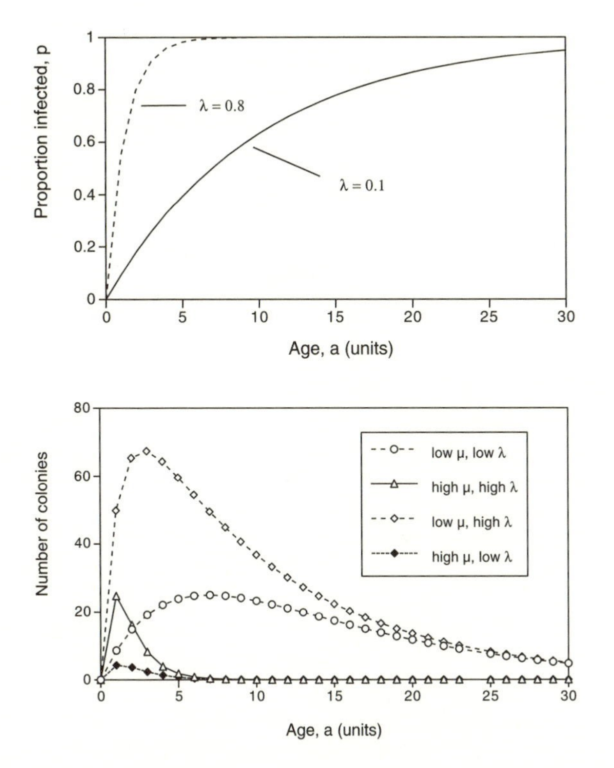

Figure 6.6 The dynamics of infection in age-structured host populations. (a) The proportion, p (ordinate), of infected colonies as a function of age (following eq. 6 in box 6.2, Type II survivorship of colonies) for low and high values of the force of infection, Λ. (b) The absolute number (ordinate) of infected colonies from an initial population of $N_0 = 100$ in relation to age. Parameter values are as in figure 6.3. Mortality rate is low ($\mu = 0.1$) or high ($\mu = 0.8$); force of infection is low ($\Lambda = 0.1$) or high ($\Lambda = 0.8$). Age in arbitrary units.

generation. Data on how and whether parasites are transmitted to the next generation of social insect hosts would thus be highly valuable.

For infections in the endemic state, the pool of susceptibles is replenished by births, immigration, or by clearance of infection. As new susceptible colonies accumulate, the ground may be prepared for the start of a new epidemic. The theory of simple dynamics predicts that the system in the endemic state eventually settles with damped oscillations (see Anderson and May 1991, p. 125). The oscillations are long-cycled when the duration of infectiousness and latency is long (D, which is likely for some species; see above), or when colonies are refractory to further infections (which is unlikely, because new susceptible workers are produced), and when infections do not occur at a young age (which is

Table 6.1

Estimates for the Average Age When Infection Is Acquired, and of R_0, the Net Reproductive Rate of the Parasite

Disease	*Age at Infection*	R_0
Chicken pox	6–8 years	7–12
Diphtheria	11 years	4–5
Measles	1–6 years	5–18
Rubella	2–10 years	6–16
Pertussis	4–5 years	10–18
Poliomyelitis	12–17 years	5–7
Crithida bombi in *Bombus*	7 days	4–12
Nosema bombi in *Bombus*	$\geq$ 20 days?	2–3

SOURCE: Data for human diseases taken from Anderson and May 1991.
NOTE: Parameters for human diseases vary considerably among geographical locations.

likely because older, larger and more active colonies are probably more prone to novel infections). Perhaps long cycles are typical in natural populations of social insect colonies. Finally, it is possible to estimate what fraction of colonies has become infected during the entire course of the epidemic process, provided the population is closed so that no new susceptibles arrive. This fraction rises steeply for small increases in R_0 near 1, but levels off for higher values of R_0 beyond values of 2 to 3 (Anderson and May 1991).

R_0 has never been estimated in social insects. From our own studies of *Bombus,* infected by *Crithidia,* the average age of the colony at infection is in the order of one week (7 days; fig. 6.4). Assuming Type III survival for the established colonies (fig. 6.2) and an average life expectancy of 2 months (60 days), this translates into $R_0 \approx 8.5$ (a likely range is $R_0 = 4 \ldots 12$, depending on assumptions about A and L; box 6.2). A similar estimate can be made for *Nosema bombi,* although the average age at infection is a much cruder guess (i.e., probably a few weeks, ca. 20–30 days). Then $R_0 \approx 2$–3. These values compare well with those calculated for a range of human diseases (table 6.1).

6.4 Seasons and Growth

Seasonal Populations

Most social wasps and bees have an annual life cycle with distinct, nonoverlapping generations. The colony grows within a season and can become infected through horizontally transmitted parasites. At the end, the colony reproduces and a new generation starts the cycle next year. A fraction of all new colonies in the next year will then be infected as a result of vertical transmission that bridges the hibernation period. The analysis of such discrete generation epidemiology can lead to quite complicated dynamics, including stable, unstable, or chaotic behaviors (Brown 1984; Régnière 1984; May 1985). The model of box 6.3 depicts a simple approach to an epidemiology with discrete generations. Box 6.3 shows how the fraction of colonies remaining susceptible at the end of the season depends on the parameters such as vertical and horizontal transmission rate and population size.

The models of boxes 6.1 and 6.3 can be slightly modified to account for the dynamics of a social insect population within a season (box 6.4). This leads to the classical epidemiological equations (Kermack and McKendrick 1927). In conventional terms, this theory is used to understand the spread of a novel infection in a large susceptible population—the epidemic part of the process. However, the endemic situation is also of interest. A full solution to these equations is difficult or impossible. It is usually assumed that the system has come to an equilibrium at the end of the season (as in box 6.3). However, the seasonal dynamics of a social insect population is almost certainly different from this scenario. For example, in annual species (many bees, most wasps), the colonies grow and eventually reproduce at the end of the cycle when the host-parasite system is very unlikely to have settled into an equilibrium. An analytical solution of this scenario is obviously quite difficult or impossible. However, the dynamics given in box 6.4 can be numerically evaluated to see how the process unfolds within a season.

Figure 6.7 shows two examples of the expected seasonal dynamics from this model. Over many generations, this pattern would of course be found only if overwinter mortality scales exactly in a way that a constant fraction of vertically infected colonies at the start of each season is found (i.e., always $f=0.1$, as in the numerical example used here). As the graphs show, prevalence steadily increases to 100% with low background mortality and low rates of recovery (fig. 6.7a), but there is a mid- to late-season peak with both high mortality and high recovery rates (fig. 6.7b). In other words, the within-season dynamics can generate different outcomes, and show patterns that are quite similar to those in an age-structured population (fig. 6.6); the two processes do in fact have some traits in common (e.g., decreasing survivorship) but are also sufficiently different in their logic.

For many of the parasites discussed in chapters 2 and 3 (e.g., *Nosema,* chalkbrood), pronounced seasonal variation in the prevalence of infection was found. The present analysis shows that seasonal variation is a likely result of the ecological dynamics of the system, often with prevalences high at intermediate times in the season, a pattern that is in fact often observed. Of course, it does not automatically follow that all of these cases can be explained by this factor alone.

BOX 6.3 DISCRETE GENERATIONS

Following May and Anderson (1983), we will analyze a situation where colonies develop within a season and reproduce to generate the next, nonoverlapping generation. With the same notations as in box 6.1, the season lasts from $t=0\ldots\infty$. All colonies reaching the end of the season reproduce. Some thereby vertically transmit the parasite into the next generation (with fraction f). Hence, the boundary conditions for the start of the season ($t=0$), given that N colonies were reproducing one generation before, are

$$X(0)=(1-f)N;\ Y(0)=fN. \tag{1}$$

Within the season, the system can be described with the usual continuous-time model as

$$\begin{aligned}\frac{dX}{dt}&=-\beta XY+vY\\ \frac{dY}{dt}&=\beta XY-(\alpha+v)Y.\end{aligned} \tag{2}$$

Note that this dynamic is subtly different from earlier formulations, because within a season colonies do not give birth to new colonies but reproduce at the end. Also, background mortality is assumed to be negligible during the season. Then, the addition of eqs. (2) gives

$$\frac{dX}{dt}+\frac{dY}{dt}=-\alpha Y, \tag{3}$$

and, per definition, let $\bar{Y}=\int_0^\infty Y(t)\,dt$, the average number of infected colonies over one season (interval $0,\infty$). Now, integrating eq. (3) in the boundaries $(0,\infty)$ and using the boundary conditions (1) yields

$$X(\infty)=N-\alpha\bar{Y}. \tag{4}$$

Note that to meet eqs. (2) we require $Y(\infty)=0$. Eq. (2) describing $X(t)$ can furthermore be transformed into

$$\ln\left\{\frac{\beta X(\infty)-v}{\beta X(0)-v}\right\}=-\beta\bar{Y}. \tag{5}$$

Let $X(\infty)=N(1-I)$, where $1-I=$ fraction of initial colonies being susceptible at the end of the season. Then, using eqs. (4) and (1), eq. (5) can be transformed into

$$\beta IN=\alpha\ln\frac{[\beta N(1-f)-v]}{[\beta N(1-I)-v]} \tag{6}$$

BOX 6.3 CONT.

to give an expression for I:

$$I=(1-\frac{v}{\beta N})-(1-f-\frac{v}{\beta N})e^{-I\frac{\beta N}{\alpha}}. \tag{7}$$

If now k_x, k_y denotes the reproductive rate of susceptible and infected colonies at the end of the season, we find, with the above assumptions, that the size of the population that is produced for the next generation is

$$N_1=k_xX(\infty)+k_yY(\infty)=k_xX(\infty)=k_xN(1-I)\,, \tag{8}$$

and hence, in terms of fitness or overall rate of increase of the population,

$$W=\frac{N_1}{N}=k_x(1-I)\,. \tag{9}$$

BOX 6.4 SEASONAL DYNAMICS

The situation described in boxes 6.1 and 6.3 is modified to account for the dynamics of the colonies (which may die) and the infection within a season. In this case, the changes in infected and susceptible colonies are

$$\begin{aligned}\frac{dX}{dt}&=-\beta XY-\mu X+vY\\ \frac{dY}{dt}&=-\beta XY-(\alpha+\mu)Y-vY\,.\end{aligned} \tag{1}$$

Contrary to the usual assumption that the host-parasite system has come to an equilibrium at the end of the season with, $\frac{dX}{dt}=\frac{dY}{dt}=0$, now the dynamics of the infection has not leveled off but is simply cut off through the end of the season. Colonies that are infected carry the disease to the next generation via the production of infected offspring (vertical transmission). Hence, the starting conditions ($t=0$) for the next season are given by

$$X(0)=(1-f)N;\ Y(0)=fN\,, \tag{2}$$

where f= fraction of vertically infected colonies and $=N$ colony number at the start of the season. This simple model is appropriate if overwinter mortality scales in a way so that on average a constant fraction, f, of infected colonies survives.

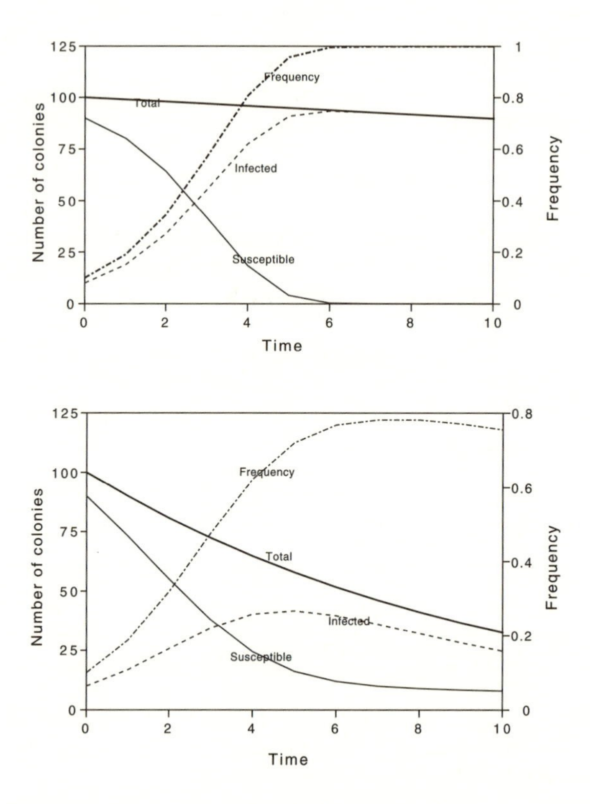

Figure 6.7 The seasonal dynamics of infection. The graphs show the numerical evaluation of the equations in box 6.4 for (a) low mortality, low recovery rates: $\mu=0.01$, $\alpha=0.001$, $v=0.001$; and for (b) high mortality, high recovery rates: $\mu=0.1$, $\alpha=0.01$, $v=0.1$. Left-hand axes: number of susceptible, infected, and total colonies in the population. Right-hand axes: prevalence (frequency of infected colonies) among surviving colonies. The season is assumed to end at $T=10$, in arbitrary units. Other parameter values are $N_0=100$, $f=0.1$, $\beta=0.01$.

In practice, differences in dynamic processes will also be difficult to distinguish with field data on prevalence alone. In other words, to understand the actual processes in a population of social insect colonies, one has no alternative but to do detailed studies on the transmission processes and study the mortality and growth patterns of colonies. Experiments with sentinel colonies or workers that are exposed to the infection at different times of the seasonal cycle should be feasible and highly desirable to increase our knowledge of the epidemiology of diseases in natural populations of social insects.

The Effect of Colony Growth

One characteristic of social insects is that the individual "hosts" (the colonies) themselves grow in worker numbers over the course of the season. A straightforward way to account for this difference is to "correct" the transmission term, βXY. This takes into account that more individual hosts can transfer the parasite and more individual hosts can receive it. For example, we could correct for colony growth as

$$x' = Xe^{c_1 t},\ y' = Ye^{c_2 t};\ \text{and replace}\ \beta XY \Rightarrow \beta x' y', \tag{6.4}$$

where c_1, c_2 describe the growth of the colonies. We require that $c_2 < c_1$, i.e., that the growth of infected colonies is slower than that of healthy ones. How does this addition affect the dynamics? As one could guess, the force of infection is gradually increased over the course of the season as colonies grow, resulting in a steadily increasing proportion of infected colonies toward the end of the season. If healthy colonies grow much better than infected ones, and if there are high rates for mortality and recovery, there may be a substantial difference in the proportion of infected colonies as compared to the standard scenario with no colony growth, leading to an unexpected increase in the number of infected colonies toward the end of the season. In this respect, the seasonal epidemiological dynamics of social insects are similar but not identical to those found in a growing host population of solitary organisms.

6.5 Regulation and Heterogeneity

Several authors have analyzed the properties of model populations with respect to whether parasites are able to control their host population at stable levels. The predictions generated by these various approaches are quite different, as the following examples show. Régnière (1984) considered a discrete generation population with vertical and horizontal transmission, nonlethal parasites, and fixed probabilities of transmission. The model predicts that regulation of hosts at stable sizes or limit cycles can be achieved over a wide range of conditions, also including parasites of low virulence, imperfect vertical transmission, vertical transmission late in the life cycle of a host, and low dispersal of the parasites themselves. The resulting dynamics should be characterized by long and large amplitudes of population cycles, especially in marginal habitats with low rates of intrinsic growth and prolonged periods of low host density. Brown (1984) also modeled seasonal host reproduction to find that high virulence and high rates of host reproduction are associated with stability of the system. In contrast, May (1985), looking at the same question, concluded that a system with nonoverlapping generations has no stable points but exhibits purely chaotic dynamics.

These conflicting predictions result from assumptions about the process to be modeled. For example, May (1985) assumed a lethal pathogen such that only the survivors reproduce. The infection follows a standard epidemiological model where the epidemic has run its course to end with a fraction of all hosts infected at the end of each generation (see the model in box 6.3). In fact, when nonlethal parasites are assumed, the system also has stable points. In Brown's (1984) scenario, a continuous formulation is still used but it is superimposed with a periodic period of host refractoriness and reproduction. Régnière (1984), finally, formulated an explicit start-to-end-of-season model where infection can be passed horizontally once during the season, and each host class (infected vertically from start, infected later, not infected, etc.) has its own, specified fecundity. He also assumed that infections of new hosts occur with a given probability, regardless of how many infected hosts exist. Hence, in this model, new infections are proportional to the number of susceptibles, something that Anderson and May (1991) call "weak homogeneous mixing." In contrast, the classical mass action principle of epidemiology discussed above assumes "strong homogeneous mixing," i.e., new infections are proportional to both the number of susceptible and infected hosts (e.g., βXY). Mixing is also affected by heterogeneity in the population.

Empirically, parasitism of honeybees by the tracheal mite *Acarapis woodi* has been analyzed with respect to host population regulation (Royce and Rossignol 1990a; Royce et al. 1991). One conclusion was that the mites can regulate the host population size at a given density, provided that scaled virulence (i.e., parasite-induced mortality rate) is larger than the rate of growth for colonies (i.e., intrinsic rate of increase for colonies). Otherwise, the colony population grows unchecked at a rate equal to the difference between the two. However, Royce et al.'s (1991) experiments with infected colonies showed that mites are rather controlled by host growth and not vice versa, because mite density is reduced with swarming when brood and young workers become rare in the colony.

Colonies vary in a number of ways. The differences are caused by differences in genotype, social structure, spatial arrangements, age structure, and behavior. To the extent that a social structure can amplify individual differences, social insect populations are clearly rather heterogeneous (see examples in table 6.2). In the present context, variability among colonies is likely to produce variation in transmission rates among given pairs of colonies. For example, large or aggressive colonies may be more likely to receive or transmit pathogens from or to other colonies. Also, spatial heterogeneities of the physical-chemical habitat should contribute to this effect. From standard theory, heterogeneities in transmission rates will, under some conditions, lead to a smaller overall epidemic,

Table 6.2

Examples of Conspicuous Variability in Important Traits among Colonies

Species	*Trait*	*References*
Ants		
Pheidole dentata	Caste ratio (major/minors)	Johnston and Wilson 1985
Formica obscuripes	Caste ratio (majors/minors)	Herbers 1980
Pogonomyrmex barbatus	Colony foraging activity	Gordon 1991
Leptothorax longispinosus	Task fidelity	Herbers 1979
Formica rufa	Task fidelity	Rosengren 1977
Pheidole hortensis	Task fidelity	Calabi et al. 1983
Solenopsis invicta	Diet preference	Glunn et al. 1981
Bees		
Apis mellifera	Pollen foraging; worker longevity; disease resistance	Rothenbuhler 1964a, Hellmich et al. 1985; Milne 1985
Apis mellifera	Swarming propensity	Winston 1980
Apis mellifera	Division of labor genetically based	Robinson and Page 1989b
Bombus terrestris	Reproductive timing	Duchateau and Velthuis 1988
Wasps		
Polistes exclamans	Worker longevity	Strassmann 1985

i.e., a smaller fraction of infected hosts. With more variable lengths of the incubation periods, the epidemic also becomes more spiky and has a longer tail (Anderson and May 1991). On the other hand, if transmission rate, β, is a variable that is nonrandomly distributed, the effective net reproductive rate of the parasite becomes (Anderson and May 1991)

$$\overline{R_0} = R_0(1 + \mathrm{var}(\beta)) , \tag{6.5}$$

where var(β) = the variance in transmission rate and $\overline{R}_0$ = the average net reproductive rate over all subgroups. Hence, an epidemic can be driven by some "supertransmitters," for example, some particularly large, active, or susceptible colonies. Contrary to intuition, heterogeneity implies that under certain conditions the fraction of resistant colonies in the population needed to eliminate the parasite is smaller than that in the homogeneous population (May and Anderson 1984). Hence, the parasite is less likely to persist despite its overall increase in R_0.

At this point, it is useful to regard colonies as "cities" and "villages" with inhabitants (the workers) that are infected and transmit an infection to other such subgroups in the population. May and Anderson (1984) have analyzed precisely this problem. In their analysis of an optimal immunization program for human diseases, the fraction of subgroups that can remain susceptible so that the parasite is still eradicated varies substantially from 5% (with high degrees of clumping; i.e., in our case a negative-binomial distribution of colony sizes) to over 50% in the case of nearly uniform distributions (all colonies of the same size). This analysis predicts that when colonies vary so that in each one a similar (absolute) number of workers remains susceptible, the parasites would more likely be eliminated due to epidemiological processes. At first sight this requirement is not very likely to be met. However, the interplay between colony organization, birthrates within colony, and division of labor may produce such relationships. The model calculations of May and Anderson (1984) also demonstrate that in such a system a parasite can persist for a long time by floating back and forth among the subgroups without any obvious large-scale epidemic. It may then suddenly break out and become visible. The case of paralysis virus in honeybees (fig. 2.3) may reflect such a phenomenon. The detailed analysis of epidemiological processes and the measurement of relevant parameters in social insects is clearly of paramount importance.

6.6 The Dynamics of Macroparasitic Infections

According to most definitions, macroparasites include the helminths, mites, and, to some extent, the parasitoids mentioned in chapter 2. The analysis of host-macroparasite dynamics requires the explicit knowledge about the distribution of parasites over their hosts. Such distributions are often approximated by a negative-binomial (e.g., Schad and Anderson 1985). The analysis shows that the threshold host population size needed to sustain a directly transmitted macroparasite is generally much lower than that needed for microparasitic infections, due to large egg production and substantial longevity of the parasite. Theory estimates the net reproductive rate (Anderson and May 1981) as

$$R_0 = \frac{m/(\alpha+\beta+v)}{\beta N/(\mu_0+\beta N)}, \tag{6.6}$$

where m = the rate of egg production by the parasite, i.e., the rate at which eggs are deposited in a pool of infective stages; μ_0 = the mortality rate of these infective stages; β = the transmission rate from the pool of infective stages to new hosts, and all other symbols as before.

The value of β is affected by the fact that many workers of a colony could carry infective stages of a macroparasite back to the nest. Such examples have been discussed in chapter 3: triungulin larvae of Strepsiptera, eggs of nematodes that are acquired with food, or proglottids of cestodes. This should be especially evident when hosts regularly congregate at localities where transmission of long-lived parasitic stages is facilitated. The extraordinary biology of the nematode *Sphaerularia bombi* parasitizing bumblebee queens (see chap. 3) is an example. At the same time, this discussion illustrates the major problem for macroparasites—the problem of how to meet and become transmitted to new hosts. There is little information on the value of the parameters in eq. (6.6). Hence, it is difficult to deduce on a priori grounds whether social insects are a favorable or difficult environment for the epidemiology of macroparasites. From the empirical evidence of appendix 2, it is at least evident that a variety of macroparasites use social insects as hosts, especially nematodes in ants and mites in bees.

6.7 A Case of Complex Dynamics

The biology of social insects is a case for complexity and generates a host-parasite dynamics that does not lend itself to easy analytical solutions of the respective differential equations. In this section, an example of a complex dynamics is discussed and analyzed with numerical simulations.

Box 6.5 describes a situation that should be appropriate for many social bees or wasps. The colonies are annual (i.e., have nonoverlapping generations) and are founded by a single queen (this latter assumption is actually not very critical and could readily be relaxed). The parasite can spread horizontally among colonies within each season. Obviously, the more colonies and the larger they are, the more intense is the transmission of the parasite. At the end of the season, the colonies reproduce and then perish. Some of the reproductives (young queens) from infected colonies will themselves become infected, go into hibernation, and thus transmit the parasite vertically to the offspring colonies next year. Reproductives from noninfected colonies go into hibernation uninfected. This assumption could be extended to include transmission during mating (e.g., for mites that transfer from males to females during copulation). In any case, a fraction of offspring will be infected when the next season starts. As the colonies grow, the parasite will again be transmitted to other colonies in the population and thus the cycle is repeated next year. In contrast to the previous analyses, it is now not required that the epidemic is at equilibrium at the end of the season (as in box 6.3) or that the fraction of infected colonies at the start of the season is constant over the years and independent of the progress of the epidemic (as in box 6.4).

This scenario, implemented in box 6.5, seems simple and is certainly very

BOX 6.5 A NUMERICAL SIMULATION

Consider a population of social insects that has discrete generations, like in annual social bees or wasps. Within a season, colonies are founded by the queens, they grow, and finally reproduce. Only the (female) reproductives overwinter and start a new generation the next year. A model similar to the dynamics described in boxes 6.1 and 6.3 can be formulated as follows. Within-season, the dynamics of susceptible (X) and infected (Y) colonies over course of time, t ($t=0, T$), is

$$\frac{dX}{dt} = -\mu X - \beta X x_t Y y_t + vY$$
$$\frac{dY}{dt} = \beta X x_t Y y_t - (\alpha + v + \mu)Y, \quad (1)$$

where X, Y, and x_t, y_t describes the number and sizes, respectively, of susceptible (X) and infected (Y) colonies at time t within a given season. Transmission follows the mass action principle as before, but also depends on both the number of host colonies and their size. More specifically, the epidemiologically effective colony size is given by

$$x_t = x_0 e^{r_x \cdot t}, \; y_t = y_0 e^{r_y \cdot t}. \quad (2)$$

This describes an exponential growth of colonies, starting with x_0 and y_0, respectively, at the beginning of the season. Susceptible colonies grow with an instantaneous rate of r_x, infected colonies with r_y; it is required that $r_x > r_y$. Eqs. (1) describe the development of susceptible and infected colonies until the end of the season, T. In contrast to the scenario discussed in box 6.3, but similar to box 6.4, the epidemic process is not so fast that the populations attain an epidemiological equilibrium at the end of the season. Rather, colony growth and parasite transmission are truncated by the end of the season, obviously leaving a nonequilibrium number of colonies infected or susceptible when the season ends. As the process unfolds over the seasons, these numbers vary over generations, in contrast to box 6.4 (where f was assumed to be constant).

Furthermore, a simple dynamics is assumed so that only the production of new queens has to be considered. We can assume that the number of healthy and infected sexuals, n_x, n_y, produced per colony depends on its size as follows:

$$n_x = a_x(x_r - b_x)^c \,; n_y = a_y(y_T - b_y)^c, \quad (3)$$

where a_x, a_y, b_x, b_y, and c are shape parameters (here $c=0.5$ is assumed) of these fitness functions (see chap. 5). Obviously, infected colonies can be assumed to produce fewer queens so that the shape parameters are chosen accordingly (i.e., $n_x > n_y$, and/or $b_x < b_y$). A fraction, f, of the queens produced by the infected colonies will themselves become infected and thus carry the disease to the next season. Fraction f thus describes the rate of vertical transmission in this system. In the next generation, a number of $y_0 = f n_y s_y$ of queens is infected, where s_x, s_y = the survival rates of susceptible and infected queens ($s_x > s_y$). These survival rates may also reflect differential success in colony foundation.

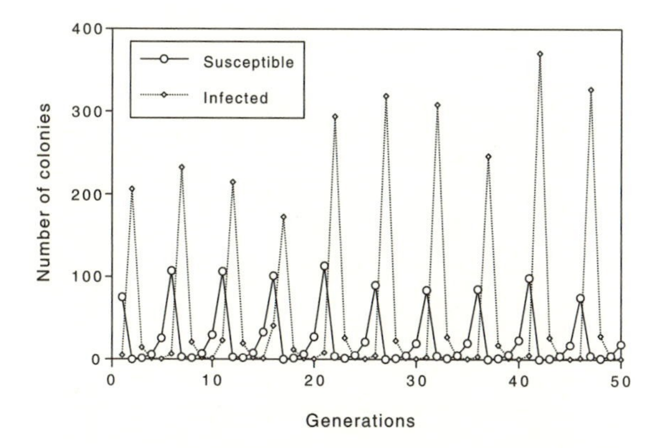

Figure 6.8 The dynamics of a host-parasite relationship with discrete generations but no endemic equilibrium at the end of the season (the model of box 6.5). The graph shows the number of susceptible and infected colonies in the population over a number of generations. The parameter values are as follows: season length = 10 (units); number of generations = 50, $X_0 = 100$, $\beta = 0.001$, $\mu = 0.02$, $v = 0.01$, $\alpha = 0.2$; r of susceptible and infected colonies respectively = 0.2, 0.1; overwinter survival of susceptible and infected colonies respectively = 0.2, 0.05; shape parameters of fitness function a, $b = 10, 0$.

close to the situation in annual social insects with nonoverlapping generations. However, it generates a wide range of dynamic behaviors in the system. Depending on the parameter combination, the system either tends to go extinct (i.e., the number of both susceptible and infected colonies falls below a critical level), oscillates for some time (as shown in fig. 6.8), runs away (either through an explosive increase in susceptible or in infected colonies or both), or the system may remain stable over time (i.e., the number of susceptible and infected colonies remains the same over time).

Figure 6.9 shows the complex landscape of the dynamic behavior of this system in a phase plane of vertical vs. horizontal transmission rates. It is readily seen that this "dynamic landscape" defies a clear picture of neatly predictable interactions. In fact, the landscape is also fine grained, i.e., small changes in the parameters may lead to runaways, oscillatory behavior, or extinctions (fig. 6.9a). This may become visible as substantial differences in the dynamics of closely neighboring localities. The situation in figure 6.9b is more predictable: for almost any domain in the parameter space, the whole host population becomes infected but continues to grow. The simulations in figure 6.9 also suggest that the rate of horizontal transmission is a very sensitive parameter in determining the dynamic outcome of the system, although the scaling of horizontal vs. vertical transmission is difficult to gauge.

In this chapter, I have discussed simple approaches to the ecological dynamics of host and parasite in social insects. From these approaches, we can formulate some expectations that should be helpful in directing further research in

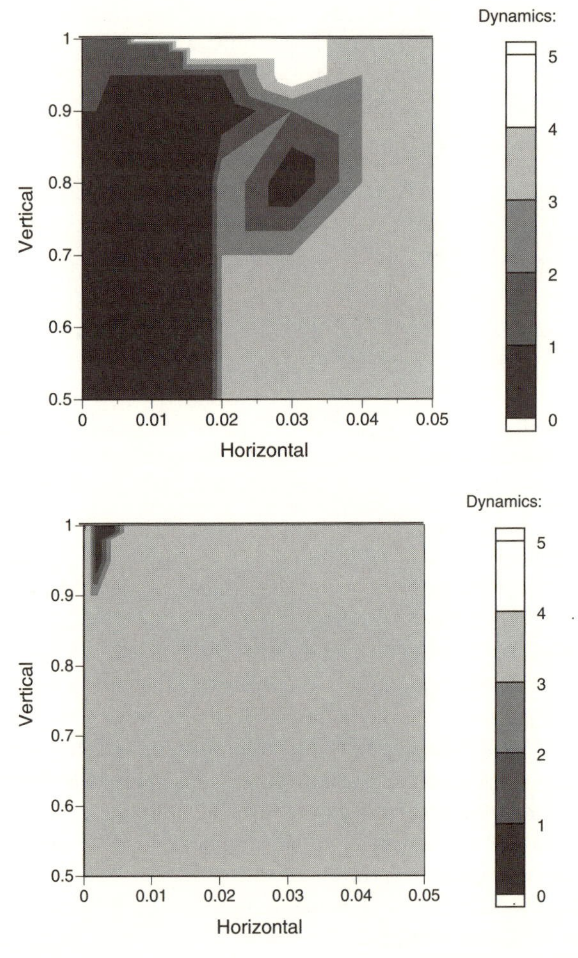

Figure 6.9 Complex dynamics in a system with discrete generations and no endemic equilibrium (model of box 6.5). The long-term dynamical behavior of the system is coded by different shadings (see panel), and defined as follows: 0 = stable (i.e., population sizes of susceptible and infected hosts remain constant); 1 = oscillating (both oscillate); 2 = exponential growth of susceptible population; 3 = exponential growth of infected population; 4 = exponential growth of both; 5 = population of hosts goes extinct before the time horizon (T_1) is reached. The phase plane is defined by the rate of horizontal transmission (β) and vertical transmission (f). The process was run over $T_1 = 500$ generations; the categories 0 and 1 were distinguished with a runs test taking into account the values of the last fifty generations of the process. Parameter values are initial colony number $X_0 = 100$, $Y_0 = 1$ (seed of infection is one colony), season length $T = 10$, $\mu = 0.02$, $v = 0.2$, $\alpha = 0.1$, $r_x = 0.2$, $r_y = 0.1$, $s_x = 0.2$, $s_y = 0.1$, $a = 10$, $b = 0$. (See box 6.5 for definitions.) The two panels show the effect of the curvature of the fitness curve (cf. fig. 5.3; parameter a, here called c) with (a) $c = 0.25$, giving rise to a highly complex dynamical landscape; (b) $c = 1.2$, where the infection cannot control the population and the host outgrows the parasite. In both cases, the same change in β has more effect than that of f.

this area. In particular, it is clear that better and more complete estimates of rates of transmission in natural populations are necessary. In addition, several of the theoretically expected outcomes have been established, e.g., long-term oscillations (cf. fig. 2.3), spontaneous and local outbreaks of diseases (e.g., fungi in tropical forests: Samson et al. 1988), and persistent infection in growing populations (see chaps. 2, 3). This chapter has served its purpose if the reader is convinced that it is both feasible and critically important to understand the ecological dynamics of social insects and their parasites, and that epidemiological considerations are ultimately necessary in understanding the ecology of the parasites of social insects.

6.8 Summary

Social insect hosts and parasites interact in ecological time. Standard epidemiological theory can be applied to elucidate some of the basic properties of this interaction. For this purpose, it is useful to consider the colonies as individual hosts in a population. From the typical biology of social insects, such hosts are characterized by long persistence, the absence of protective immunity, low rates of recovery from infection, approximately constant population sizes, and population regulation by factors other than parasites. But considerable variation exists among different species. For example, annual colonies and nonoverlapping generations (as in many bees and wasps), and perennial colonies with overlapping generations (as in most ants and termites) must be considered differently. Some important assumptions of standard models are certainly violated, e.g., multiple infections should be common for entire colonies.

Most of the parameters that affect the ecological dynamics of parasites in social insects are poorly known or unknown. In addition, different reproductive systems, e.g., mating by dispersing sexuals, fission, or hibernation periods, vertically transmit different parasites in different ways. Theory suggests that a major determinant of the dynamics is the rate of horizontal transmission of parasites between colonies. Such transmission should occur in a variety of ways, including territorial disputes, contact at feeding sites, or specialized dispersal stages of the parasites. For transmission occurring with daily activities, such as nectar collection on flowers, the horizontal transmission rates are likely to be substantial. A range of parasites is vectored by mites or transmitted by drifting workers. In addition, it is possible though not yet known that inquilines or habitual robbers can act as vectors of disease. Since a parasite gains transmission from these behaviors, at least some of these activities may be rigged by parasites. Overall, the parameter values for the standard epidemiological models that are typical for social insects suggest that the epidemiological potential for many social insect parasites should be rather high.

The age structure of a population of colonies leads to an ecological dynamics that produces age-related prevalence patterns. Particularly, in the long-lived perennial ants and termites, these processes have to be taken into account when prevalences or infection patterns must be inferred from natural populations. The estimates may typically be inflated by intermediate age classes. In addition, the models also show that the force of infection is expected to be lower for the long-lived ants and termites than for short-lived species. Age-dependent dynamics can therefore produce patterns that can be easily mistaken for variation in host resistance. Calculations based on age dependency show that the net reproductive rate of two social insect parasites probably is in the range of parasites of humans, some with considerable epidemic potential.

Populations of social insect colonies also have pronounced seasonal cycles in worker numbers, activities, and so forth. The analysis of such situations is potentially very involved. Typically though, its is expected that parasite prevalence increases with season, but it may also show a midseason peak if background mortality rates of colonies are high. In addition, growing worker numbers within colonies also boost the epidemics, especially toward the end of the season. In many species, seasons are separated and hosts have discrete generations. Such problems can be analyzed in different ways. However, any realistic scenario, even for simple cases, results in rather complex dynamical landscapes that are primarily dominated by the rate of horizontal transmission. In fact, with the many nonlinearities inherent in social insect-parasite interactions, useful models will have to rely on numerical simulations of precisely defined situations. Some of the nonintuitive properties of host-parasite dynamics relate to variation in transmission rate among hosts, or to variation in local host numbers that can both increase the net reproductive rate of the parasite, at the same time making it less likely to persist. Although certain diseases can kill entire colonies, it is unknown to what extent parasites are able to regulate populations of social insects in ecological time. Theory also suggests that macroparasites can be more readily maintained in a host population. Social insects do in fact have a large number of macroparasites, such as nematodes, mites, and helminths. To date, too few data are available to estimate some of the critical parameters that determine the ecological dynamics of social insect parasites.

7 Virulence and Resistance

Host resistance and parasite virulence determine how host and parasite interact in ecological time and how they both coevolve. This chapter is introduced by some general remarks on host resistance and parasite virulence. The situation in social insects will then be discussed in an attempt to identify promising questions or future research.

7.1 The Nature of Resistance and Virulence

The success of a parasite depends on its ability to find a host, infect, multiply, and eventually be transmitted to a new host. The broad range of parasite life cycles leads to very different ways of achieving this result. However, one critical element is generally found: the production of large numbers of transmissible forms of the parasite. Microparasites rely on extensive multiplication within their host to increase density and thus to enhance the probability of being transmitted. Macroparasites, in contrast, rely on individual growth and survival within their host and on the production of a large number of offspring that leave the host to infect others. An increased success of the parasite by definition implies a reduction in host fitness and hence will be opposed by the host.

Resistance normally refers to the results of processes on the part of the host that reduce the number or survival of parasites and thus the negative impact on host fitness. This widely used concept normally excludes aspects of behavior and ecology that reduce the chances of contracting the parasite, or that affect the probability of transmission. Such aspects include the selection of nesting sites with a low infection risk, or the choice of food items that are unlikely to be intermediate hosts of parasites (Moore 1984). Defense against parasites may also involve the modification of individual life histories. For example, reproduction at an earlier age increases the chances of leaving offspring before sterilization by the parasite occurs, as shown for snails infected by schistosomes (Minchella 1985). Also, the reallocation by the host of energy and nutrients from growth or reproduction to defense to increase the chances of host survival may reduce the parasite's impact. In some cases, however, for example gigantism in tapeworm-infected rats (e.g., Mueller 1974), such reallocation seems to benefit the survival of the parasite too. In the case of social insects, these additional strategies must be kept in mind, because individual behaviors and the dynamics of the social structure allow a wide range of possibilities to combat parasites. On the whole, comparatively little is known about resistance mechanisms in insects when compared to the better-studied plants or vertebrate immune systems. In insects, most of the knowledge comes from two commercially important species, the silkworm and the honeybee. Among the invertebrates mollusks and crustaceans have also been studied in more detail.

The evolutionary biology of virulence has recently become a matter of vivid debate. Virulence is often defined as an increase in the host mortality rate as a result of the parasite's presence. But reduced host fecundity, parasite replication rate within the host, and several other measures have also been used (Bull 1994). Here, the term is used in a broad sense that encompasses these meanings. It will refer to processes that are caused by the parasite and which lead to a reduction in some component of the host's fitness. More recent discussions, for example, consider the possibility that some of these negative effects are not the result of parasite action but an accidental by-product of the infection (e.g., Levin and Svanborg Eden 1990), or a defense strategy of the host itself (e.g., Williams and Nesse 1991).

Virulence is associated with a number of processes at various stages of the infection. For example, harm could be caused through the penetration of the host body wall, with the depletion of host resources when the parasite feeds, with damage or destruction of host tissue, in the production of toxic substances by the parasite, or by the suppression of the host's immune response. Consequently, virulence should in principle also include instances where the behavior of the host is manipulated by the parasite to increase the probability of its successful transmission and where it places the individual host at greater risks. Acantocephalan parasites, for example, change the behavior of intermediate hosts, e.g., wood lice, in a way that makes them more likely to be eaten by the final host, a songbird (Moore 1984) (table 3.4).

Resistance and virulence are therefore properties that emerge as a result of host-parasite interaction in a given environment. Their expression is as diverse as the lifestyles and characteristics of hosts and parasites themselves. Also, our knowledge is very far from complete or even satisfactory on most accounts.

7.2 The Biology of Resistance in Insects

The very sophisticated immune system of vertebrates is generally not found in insects. However, some seemingly very old elements, such as the ability to discriminate self from non-self, or phagocytosis, are found in invertebrates and insects, too. Moreover, the insect defense system is overall similar in its structure to the vertebrate system; for example, it has a humoral and cellular component. The humoral component involves the action of lectins, specific carbohydrate-binding proteins or glycoproteins, that recognize and likely bind to antigens.

They can be seen as "natural" antibodies (Pathak 1993; Beck and Habicht 1996). In insects, the fat body is perhaps the major source of lectins, but they may also be produced by the epidermal membranes (as in the butterfly *Pieris:* Pathak 1993). Antibodies against bacteria and toxins have been found in cockroaches (Dunn 1990; Karp and Duwel-Eby 1991). Since cockroaches are assumed to be phylogenetically close to termites as well as primitive with respect to most other insect orders, antibodies could also be present in the socially living termites or in other, more advanced insects. In addition, cytokines help to organize the immune process in ways similar to that in vertebrates. In the process, iron is sequestered and it functions in bindings to proteins. This observation may be significant for social insects living in the tropics where iron-enriched clay soils are common. Cytokines are known from vertebrates as interferons, interleukins, and tumor necrosis factors but are present in almost all living organisms. Invertebrates have a number of elements that are more specific to their immune system. For example, antibacterial peptides and proteins (lysozymes) are part of their humoral defense system (in mammals they function to digest bacteria in the oral cavity). Some of these molecules have been more intensively studied, such as cercopins that kill bacteria at low concentrations (in the silk moth), or defensins in many insect orders (Pathak 1993). Several insects also produce proteins that belong to the very old immunoglobulin superfamily (also found in vertebrates; Beck and Habicht 1996).

The cellular component is interrelated with the humoral system as certain hemocyte cells (blood cells) secrete humoral factors, while humoral factors are involved in regulating cellular activities (e.g., Tanada and Kaya 1993). Many different hemocyte cell types have in fact been found and characterized. But for most of these, their function in the immune system is not clear. In general, relatively small particles are directly phagocytosed by some classes of granular hemocytes within hours. Phagocytosis is ineffective if a high load of foreign objects has invaded the host. In this case, granular hemocytes discharge a coagulum that becomes centrally melanized. This coagulum traps the foreign objects within minutes (nodule formation). Objects that are too large to be engulfed or trapped, such as parasitoid eggs or larvae, are encapsulated (Drif and Brehélin 1993). This process takes place within minutes to hours in two distinct phases. Hemocytes first aggregate around the foreign object, then lyse and release substances into the hemolymph to attract other hemocytes. This eventually leads to the formation of a tight capsule around the antigen that becomes melanized and kills the parasite (Ratcliffe et al. 1976; Schmit and Ratcliffe 1978; Ratner and Vinson 1983; Ennesser and Nappi 1984; Gupta 1985; Drif and Brehélin 1993). Granular hemocytes (granulocytes) (absent in Diptera and most Lepidoptera), plasmatocytes (in Lepidoptera and Diptera), and thrombocytoids (in Diptera) are involved in these processes. In many insects, specialized tissue during postembryonic development is the source of hemcoytes, but it is not known for certain which cell types originate from which tissue. Cell counts in the hemolymph often change during development—with the condition, with the action of hormones, and with the presence or absence of parasites—although in nontrivial and different ways (Drif and Brehélin 1993).

Often, the insect immune system is in fact challenged by other insects, e.g., when a parasitoid develops inside the hemocoel of its host. In such cases, the parasite is typically exposed to the cellular defense where hemocytes attack the invader. Immune reactions are also known to occur against protozoan parasites. For example, cysts are formed in the host *Anagasta kuehniella* against the gregarine *Mattesia dispora,* or in *Heliothes virescens* to contain the microsporidium *Nosema heliothides* (Brooks 1988). Some recent work has also found evidence for an immunological memory in insects, i.e., that a second infection by the same parasite is countered more rapidly (Faye and Hultmark 1993). The memory appears to be based on cecropines and related proteins that mostly act against gram-negative bacteria (Faye and Hultmark 1993; Hoffmann et al. 1994). So there is no intrinsic reason to expect that insect defense is less effective than vertebrate defense, although the sophisticated immunological memory of vertebrates is typically absent (for a review, see Gupta 1986).

In contrast to other organisms, effects resulting from defense against parasites can benefit the colony at the expense of the individual member. For example, control of infestation of the brood mite *Tropilaelaps clareae* in the Asian honeybee *Apis dorsata* occurs by workers that remove larvae infested with mites; this clearly is to the advantage of the colony but not the individual larva. In *Apis cerana,* in contrast, workers selectively remove the brood mites *Varroa jacobsoni* and kill them but do not necessarily kill the larva (Burgett et al. 1990; Peng et al. 1987a,b).

Genotypic Variation

Genotypic variation underlies host-parasite interactions in almost any insect host that has been investigated so far. For example, in the well-studied silkworms *Bombyx mori,* where the strains are in almost homozygous condition, variation in resistance has been found for infections by a number of different viruses (Watanabe and Maeda 1981). However, a simple gene-for-gene system, such that host resistance is coded by single genes that have counterparts in the form of corresponding virulence alleles in the pathogen (e.g., Thompson and Burdon 1992), is not the rule in insects. Except in a few cases (Watanabe 1967; Briese 1982), resistance is typically controlled by several genes through polygenic inheritance.

Genetically based resistance against parasites includes examples of parasitoids (in *Drosophila:* Hadorn and Walker 1960; Walker 1962; Carton and Boulétreau 1985; Boulétreau 1986; Nappi and Carton 1986; Carton et al. 1992), bacteria (*Bacillus thuringensis* in houseflies: Harvey and Howell 1965), nematodes (in mosquitoes: Goldman et al. 1986), and microsporidia (in moths:

Weiser 1969). Genotypic variability in susceptibility to fungal pathogens (chalkbrood, *Ascosphaera aggregata*) is known from the solitary leaf-cutter bee *Megachile rotunda* (Stephen and Fichter 1990a,b).

Lines of the honeybee also differ in their resistance to viral (Bailey and Ball 1991, CPV = hairless black syndrome; Kulincevic and Rothenbuhler 1975) and a variety of nonviral diseases (table 7.1), for example, bacteria (*Paenibacillus larvae:* Rothenbuhler and Thompson 1956; Rothenbuhler 1964a), fungus (chalkbrood, *Ascosphera apis:* Milne 1983), or protozoa (*Nosema apis:* Sidorov et al. 1975). Resistance to American foulbrood (the bacterium *P. larvae*) in the honeybee is one of the best-understood cases. Differences in the susceptibility of lines has been utilized. For example, the replacement of susceptible German lines by Italian races has contributed much to the control and elimination of foulbrood in North America (e.g., Krieg 1987) (see also reviews in Cale and Rothenbuhler 1975 and Taber 1982).

If more than one parasite infects a host, resistance against each may be genetically covarying. For example, there is a negative correlation between the capability of encapsulating eggs of parasitic wasps and the defense against a fungal disease among aphid clones (Read et al. 1995). In contrast, immunity against trypanosomes in cockroaches also increases susceptibility to infection by mermithid nematodes (Molyneux et al. 1986). The occurrences of several of the parasites discussed in chapters 2 and 3 show negative or positive associations. This could be due to covariation in resistance against the different kinds of parasites, but alternative explanations are of course possible.

When pathogens cause substantial host mortality, host resistance has been found to rapidly evolve within a period encompassing five (e.g., flacherie virus in silkworm: Watanabe 1967) to fifty generations (e.g., *Bacillus thuringensis* in houseflies: Harvey and Howell 1965). In many cases, though, there is considerable individual variability in this dosage-mortality relationship. Because of this, the response to selection is often rather shallow. This seems especially true for bacterial infections (Krieg 1987). For example, Rothenbuhler and Thompson (1956) found only a slight increase in resistance of honeybees to *Paenibacillus larvae* after selection.

Nongenetic and Environmental Effects

The efficiency of encapsulation in *Drosophila* against parasitoid eggs varies not only with genotype but also with age and health status, general host condition, and many environmental factors such as temperature (Muldrew 1953; Bosch 1964; Salt and van den Bosch 1967; Morris 1976; Blumberg and DeBach 1981; Blumberg 1988; Nappi and Silvers 1984). As a simplifying metaphor, we may assume that an overall level of immunocompetence varies with genotype and environmental conditions. Here, the word "immunocompetence" is used to refer to a general capability to resist parasites (e.g., Siva-Jothy 1995), although

Table 7.1

Examples of Causes for Resistance against Parasites in Social Insects

Host	*Parasite, Disease*	*Finding*	*References*
		INNATE FACTORS	
Apis mellifera	*Paenibacillus larvae* (American foulbrood)	Natural variation in resistance. Selection for increased resistance possible.	Rothenbuhler and Thompson 1956; Rothenbuhler 1964a; Hoage and Rothenbuhler 1966; Cale and Rothenbuhler 1975
Apis mellifera	Hairless-black syndrome (= chronic bee paralysis virus: Rinderer and Green 1976)	Breeding for resistance possible.	Kulincevic and Rothenbuhler 1975
Apis mellifera	*Acarapis woodi* (Acarosis)	Natural variation in susceptibility among different lines and stocks.	Rothenbuhler 1958; Bailey 1967b; Alexejenko and Wowk 1971; Gary and Page 1987; Page and Gary 1990; but see Taber 1984
Apis mellifera	*Nosema apis*	Natural variation in suceptibility among lines; in queens.	Furgala 1962
Bombus terrestris	*Crithidia bombi*	Susceptibility to infection varies among colonies kept in identical environments. Also probability of transmitting disease at different times varies among colonies.	Shykoff and Schmid-Hempel 1991b; Schmid-Hempel and Schmid-Hempel 1993
Bombus terrestris	Parasitoids	Intensity of encapsulation varies among colonies.	Schmid-Hempel 1994b; König and Schmid-Hempel 1995
		ENVIRONMENTAL FACTORS	
Apis mellifera	*Paenibacillus larvae* (American foulbrood)	Pollen feeding reduces mortality rate of workers.	Rinderer et al. 1974
Apis mellifera	*Paenibacillus larvae* (American foulbrood)	Food quality (pollen composition) affects resistance.	Rose and Briggs 1969

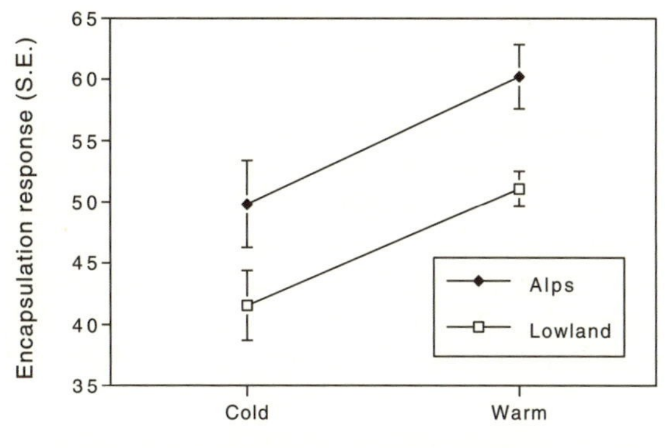

Figure 7.1 Encapsulation response (in arbitrary units; see Schmid-Hempel and Schmid-Hempel 1998) against an artificial parasite for workers of the bumblebee *Bombus lucorum*, depending on origin of the colony queen (Alps, Lowlands) and on the experimentally induced temperature regime under which the colony developed (low: 18°C, high: 28°C). Both main effects are significant while the interaction is not (ANOVA). (Benelli and Schmid-Hempel 1998.)

one should keep in mind that the relationships among different functions of the immune system are far from simple or completely known.

With these remarks in mind, we observe that a reduction in immunocompetence is often associated with environmental stress (Steinhaus 1958). For example, *Nosema apis* infections of the honeybee break out only when the colony is stressed (Doull 1961). Typically, prolonged starvation reduces the hemocyte count (i.e., the concentration of hemocytes in the hemolymph), which has often been used as a measure for the level of cellular immunocompetence (Jones and Tauber 1952; Shapiro 1967). Also, diet affects resistance, e.g., in moths against viruses (NPV) (Keating and Yendol 1987; Pimentel and Shapiro 1962) and in spruce budworms infected by *Nosema* (Bauer and Nordin 1988). In the bumblebee *B. lucorum*, the degree of encapsulation of a novel parasite depends both on genotype and on thermal environment (Benelli and Schmid-Hempel 1998) (fig. 7.1), but it is not affected by starvation (Schmid-Hempel and Schmid-Hempel 1998).

Quite often, younger larval instars are more susceptible to disease than later stages or adults. For example, early instar larvae of the silkworm are more susceptible to viral infection (CPV) (Aruga and Watanabe 1964). Similar findings have been made in forest tent caterpillars (Stairs 1965), the moth *Heliothis zea* (Allen and Ignoffo 1969), and flour beetles (*Tribolium castaneum:* Milner 1973). In the honeybee, this is the case for sacbrood virus (Bailey 1960) and European foulbrood (*Melissococcus pluton:* Bailey 1963). However, younger stages can also be more resistant than adults. Examples include the codling moth infected by the granulosis virus (Camponovo and Benz 1984) and by some fungal infections (Mohamed et al. 1977).

Infections in insects are normally favored by high temperatures and humidity. There are, however, a number of cases where higher temperature actually reduces the success of parasites. For example, the cabbage worm recovers from fungal infection when transferred to warm temperatures, although it is not precisely known whether decreased parasite virulence or increased host resistance contributes to this effect (Tanada 1967). Also the rate of development and multiplication of viruses (e.g., Watanabe and Tanada 1972), microsporidia (Weiser 1961; Wilson 1979), some fungi (Gardner 1985; Seeley 1985) and perhaps nematodes (Kaya et al. 1982) is reduced at elevated temperatures. The implication of this resistance mechanism for the organization of the colony was already discussed in chapter 4.6.

Behavioral Defense

Besides physiological components, parasite resistance in social insects also involves behavioral strategies. For example, workers of colonies of the termite *Coptotermes* wall off nestmates infected by nematodes (Fuji, in Klein 1990). Guard bees of the honeybee attack workers infected with chronic bee paralysis virus more aggressively than healthy bees, especially during times of nectar flow (May–July in the Northern Hemisphere) (Drum and Rothenbuhler 1985). Colonies of fire ants that are infected by the microsporidium *Thelohania,* in contrast, defend their nests less vigorously (Allen and Buren 1974).

There are also more subtle behaviors. Workers of the bumblebee *Bombus terrestris* parasitized by larvae of conopid flies are more likely to stay outside the nest at night rather than return home. Experiments have shown that this behavior prolongs the life span of the worker because the parasite larva develops more slowly and with less success in a cold environment. Furthermore, the workers, when given the choice, actively seek out the cold (Müller and Schmid-Hempel 1993b). The opposite choice, i.e., preference of warm temperatures, has been found in the grasshopper *Melanoplus sanguinipes* infected by *Nosema acridophagus* (Boorstein and Ewald 1987). In the honeybee, resistance against *Ascosphaera* (chalkbrood) or *Melissococcus* and *Paenobacillus* (foulbrood) depends largely on hygienic behavior (Milne 1983; Bailey and Ball 1991). In the case of American foulbrood, the relevant behavioral elements consist of uncapping the diseased brood cells and removal of the larvae by workers. These behaviors are genetically based on two loci and two recessive genes (Rothenbuhler 1964a). With chalkbrood, only uncapping seems to be affected genetically (Milne 1983). As discussed in chapter 4, additional factors, such as division of labor, resilience associated with colony organization, and frequent nest relocation, can add to "resistance" against parasites. Loehle (1995) proposed a concept of social barriers against pathogen transmission, involving mating strategies, social avoidance, group isolation, and other relevant behavioral mechanisms. Unfortunately, his review completely ignores an obvious case—the social insects.

Nutrition and Feeding Mechanisms

Choice of diet can help to combat parasites, too. For example, elevated nitrogen content in the diet of the spruce budworm reduces the incidence of infections by *Nosema* (Bauer and Nordin 1988). Similarly, nutrition renders the leaf-cutter bee *Megachile* more refractory to chalkbrood disease (Goettel et al. 1993). In resistant lines of *Apis mellifera,* the larval food is more effective in inhibiting the germination of spores and in reducing the number of vegetative cells of *Paenibacillus larvae* than it is in susceptible lines (Rose and Briggs 1969). A high-protein diet (pollen) reduces worker mortality in foulbrood colonies (Rinderer et al. 1974) and in caged bees infected by *Nosema* (Hirschfelder 1964); but at the same time, the pollen diet promotes spore development of *Nosema* (Rinderer and Elliott 1977). In some cases, therefore, nutrition can have a major effect on the resistance against disease, e.g., foulbrood in the honeybee (Bamrick 1964; Davidson 1973). The case of foulbrood also illustrates that defense often rests on several elements, e.g., hygienic behavior, larval resistance, efficiency of spore elimination by the proventriculus, and nutrition, including the presence of antibacterial substances in food.

As discussed in chapter 3, many parasites are contracted by ingestion of spores and cells. In bees, the proventriculus forms a filtering device at the posterior end of the crop just ahead of the midgut. Its hair-lined lips project into the crop (the "social stomach") and are able to collect particles out of the fluid. Its main function is the collection of pollen grains (5–100 μm) for later digestion in the ventriculus (the digestive chamber of the gut), but smaller particles, such as parasite spores, are also removed (Seeley 1985). In ants, larger particles are retained by the infrabuccal cavity. The size of particles that are effectively removed varies considerably. In the leaf-cutter ant, *Acromyrmex,* particles of 10 μm or larger are kept back. This would not be sufficient to prevent microsporidia or bacteria that have dimensions of a few micrometers from passing. In the carpenter ant *Camponotus,* this limit is as large as 150 μm, but in the fire ant *Solenopsis,* an effective filter size of 0.88 μm was reported. These values correspond to the differences in worker sizes of these two species, which are 6.0 mm, 9.5 mm, and 4.4 mm, respectively (Eisner and Happ 1962; Glancey et al. 1981; Sanchez-Pena et al. 1993). Perhaps larger castes in polymorphic ants are more parasitized by certain microparasites as a consequence of size differences in these filtering devices. Seeley (1985) indicates that in honeybees spores of *Paenibacillus larvae* and *Nosema apis* (1–5 μm) are efficiently removed by the action of the proventriculus.

If the infrabuccal cavity and proventriculus have evolved mainly in response to parasite pressure, a number of expectations could be formulated. For example, species exposed to a large number of parasites that are infective by os should have well-developed structures. Microsporidia and bacteria probably are the major parasite types that use this pathway (chap. 3). Plurad and Hartmann (1965) report that resistant lines of honeybees are more efficient in filtering out particles of *Paenibacillus larvae.* Similarly, the size of parasite spores should be under selection to coevolve with more efficient filtering devices, for example, to become smaller. A negative correlation between spore sizes, microparasite prevalence, and/or virulence could thus be expected.

Defensive Secretions

The rich variety of chemical compounds that are produced by the glands of social insects are not only important for communication (see Hölldobler and Wilson 1990 for a review) but also for defense against parasites. These glands constitute a highly sophisticated and diversified chemical factory. Chemical defense is actually widespread and not restricted to social insects. For example, in the solitary Alkali bee *Nomia melanderi,* the integument of prepupae contains substances that inhibit the growth of bacteria (Bienvenu et al. 1968). Many tropical bees (most of them solitary species) produce biocidal products in Dufour's gland and in the cephalic glands of adult bees, e.g., volatile acylic terpenoids and derivatives of fatty acids (Hefetz 1987; Roubik 1989). In fact, fungicides and bactericides are obviously found in almost any mandibular gland secretion of bees (e.g., Cane et al. 1983). In addition, plant resins that are typically collected by bees contain triterpenes and isoprenoids with antimicrobial properties. The production of antimicrobial substances is also known from the honeybee, where it is present in honey (White et al. 1963), larval food (Rose and Briggs 1969), and in the royal jelly that is fed to larvae destined to become queens (McCleskey and Melampy 1939). The mandibular glands of ants produce organic sulfides, ketones, pyrazines, and salicylate esters (e.g., Hölldobler and Wilson 1990).

The metapleural glands of ants were once thought to produce secretions involved in nestmate recognition. But it has been demonstrated that these substances have antiseptic effects (Maschwitz 1974; Hölldobler and Engel-Siegel 1984; Attygalle et al. 1989; Veal et al. 1992), and in particular also eliminate fungi (Beattie et al. 1985, 1986). One component, myrmicacin, probably has evolved as a highly potent defense against soil pathogens. Recall from chapter 2 that many parasites of social insects are in fact residents in the soil and are contracted as opportunistic infections. A homologue of myrmicacin is also secreted by bees (Beattie 1986; Iwanami in Cherrett et al. 1989). Obin and Vander Meer (1985) found that gaster flagging in fire ants is associated with spraying venoms that contain antimicrobial and antifungal compounds (Jouvenaz et al. 1972; Cole 1974). In his review, Jouvenaz (1986) refers to this chemical defense to explain why ants are often refractory to infections by protozoa, viruses, and bacteria, but typically become infected after an unrelated injury occurs. This would support the view by Hölldobler and Wilson (1990) mentioned before, i.e., that ants have relatively few diseases because of effective defenses. Figure 2.5 does not, however, support this simple expectation but suggests that perhaps for fungal disease, this argument may hold. The fact that such activity in the secretions of metapleural glands is even found among the most primitive ants extant today (*Myrmecia:* Veal et al. 1992), and possibly in the extinct *Sphecomyrma* (Hölldobler and Wilson 1990), suggests that microorganisms attacking in this way have been a very serious problem for social insects for a long time, and that there is a long history of coevolution. Some isolated genera of social insects, on the other hand, have secondarily lost this gland (Hölldobler and Wilson 1990) and it would be interesting to understand the reasons for this. The secretions can also serve dual functions. In the ant *Crematogaster deformis* the metapleural glands are hypertrophic and regularly release small amounts of secretions that have antiseptic effects. If the worker is attacked, large amounts are discharged, serving as defense against predators (Attygalle et al. 1989).

Unwanted fungal infections must pose a special problem for species such as leaf-cutter ants and fungus-growing termites that rely on a fungus for their nutrition (e.g., Martin 1970; Maschwitz et al. 1970). In the Attini, the fungus garden does in fact overgrow quickly with undesired fungi (Schildknecht et al. 1973), which must then be kept controlled under normal conditions. In fact, Bot and Boomsma (1996) observed considerable variation in the individual size of the metapleural gland (the bulla) between two species of leaf-cutter ants, *Acromyrmex*. Moreover, a large and significant proportion was due to colony variation within species, suggesting a selection potential. Termites are known to release secretions that act as a fungistat and increase the competitiveness of the symbiont fungus over its undesired competitors (Thomas 1981 in Cherrett et al. 1989).

Costs of Immune Defense

Resistance reduces the effect of a particular parasitic infection. But the development and maintenance of the immune system itself is costly and may show up as negative effects on other fitness components. Indeed, the immune response itself may be dangerous. For instance, the severe liver damage after an infection by human hepatitis B viruses seems to be caused by the immune response of the host itself rather than the parasite (Ganem and Varmus 1987). This is also demonstrated by autoimmune diseases, where the immune system turns itself against other body cells. Moreover, structures that are necessary to produce a facultative immune response are probably expensive to produce and maintain. Finally, mounting the immune response against a parasite involves direct cost in terms of the necessary energy and nutrients.

So far, the cost of immunity has been analyzed in only a few cases, mostly in birds. For example, clutch size manipulations have shown that an increased parental effort is associated with larger parasite loads and lower titers of defense proteins (immunoglobins) (e.g., Norris et al. 1994; Apanius 1991). A corresponding relationship between MHC-haplotype (a vertebrate locus complex that codes for proteins involved in antigenic recognition) and growth/fecundity parameters is known for chickens (Dietert et al. 1990). In insects, the question of the costs of immunity has rarely been addressed from the same point of view. Nevertheless, in a review on cellular immunity in insects, Lackie (1988) suggested that reduced immunocompetence in stress situations such as starvation, crowding, or nonphysiological temperatures is the result of a cost of immunity. A study of Carton and David (1983) on *Drosophila* showed that increased encapsulation responses are associated with a loss in other fitness components, i.e., with fecundity of the female. Similarly, Boots and Begon (1993) selected lines of the Indian meal moth, *Plodia interpunctella,* for resistance against granulosis virus. As a result, increased resistance became associated with longer larval development (reduced growth rates) and reduced egg viability, thus with lower fitness. As an aside, the benefits of resistance were not large enough to outweigh the fitness losses associated with slower growth whenever infection levels were either too low or too high.

For social insects, a cost of resistance was demonstrated in honeybees. Lines of larvae resistant to American foulbrood have reduced growth rates when compared to lines susceptible to the disease (Sutter et al. 1968). Skou and Holm (1980) found that honeybees treated with herbicides or antibiotics—thought to be stress factors—are infected more by the yeast *Candida,* i.e., are more susceptible to fungal infections. In an explicit experiment, König and Schmid-Hempel (1995) either prevented or permitted foraging activity for workers of the bumblebee *Bombus terrestris.* Foraging workers were more active and worked more during the experimental period. An experimental "test parasite" was implanted into the workers at the same time. The experiment showed that foraging activity was associated with reduced levels of encapsulation (fig. 7.2). Bot and Boomsma (1996), analyzing the metapleural glands of leafcutter ants, noted that gland secretions, supposed to have antibiotic effects, are no longer produced when workers are food deprived. They accounted for this fact with the cost of defense.

Interactions among Parasites

Resistance is also influenced by the presence of other parasites in the host. For example, larvae of *Trichoplusia ni* parasitized by the parasitoid *Hyposoter* are less prone to infections by a virus (NPV) (Beegle and Oatman 1974). Also, honeybees parasitized by the mite *Acarapis woodi* seem less susceptible to infection by airborne bacteria (*Pseudomonas*) and paralysis virus (Bailey 1965). The

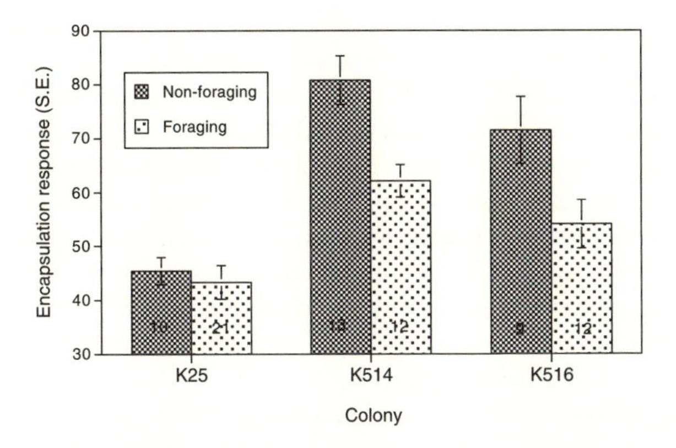

Figure 7.2 Encapsulation response (in arbitrary units) against a novel antigen in workers of the bumblebee *Bombus terrestris,* either allowed (foraging) or restrained (nonforaging) from foraging. Data from three colonies (K25, K514, K516) are shown. Small figures refer to sample sizes (number of bees). (ANOVA: colony effect: $F_{2,71}=8.36$, $P<0.001$, treatment; $F_{1,71}=4.87$, $P=0.03$, interaction: N.S.). (Reproduced from König and Schmid-Hempel 1995 by permission of The Royal Society.)

etiologic agent of (American) foulbrood, *Paenibacillus larvae,* is in fact suspected to produce "antibiotics," so that secondary infection by further pathogens is less likely (Holst 1945). A somewhat similar situation may exist in the stingless bee *Melipona fasciata* in Panama. The nests contain a variety of bacteria, *Bacillus* spp., in brood cells and storage pots of honey and pollen. These bacteria apparently produce a rich variety of enzymes, and perhaps also antibiotics (although this is not known). The provisions of these nests are typically much less infected by microbial infections than those of other species, perhaps because of the bacteria. The microbes in turn may gain an advantage by keeping away competitors from their food source (Gilliam et al. 1990).

In many cases, parasites act synergistically so that infection by one increases susceptibility to another, as shown in a number of cases. For example, two different viruses of the army worm interact in this manner (Tanada 1959). Bacterial infections make silkworms more susceptible to viruses (Ishikawa and Miyajima 1968). Larvae of the rice moth parasitized by the wasp *Bracon* are more susceptible to *Bacillus thuringensis* than unparasitized larvae (Tamashiro 1960). Similarly, infection by the amoeba *Malpighamoeba mellificae* in the honeybee correlates positively with infections by the microsporidian *Nosema apis.* Fyg (1964) found that in cases where the colony is doubly infected by *Nosema* and *Malpighamoeba,* the queens are also more often infected by *Nosema* (61%, $N=164$ colonies) than queens from colonies that contain only *Nosema* infections (41%, $N=310$ colonies). This is attributed to the fact that *Malpighamoeba* can cause dysentery and thus spread the other parasite more readily in the colony too. A similar positive association was found for *Nosema bombi* and *Crithidia bombi* infecting their common host, *Bombus terrestris* (Müller and Schmid-Hempel, unpubl. data). However, only the net effect of associations of different parasites is usually observed. So, in most cases it remains untested whether this association is based on genotypic covariance among loci coding for resistance to different parasites, on interindividual variability in general susceptibility to disease, on interactions among the parasites themselves, on epidemiological similarities, or on any combination of these.

As the examples make clear, our knowledge on resistance mechanisms in social insects is to a large extent restricted to the honeybee (table 7.1). However, we can safely assume that many of the relevant characteristics of other social insects are not different from *Apis* or, for that matter, from insects in general.

7.3 Evolutionary Ecology of Virulence

Host and parasite are typically linked in a coevolutionary arms race, with the parasite sometimes evolving toward higher virulence and the host evolving to reduce it. A good indication for this hypothesis is that parasite virulence is typically reduced (Olmsted et al. 1984; Kilbourne 1994) when the parasite is passed through different hosts and it is often increased when passing through similar hosts (Black 1992). The idea is also supported by the observation of local adaptation, such that parasite virulence on hosts often decreases with geographical distance between the two, or with a larger genetic difference (Anderson and May 1982; Lively 1987; Ebert 1994b; see also Gandon et al. 1996; Morand et al. 1996). Furthermore, virtually all entomopathogenic microorganisms have strain-specific variation in virulence (Tanada and Kaya 1993). Hence, this trait is likely to be subject to evolutionary changes. In the following discussion, parasitoids are not considered. Their lifestyle usually requires that the host is killed, i.e., a nearly maximum virulence is adopted. Consequently, there is very little meaningful variation in this pattern (Godfray 1994).

The above pattern is not universal, though. For example, there are also cases where a parasite infects the "wrong" host and is very virulent at the same time, e.g., Ebola virus from an animal reservoir infecting humans. In addition, parasites typically show tissue tropisms, i.e., they normally reach a specific tissue inside the host, where they establish, multiply, or reproduce. Sometimes, the parasite's presence in tissues other than the "normal" one is associated with high virulence, although the parasite cannot be transmitted from there to new hosts and is therefore locked into a dead end. Low virulence, on the other hand, may be associated with a reduced ability to replicate in a critical tissue (e.g., in Venezuela equine encephalitis where avirulence is linked to a reduced ability to replicate within the central nervous system) (Svanborg-Eden and Levin 1991; Bull 1994). Furthermore, the parasite may not be able to penetrate and cross the peripheral tissue surrounding the target tissue. Here, avirulence follows from the failure to reach the tissue where virulent effects would become manifest. These examples demonstrate that virulence does not always reflect a tight, common coevolutionary history with an adaptive outcome. In most cases, however, the adaptive view (e.g., Ewald 1994) seems to be the most appropriate one in trying to understand host-parasite interactions. It will be adopted for the remainder of this chapter.

The classical adaptive view is that parasites should inevitably evolve toward lower levels of virulence; otherwise, they would eliminate their hosts and go extinct themselves (e.g., Palmieri 1982; see also Svanborg-Eden and Levin 1991). There are in fact cases where parasites with a long coevolutionary history have low levels of virulence. For example, trypanosomes that commonly infect native ungulates in Africa (e.g., the bushbuck) produce mild symptoms, and cattle breeds that have been in Africa for a long time suffer little from these parasites. Yet the newly arrived European or zebu cattle breeds show severe symptoms if infected (Allison 1982). It is not clear, however, whether these differences reflect evolutionary changes of parasite virulence or of host resistance. Also, in the well-studied case of myxoma virus that was introduced to Australia to control the rabbit populations, a decrease of virulence over time has occurred (Fenner and Ratcliffe 1965).

Unfortunately, the classical view presented above rests on group selection, usually considered to be a weak evolutionary force. Models for virulence evolution based on individual selection, in contrast, consider the individual parasite's fitness advantage to be associated with virulence. For example, it matters how many offspring per individual parasite are successfully transmitted and established in a new host. This quantity is captured in the concept of R_0 considered in chapter 6. It is often reformulated as the number of newly infected hosts (secondary infections) created by each currently infected host (primary infection). Obviously, parasite fitness and R_0 must not always be the same. For example, calculations of R_0 assume that the uninfected population is at equilibrium and regulated by something else than parasites, which is clearly not always the case. For social insects, an additional complexity sets in because the rate of novel infections can refer to individual hosts as well as to entire colonies.

Selection within the Host

For the discussion of parasite virulence, it is useful to consider the different selection regimes separately. Within-host dynamics is one of the important domains. Our empirical knowledge is scant, but a range of theoretical models exists. For example, Sasaki and Iwasa (1991) assumed exponential growth of a parasite within its host. As the number of parasites within the host increases, host survival decreases. Therefore, this simple scenario should select for intermediate levels of virulence—if the parasite grows too rapidly and thus becomes too virulent, the host is killed too quickly. If growth is too slow, too few propagules are produced before the host dies for other reasons. Therefore, parasites in host populations subject to a high background mortality rate should evolve toward higher virulence. In social insects, this could be relevant for short-lived, annual type of societies with relatively high worker turnovers (bees, wasps), in contrast to the ant-termite type with relatively long-lived workers. Similarly, this hypothesis predicts that parasites should generally become more virulent in species with typically large colonies, as worker life span generally decreases with colony size (see fig. 1.3).

Antia et al. (1994) assumed that the growth of a parasite within the host is determined by its virulence (assumed to be identical to growth rate) and by the immune system that eliminates the parasites. The host dies when a critical, lethal threshold number of parasites is reached. The optimal strategy would be to choose a growth rate at which the lethal threshold is reached exactly when the action of the immune system starts to decrease parasite numbers by clearing the host body of foreign cells. In this case, the total number of propagules produced by the parasite is maximized. It is assumed that the parasite's success of transmission to new hosts, and thus its fitness, is positively correlated with the number of propagules produced over the course of infection. Again, this process would select for intermediate levels of virulence. In this case, parasites that grow too rapidly would outrun the immune attack and kill the host too early. Those that grow too slowly would stimulate the immune system but not be able to multiply much before they are eliminated by the host's immune defense. This view is supported by the observation that parasites have evolved a number of strategies to evade the host's immune system. For example, some bacteria (*Neisseria, Klebsiella*) produce capsules of polysaccharides that counter the immune system's attack (Finlay and Falkow 1989). Others produce proteases that are capable of cutting IgA molecules (immunoglobulins) that are important in vertebrate immune defense. A particularly sophisticated method is antigenic variation, as, for example, is found in many *Trypanosoma* species. In this case, the outer surface of the parasite, carrying polysaccharides that reveal the cell's identity, is frequently replaced by new surface antigens. Hence, the host defense system must activate antibodies against ever changing variants of the antigenic signature; an estimated 100 to 1,000 variants can be produced by each infecting trypanosome antigenic group (e.g., Wakelin 1984). Trypanosomes are also known from social insects (appendix 2.4.), but it is not known whether the immune system has to cope with similar antigenic variation. Perhaps, though, such variation is more frequent in social insects with long-lived workers (as in many ants), where the parasite can persist in an individual for longer periods and has fewer opportunities to transmit to a new host.

Selection between Hosts

Virulence may be adaptive if it increases the chances of transmission to new hosts. Some examples include the observation that coughing leads to transmission of viral particles in human influenza, that sores around genitalia ease transmission of sexually transmitted diseases (gonorrhea), or that open lesions of myxomatosa-infected rabbits make it more likely that vectors, e.g., flies, pick up the virus. In all of these cases at least one of the manifestations of virulence increases transmission. This logic obviously fails when the effects of virulence occur after the time when transmission normally occurs. However, in most cases, "virulence" does build up when transmission also is possible. Furthermore, increased parasite loads (e.g., expressed as the number of spores or cells of the parasite per host) are correlated with increased transmission, e.g., in myxoma virus (Fenner et al. 1956), bacteriophages (Bull and Molineux 1992), malaria (*Plasmodium:* Day et al. 1993), or microsporidia (Ebert 1994b). At the same time, however, there is not necessarily a positive relationship between parasite multiplication rate within the host and pathogenic effects (Ebert 1994a) or transmission rate to new hosts (Schmid-Hempel and Schmid-Hempel 1993).

Ewald (1994) suggested that where parasites are picked up by a vector, high levels of parasitemia (i.e., high cell numbers in the host) would increase the chances of getting into the vector. At the same time, the resulting high virulence could immobilize the host so that it is more likely to be found by the vector in the first place. As a consequence, vector-transmitted parasites should generally be more virulent than non-vector-transmitted ones. Chapter 3 showed that there are few parasites in social insects that are vector-transmitted in this sense. The carryover of parasites by drifting workers, robbing and raiding behavior, or "sampling" inquilines has been suggested to be analogous. In addition, viral infections activated vectored by mites may be an important exception (appendix 2.1). Hence, this hypothesis predicts that parasites would evolve toward higher levels of virulence in species where such transmission pathways are frequent, e.g., in many victims of raiders or in mite-infested populations.

Many studies have shown that the difference between vertical (to offspring) and horizontal transmission (to other hosts in the population) is important for the evolution of virulence. Theory predicts that an increase in horizontal transmission will generally lead to increased virulence (Ewald 1987; Nowak 1991). This pattern has been found, for example, in nematodes infecting fig wasps (Herre 1993) and in phages of bacteria (Bull et al. 1991). However, this scenario assumes that parasites are either vertically or horizontally transmitted but not in both ways at the same time. As summarized in chapter 3, few parasites of social insects appear to meet this strict assumption, since most can as readily be transmitted to sexuals (or workers) within the colony as they are to a worker of a different colony.

Where vertical and horizontal transmission are thus correlated, as is the case for many typical social insect parasites such as protozoa and bacteria, Lipsitch et al. (1996) have shown that an increase in the frequency of horizontal transmissions will actually select for lower virulence. For example, in the ant-termite type of species with perennial, long-lived colonies (e.g., also in honeybees), horizontal transmission should be frequent relative to vertical transmission (i.e., to newly founded colonies). In the bee-wasp type with annual short-lived colonies, in contrast, vertical transmission should be relatively more frequent. Hence, these two transmission patterns should select for different outcomes—with this argument, generally lower virulence in the ant-termite type but higher virulence in the bee-wasp type. Similarly, any structure in populations shifts the balance from horizontal to vertical transmission because more parasites will be transmitted within local groups. Generally, increased structure in populations should favor less virulent parasites (e.g., Lipsitch et al. 1995a), a situation with obvious relevance to social insects (see box 7.1; but see Wood and Thomas 1996).

In addition, parasites in social insects that reproduce by budding or fission (such as many polygynous ants and stingless bees) would probably fare better when colony growth rather than the actual production of sexuals (of which not too many are needed) is maximized. This is because they could pass vertically to the daughter colonies with the respective worker forces that become a part of the new colony. Parasites in host species that depend on sexuals for reproduction should fare better when sexual production is maximum, especially when this is the only way to bridge an important diapause, such as hibernation in temperate habitats. Clearly, quite different predictions would follow from these different scenarios. These possibilities also raise intriguing questions about the relationship of social insect life histories and parasitism.

Models that address the problem of multiple infections of a host by the same parasites often assume that the more virulent strain entirely takes over the host from a less virulent strain, e.g., as it outgrows its competitor. This puts virulence back in the ecological framework of competition theory (e.g., Bremerman and Thieme 1989; Lipsitch and Nowak 1995). Theory suggests that only a limited number of parasites can coexist stably. Generally, the models also predict an increase in virulence under multiple infection (e.g., Bremerman and Pickering 1983; Nowak and May 1994). Multiple infection can also lead to very complicated dynamics (e.g., Nowak and May 1994). Again, this is a situation that should be fairly typical for social insects where the colony is a large and persistent target that is able to collect a large number of infections. As we have already seen earlier, next to nothing is known as to whether multiple infections by several variants or strains of a parasite is common in social insects. Moreover, different parasite strains can infect the same or different workers of a colony; additional complexity is thus introduced.

Several diseases of social insects are known to produce long-lived spores that can infect new occupants of an abandoned nest, e.g., *Paenibacillus larvae* (foulbrood) in the honeybee (chap. 2). Interestingly, from a theoretical analysis of the problem, Bonhoeffer et al. (1996), contrary to Ewald (1994), have concluded that parasite virulence should be independent of the longevity of the propagule. However, if the host-parasite system is not in an equilibrium, then increased longevity of the parasite propagule should indeed favor higher levels of

BOX 7.1 VIRULENCE IN SOCIAL INSECT POPULATIONS

A host population consists of a number of groups each of which contains n host individuals; for example, colonies of social insects of size n. The parasite spreads from host to host within the colony, and eventually to other colonies. Let c denote the rate of contact among colonies. If a colony is infected, a contact leads to the transmission of the parasite with probability β (rate of transmission) (see figure). Single hosts and colonies are assumed to be singly infected and host density does not depend on presence of the parasite.

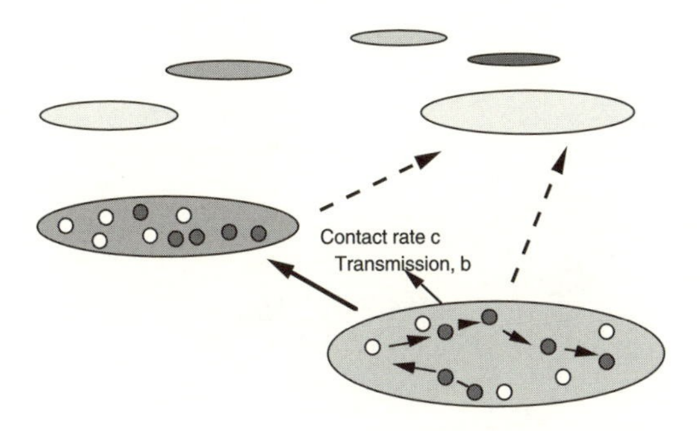

Lipsitch et al. (1995a) have analyzed this situation by assuming that transmission per contact is

$$\beta(\alpha) = 1 - q^n(\alpha)\,, \text{ where } q(\alpha) = 1 - p(\alpha)\,, \tag{1}$$

and where $p(\alpha)$ describes the probability of infection per host-host contact within a colony as a function of virulence α. Transmission rate between colonies, $\beta(\alpha)$, is assumed to depend on α and is proportional to the probability that a given colony member is infected. Therefore, the overall transmission between colonies depends on the number of individual hosts per colony (e.g., worker number, n). With standard epidemiology (see chap. 6, box 6.1),

$$R_0 = \frac{c\beta X}{\mu + \alpha + \gamma}\,, \tag{2}$$

virulence. Again, data on this point are unfortunately missing. However, it is likely that many social insect hosts, due to their biology, are indeed not in equilibrium with their parasites (see also chap. 6). Therefore, it is likely that parasites with long-lived propagules are typically virulent parasites of social insects, e.g., as with American foulbrood.

Finally, when these normal scenarios are enriched with the ecological dynamics, so that infection results in a decrease of the host population, parasite

BOX 7.1 CONT.

where X = number of susceptible (not infected) host (groups), μ = background mortality rate, α = virulence (i.e., parasite-induced mortality), and γ = recovery rate of hosts. The accompanying graph shows how R_0 depends on virulence, α, in groups of different sizes, n. Evolution will tend to favor parasite types that maximize R_0 for a given n and α. In the graph, the points on the surface that indicate maximum R_0 form a trajectory from the lower right hand corner to the upper left one, as colony size, n, increases. Hence, the model predicts that virulence should decrease as colony size increases (in the graph from $\alpha \approx 0.3$ to $\alpha \approx 0.02$). Parameters are $\mu = 0.05$, $c = X = 1$, $\gamma = 0$; $p = 1 - 10/(1 + 10\alpha)$.

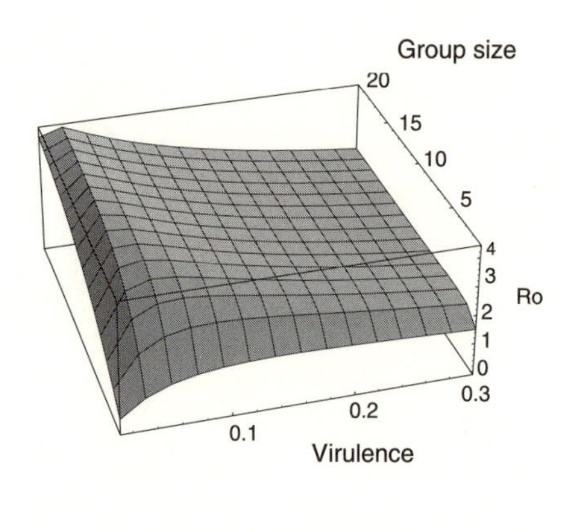

variants with high virulence are bound to become successively lost from the population, because they generally need a higher host population size to be maintained. Consequently, this process leads to a decrease in parasite virulence over coevolutionary time, but actually not to zero virulence (Lenski and May 1994). To some degree, this reconciles the classical view with the modern, individual-based approach. Table 7.2 summarizes these predictions.

7.4 Host-Parasite Coevolution in Social Insects

With the theoretical foundations just discussed, it becomes possible at least to ask how host and parasites should coevolve in social insects; in particular, how host resistance and parasite virulence should evolve in these systems. The existing studies are still a far cry from reliable empirical evidence, but social insects may in fact provide good opportunities to study these questions.

Table 7.2

Synopsis of Adaptive Hypotheses for Parasite Virulence

Scenario	*Predictions*	*Examples, References*
	WITHIN-HOST SELECTION	
Faster strains are more virulent	Higher virulence evolves as a consequence of competition for hosts.	Bremerman and Thieme 1989; Sasaki and Iwasa 1991
Multiple infection	Multiple infection increases virulence because more virulent strains can take over. Stable coexistence of many strains with variable virulence.	Nowak and May 1994
Balance with immune system	Virulence given by optimal growth rate that balances transmission and clearance by immune system.	Antia et al. 1994
Short-sighted selection	Winners of within-host competition are poorer at transmission to new hosts.	Not demonstrated, but perhaps HIV, polio, bacterial meningitis
Tissue tropism	Virulence depends on ability to infect and replicate in specific tissues.	Many viruses
Diversity	Virulence is a consequence of overloading the immune system with too many antigenic variants.	Antigenic variation in *Trypanosoma* HIV-virus; Nowak et al. 1991

Expectations for Virulence

Our empirical knowledge of the virulence levels and its variation in parasites of social insects is rather limited. Table 7.3 gives a summary with a crude arrangement according to virulence levels. Very few of the studies are detailed enough to actually test the predictions discussed above, so any conclusion must remain tentative for the time being.

As stated before, theory suggests that parasites transmitted horizontally or vectored between colonies should more likely evolve toward high virulence, if at the same time they do not depend on vertical transmission (Ewald 1987). This is almost by definition true for parasitoids (Strepsiptera, parasitic wasps and flies) and most nematodes. Parasitoids are lethal as a rule, and many nematodes often have detrimental effects, though not always. In most other parasites, both vertical and horizontal transmission occurs, but the relative rates are generally not known. As a crude approximation, it can be assumed that horizontal transmission may be relatively more frequent where the parasite has been found at

Table 7.2, continued

Synopsis of Adaptive Hypotheses for Parasite Virulence

Scenario	*Predictions*	*Examples, References*
	BETWEEN-HOST SELECTION	
More virulent strains have more transmission	High virulence corresponds to high parasite loads, therefore increasing chance of transmission.	Perhaps many parasites; Ebert 1994b
Vector-borne transmission	Transmission to vector increased with high virulence (parasite load) and with host immobility.	Malaria, Myxoma; Ewald 1993, 1994
Host population size, density	If parasites reduce host population, virulence decreases over time.	Lenski and May 1994
Vertical transmission	Vertical transmission leads to lower virulence, because parasite success is closely correlated to host fitness.	Nowak 1991; Herre 1993
Coupled vertical and horizontal transmission	Decreased virulence with increasing horizontal transmission frequency.	Lipsitch et al. 1995b
Long-lived propagules	Virulence should be independent of longevity in equilibrium, but select for increased virulence when parasite invades.	Bonhoeffer et al. 1996
Spatial distribution	High virulence maintained if host population is spatially distributed over wide area.	Wood and Thomas 1996
Structured populations	Population structure is associated with frequent transmission within groups; decreases virulence.	Lipsitch et al. 1995a

SOURCE: After Bull 1994 and modified.

food sources, at sites where hosts congregate, and where vectors are known (see chaps. 2, 3). Similarly, horizontal transmission should be relatively more important in host populations of the long-lived, perennial type. In table 7.3, these conditions apply, for example, to chronic bee paralysis virus and other viruses that are vectored by mites, but also for *Spiroplasma* or *Nosema*. According to Lipsitch et al. (1996), low virulence is expected under these conditions. In fact, none of them are typically very virulent parasites. Theory also suggests that in social insects parasites with long-lived propagules should have higher levels of virulence (Ewald 1987; and Bonhoeffer et al. 1996 under certain conditions).

Table 7.3

Virulence Levels of Social Insect Parasites

Parasite	*Host*	*Virulence Correlates, Transmission**
BENIGN		
FV, BVY viruses	*Apis mellifera*	Mite-associated infections (vectored?).
Malpighamoeba mellificae	*Apis mellifera*	
Leidyana gregarines	*Apis mellifera*	
Crithidia mellificae, Crithidia bombi	*Apis mellifera, Bombus* spp.	Horizontal transmission via flowers known. Multiple infection occurs.
Kuzinia, Parasitus, Scutacarus	*Bombus* spp.	Horizontal transmission by phoresy.
Myrmicinosporidium durum	Many ants	
Myrmecomyces annelisae and several yeasts	*Solenopsis* spp.	
Diploscapter lycostoma	*Formica* spp. *Lasius* spp. *Myrmica* *Iridomyrmex humilis*	
MODERATE		
Spiroplasma (severe?)	*Apis mellifera, Bombus*	Horizontal transmission via flowers.
Cephalosporium, Cladosporium, and other fungi	*Apis mellifera, Bombus* spp.	
Nosema apis (severe?)	*Apis mellifera*	Horizontal transmission via flowers?
Acarapis woodi	*Apis*	Multiple infection by females.
Tropilaelaps clareae	*Apis dorsata*	Horizontal transmission by phoresy; multiple infection by females.
Gregarines	*Myrmecia*	
Anomotaena brevis	*Leptothorax* spp.	
Allomermis myrmecophila	*Lasius* spp.	
Mermis (severe?)	*Apis,* many ants	Horizontal transmission via contaminated surfaces.
Pheromermis villosa	*Lasius* spp.	Horizontal transmission via infected food.

Table 7.3, continued

Virulence Levels of Social Insect Parasites

Parasite	*Host*	*Virulence Correlates, Transmission**
SEVERE		
BQCV, BVX, KBV, AIV, EBV, Kashmir bee virus	*Apis mellifera, A. cerana*	BVX with winter epizootics.
Deformed wing virus	*Apis mellifera, A. cerana*	Vectored by mites.
Paralysis virus	*Apis mellifera, Bombus* spp.	Horizontal transmission via flowers; vectored by mites.
Sacbrood virus	*Apis mellifera, A. cerana*	
Paenibacillus larvae, Melissococcus pluton	*Apis mellifera*	Long-lived spores.
Pseudomonas antiseptica, Serratia marcescens	*Apis mellifera*	Opportunistic soil organism. Transmitted horizontally by air.
Ascosphaera apis	*Apis mellifera*	Transmission from reservoir (other bee species). Long-lived spores.
Aspergillus flavus	*Apis mellifera*	
Nosema bombi, Apicystis (Mattesia) bombi	*Bombus* spp.	
Varroa jacobsoni	*Apis*	Horizontal transmission by phoresy, e.g., during robbing. Multiple infection by females.
Sphaerularia bombi	*Bombus* spp.	Transmission from reservoir (soil).
Beauveria bassiana	*Solenopsis richteri, Vespula germanica*	Horizontal transmission by air?
Entomophtora spp.	Many ants	Horizontal transmission by air?
Burenella dimorpha	*Solenopsis geminata*	
Mattesia geminata	*Solenopsis geminata*	
Thelohania solenopsae	*Solenopsis* spp.	
Mermithid nematodes, *Tetradonema solenopis*	*Solenopsis* spp.	

Table 7.3, continued
Virulence Levels of Social Insect Parasites

Parasite	*Host*	*Virulence Correlates, Transmission**
SEVERE, continued		
Brachylethicium, Dicrocoelium, Anomotaenia, Choanotaenia, Protocephalus, Raillietina	*Formica, Camponotus, Leptothorax, Euponera,* and many other ants	Intermediate hosts for cestodes and trematodes.
Cordyceps spp.	Many ants, *Vespa, Vespula*	Horizontal transmission by scavenging?
Metarhizium anisopliae	*Cryptotermes brevis*	
Diplogaster aerivora	*Leucotermes lucifugus*	

NOTES: Categories: "Benign" = few if any known effects. "Moderate" = small reduction in life span or fecundity with moderate pathological effects; hosts often survive; or color aberrations with unclear role. "Severe" = drastic reduction in life span or fecundity of individual host; host death, behavioral changes with serious consequences; death of colony may follow. Symptoms may differ substantially among geographical regions. Examples taken from appendix 2 and tables 3.1, 3.2. Parasitoids not listed.

*Correlates are usually not exactly known but inferred from reports on occurrence and distribution of the parasite.

Although table 7.3 lists some examples in the predicted direction (e.g., *Paenibacillus, Ascosphaera*), no striking relationship stands out. Table 7.3 also shows that virulent parasites have perhaps more often been found in species with large-sized colonies (e.g., the honeybees *Apis,* fire ants *Solenopsis,* or wood ants *Formica*). This is not expected from the considerations in box 7.1. On the other hand, the pattern is heavily biased by the particular species that fall into the large-size category and which have been well investigated.

For the trypanosome *Crithidia bombi* in bumblebees a more detailed knowledge is available. This parasite is readily transmitted between and within colonies, to workers and to sexuals. In such a case of coupling of vertical and horizontal transmission events, low virulence is expected (Lipsitch et al. 1996). This is indeed the case. In addition, parasites of annual, diapausing hosts, such as *Crithidia* in bumblebees, must transmit vertically through a period when their host is severely stressed, e.g., has to survive hibernation. Hence, this particular selection episode may additionally select for low levels of virulence. We would thus also expect that virulence should decrease with a long diapause, for example, in northern or high-alpine populations (e.g., Rennie 1992).

Experimental evidence for the potential effect of within-host selection on parasite virulence in social insects is available from studies where pathogens have been passed through a series of hosts. Usually this has been done in an attempt to modify virulence for biocontrol programs. For example, Clark (1977) found that *Spiroplasma* remains infective after more than twenty passages in culture. But it loses infectivity when it is passed alternatively through culture media and insect host. It is also less pathogenic when transferred to another host, for example, from the honeybee *Apis* to the moth *Galleria* (Dowell et al. 1981). *Paenibacillus larvae* also loses infectivity when cultivated on media (Bailey and Ball 1991). In contrast, serial passage of *P. larvae* through two lines of the honeybee increased virulence (Krieg 1987). This is also the case for the fungus *Aspergillus flavus* when it is passed several times through its host (honeybee larvae) and for *Ascospharea* (Bailey and Ball 1991). Therefore, within-colony passage in the same host typically increases virulence. However, it is not known whether in the cited cases, the situation is frequent enough to drive the evolution of virulence under field conditions, too.

An important deficiency in the models discussed before is the neglect of genotypic host variability which must be particularly relevant for the horizontally transmitted and cross-species infecting parasites (e.g., Jaenike 1993). For the trypanosome *Crithidia* in bumblebees, there is now good evidence that genotype-genotype interactions are of overriding importance (Shykoff and Schmid-Hempel 1991b; Schmid-Hempel and Schmid-Hempel 1993; Schmid-Hempel et al., unpubl. data). Perhaps genotypic variation of the host is maintained by a number of different processes, and that *Crithidia* is exposed to this variation rather than affecting it. Although never explicitly tested, this should also imply that frequent passage between colonies (with different genotypes) selects for lower levels of virulence, as the results of cross-infection experiments also suggest where the parasites are usually less pathogenic when forced onto foreign hosts (including culture media). Similarly, in their extensive review, Tanada and Kaya (1993; also Samson et al. 1988) conclude that entomopathogenic fungi typically have many strains that are fairly host-specific. They are usually virulent in the host from where they are isolated but not in other hosts. Note that this does not invalidate the arguments presented before. In fact, the foregoing models should actually be tested with populations of the same parasite that are subject to different ecological selection regimes and where the hosts do not coevolve.

With respect to within-host selection, some characteristics of social insect biology may be particularly interesting. First, for example, the life span of workers varies widely among species of social insects (table 1.3) and is generally shorter in species that typically attain large colony sizes (fig. 1.3). However, life spans also vary among colonies in a population (e.g., Müller and Schmid-Hempel 1993a; Schmid-Hempel 1994b). As argued above, shorter life span of the hosts should select for faster multiplication rates of infectious parasites.

Hence, we should expect that the virulence of parasites that infect adults should be higher in species or populations with high turnover rates of workers, i.e., shorter life spans. This has not been tested yet but it would seem to be straightforward to do (see table 7.4). Second, parasites that infect species with genetically highly variable colonies should—*ceteris paribus*—on average be less virulent. Obviously, we would then require that genetic variability of the hosts is maintained over time. This would predict that virulence levels are generally lower in polyandrous or polygynous species. The prediction depends somewhat on how the different mating strategies distribute the available genetic variance among and within colonies. With polyandry, each colony becomes more heterogeneous but the variance among colonies decreases. So, if transmission between colonies is frequent, parasites could actually adapt to this background more quickly and increase rather than decrease in virulence.

There is also a note of caution. If virulence is an adaptive trait and coevolves with the host, then the relevant consequences of parasitism are at the level of the inclusive fitness of the affected host individual. This will often be better captured by measures of colony success (as in Sherman et al. 1988) than by measures of individual welfare. Hence, the evidence summarized in table 7.3 is only partly useful. Unfortunately, virtually all studies concentrate on the harm done to individual workers or brood and rarely measure parasite effect on colony survival and reproduction. Obvious exceptions are investigations in connection with commercial breeding, as in the honeybee (also Müller and Schmid-Hempel 1992b in bumblebees).

Expectations for Resistance

Unlike virulence, the evolutionary ecology of resistance has not received the same recent amount of attention. Therefore, this section proceeds by developing some ideas that could be useful in organizing the field and in highlighting some of the intrinsic problems surrounding defense in social insects.

In chapter 4 I discussed the relationship of colony organization and parasites. Here, one further step can be taken. For example, highly resistant workers have a higher value for the colony under parasitism and thus make the system more reliable in the sense discussed in chapter 4. It can also be assumed that defense is costly. Hence, with a given amount of resources, only a limited number of resistant (high quality) workers can be produced but more of the susceptible (low quality) ones. More susceptible workers will be advantageous if parasitism is absent, since they retrieve more resources, but this is different if parasites abound. Box 7.2 develops these ideas and calculates the optimum number of workers under this scenario. It is shown that colonies living in good habitats (i.e., with high levels of resource availability) should produce many workers but not invest much in defense against parasites. Furthermore, variation in parasite

Table 7.4

Predictions for the Evolution of Virulence in Relation to Characteristic Elements of Social Insect Biology (see text for further discussion)

Trait	*Virulence Should...*	*Justification*
Polyandry, polygyny	Decrease . . .	Genetically variable hosts more difficult to adapt to. Applies if within-colony selection is prevalent.
	. . . or increase	High degrees of polyandry and polygyny reduces genetic variance among colonies. Hence, population more uniform at large. If between-colony transmission prevalent, may select for higher virulence.
Large colony sizes	Decrease . . .	Transmission within-colony becomes more frequent relative to between-colonies.
	. . . or increase	Under same conditions, but if horizontal and vertical transmission coupled. Large colonies tend to have short-lived workers. High turnover sets within-host competition and higher virulence.
High worker turnover (short life spans)	Increase	Short-lived hosts select for high multiplication rates of parasites.
High degree of sociality	Increase . . .	More social contacts promote more transmission.
	. . . or decrease	Tighter social integration contains parasite within colony and thus gives less weight to horizontal transmission.
Sophisticated caste structure, division of labor	Decrease	Relatively more local transmission within colony-subgroup (caste). Also higher degree of redundancy selects for lower host defenses.
Inquilinism, raiding, robbing	Increase	Transmission to new colonies frequent.
Iteroparity vs. semelparity	Decrease	Transmission within-colony becomes more frequent relative to between-colonies. If coupled, may increase.
Ephemeral habitats	Increase	Transmission to new colonies more frequent as compared to within-colony or to local clusters of colonies.

Table 7.4, continued

Predictions for the Evolution of Virulence in Relation to Characteristic Elements of Social Insect Biology (see text for further discussion)

Trait	*Virulence Should...*	*Justification*
Flying vs. walking foragers	Increase . . .	Parasites are spread more widely (i.e., less locally) in population.
	. . . or decrease	Flight energetically more demanding and hence less defense by host.
Cavity nesting	Increase	More likely to encounter parasites with long-lasting spores when nest sites are limited. Only if populations always out of equilibrium.

pressure has a decelerating (logarithmic) effect on the predicted outcome. This result is rooted in the fact that workers of social insects essentially are disposable elements. In other words, a colony can choose to produce many low-quality workers or a few high-quality ones and still achieve the same success. This option would not be available for solitary organisms, at least not in the same sense. Modular organisms in general would follow a similar logic. Note also that the model of box 7.2 does not explicitly take into account the chances of escaping parasitism. In fact, it is readily appreciated that below a certain parasite prevalence in the habitat, defense does not pay because the expected costs exceed the expected benefits. Above a certain threshold, some defense should be employed—which is the problem in box 7.2.

It is easily shown that redundancy in colony organization drastically reduces the need to have well-defended workers (see chap. 4). On the other hand, where colonies are well buffered against negative effects, parasites will be under lower levels of selection by the host and also attack less well defended workers. With the standard arguments presented above, parasites should thus evolve toward higher levels of virulence for the same cost to them. In principle then, this coevolutionary effect may cancel the advantages of colony organization and lead to an outcome where host defense and parasite virulence level off with the same net result, but at a higher cost to each of the parties. This could be tested by comparing the level of parasite virulence in relation to colony organization among different species of social insects. Alternatively, the evolution of virulence on hosts that differ in resistance (e.g., different lines of the honeybee) could be tested experimentally to check the assumptions and some of the predicted results.

Consider the case where the risk for the colony of acquiring an infection that

BOX 7.2 RELIABILITY AND PARASITE RESISTANCE

Colony organization of social insects generates resilience against external perturbations (chap. 4). The performance of the whole system (the colony) depends, in a highly nonlinear manner, on how well each individual performs (see eqs. 4.1). Suppose that individual reliability, p $(0 \leq p \leq 1)$, depends on whether or not the individual worker is infected by a parasite and how resistant the worker is. A large value of p indicates that each individual worker is resistant against parasite attack and therefore performs well under parasitism. High values of p correlate with high cost, $C(p)$, for the colony to produce and maintain high-quality workers. For example, the cost per worker is

$$C(p) = c_0 + c_1 p^{c2}. \tag{1}$$

Here, "costs" are not further specified but refer to resources that the colony acquires from the environment and can invest into workers or something else (e.g., sexuals at the end of the cycle); c_0, c_1, and c_2 are shape parameters that determine how defense level, p, and costs relate to each other. Generally, $c_0 > 0$, $c_1 > 0$, and $c_2 > 1$, for an accelerating cost with increasing defense and a positive intercept of this function. The value of c_2 (and also c_1), which describes how defense level and cost relate to each other, could also be interpreted as a crude measure of parasite pressure in the environment. With high values of c_2 (c_1), a given level of defense is costly to achieve, for example, due to the presence of many or virulent parasites that require costly defenses.

In addition, it is the workers that acquire the necessary resources. If there are many workers per colony (n), and each one of them is well defended (p), a large amount of resources, B, can be collected per unit time. For example, the rate of resource collection with n workers at defense level p, where parasites are present, is

$$B(n,p) = b_0(1 - e^{-b_1 np}), \tag{2}$$

where b_0, b_1 again are shape parameters. The value of b_0 is the asymptotic value of $B(n, p)$ and therefore reflects the quality of the habitat.

We assume that maximum performance of the colony is achieved when a maximum number of workers is produced. Consulting chapter 4 and Oster and Wilson (1978), it is readily seen that a large number of workers also increases the overall reliability of the system and is thus a useful fitness token. Since workers are costly but also retrieve resources and since the colony must grow, we require that $B(n, p) \geq n\, C(p)$. Assuming equality, we have

$$b_0(1 - e^{-b_1 np}) - n(c_0 + c_1 p^{c_2} = 0.$$

$$\text{Thus } n^* = \frac{1}{b_1 p} \ln\left(\frac{b_0 b_1 p}{c_0 + c_1 p^{c_2}}\right), \tag{3}$$

BOX 7.2 CONT.

where n^* is the maximum number of workers that are in a colony under this scenario. The accompanying graph shows how n^* relates to p and habitat quality, expressed by the value of b_0 (other parameter values are $c_0 = 1$, $c_1 = 1$, $c_2 = 3$, $b_1 = 1$). The graph shows that the maximum value of n^* (i.e., the ridge of the surface) increases while habitat quality increases. At the same time, p decreases. Hence, in good habitats the colonies should produce more but lower quality workers. The calculations show that parasite pressure, as expressed by the value of c_2, has relatively little effect (eq. 3).

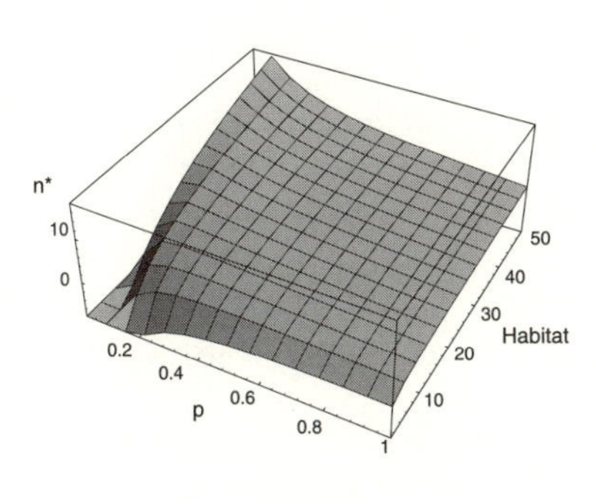

causes a disease severity, X, is a random variate. Box 7.3 explains how this situation can be analyzed. In a simple success vs. failure scheme, the colony needs to adopt a level of defense, D, that gives a maximum chance to survive the event X. The result of the analysis shows that regardless of the exact details of the distribution or of the cost of defense, it always pays more to invest into defense in a relatively predictable habitat (i.e., low variance in the distribution of X). This is simply because in a low-variance habitat any increase in defense will make it more certain that a given disease severity can be controlled. The reasoning is very similar to standard arguments in life history theory about resource allocation to survival and maintenance in relation to age and size at maturity (e.g., Stearns 1992).

Phenology and Life History

Variation in the temporal pattern of how the colony cycle and the activities of the workers unfold is the realm of life history theory and clearly important for

BOX 7.3 VARIABLE ENVIRONMENT AND DEFENSE

In a given habitat, there is variation in the risk of acquiring an infection and disease severity independent of defense levels and resource allocation by the colony. Such variation may result from variation in microclimatic factors, behavior of the animals, heterogeneity in habitat structure, the presence of competitors, variable parasite strains, chance events, and so forth. Suppose that with many contributing factors, the chances of acquiring an infection and experiencing a given level, X, of disease severity during the life cycle of a colony is normally distributed with mean μ, and variance σ^2. On the other hand, the colony is assumed to adopt a given defense level, D. We assume that the chances that the colony survives to reproduce is given by

$$X \leq D \text{ for colony success },$$
$$X > D \text{ for colony failure }.$$

With X being a normal deviate, the chances of colony success are

$$P(X \leq D) = \phi(z) \text{ , the cumulative probability distribution,}$$
$$\text{where } z = \frac{D - \mu}{\sigma} \text{ , the standard normal deviate.}$$

The best strategy is to maximize $P(X \leq D)$, hence to maximize $\phi(z)$. Fortunately, the details of z are not particularly important, since any increase in z will increase the probability of success, $\phi(z)$. Now consider the relationship of habitat variance, σ, with defense level:

$$\frac{dz}{dD} = \frac{1}{\sigma}.$$

Hence, an increase in defense level is most effective when habitat variance is small. Therefore, colonies living in relatively predictable habitats (low variance) should invest more in defense and, vice versa, those in variable habitats should reduce defense. Obviously, if defense is costly, the actually adopted level should become generally lower, but the logic of the argument is not reversed.

the relationship with parasites. Throughout this book, several instances of life history approaches have been used, both for the individual worker within the colony as well as for colonies within their populations. Here, a brief explicit section on how parasitism could enter the picture seems warranted, though the field has to acknowledge its ignorance on this subject.

According to Bourke and Franks (1995), gap dynamics drives the evolution toward iteroparity and high-cost offspring in ants. Because the population of adult colonies is dense and persistent, offspring compete for gaps left by perished colonies, and they need to be well equipped to be successful. Iteroparity, i.e., several reproductive stages, reduces the risk of producing offspring during adverse conditions. The variety of life styles of ants is in fact accompanied by a suite of adaptations to habitat characteristics that determine the success of dispersal and establishment of offspring (Bourke and Franks 1995). But variation in important characteristics, such as the length of the prereproductive phase (the initial ergonomic phase sensu Oster and Wilson 1978), the length of the reproductive life span, age at maturity, and semel- or iteroparous lifestyles, should also affect how host and parasite interact. This is nicely borne out by the ecological dynamics discussed in theory in chapter 6, showing how age-dependent changes or life cycle lengths can affect the persistence and prevalence of infections in social insect populations.

To date, we lack well-founded empirical knowledge on how host life history and parasitism interact in social insects. Some hints come from studies in other taxa or from theoretical analyses. For example, in Minchella's (1985) classical studies on snails, earlier reproduction reduced the expected fitness loss due to parasitic castration. Similarly, Hochberg et al. (1992) suggested that host and parasite can coevolve via life-history traits. Hosts are predicted to reproduce sooner when more virulent parasites are present, and to become shorter lived when semelparous. Quite generally, an increase in mean and variance of adult mortality should select for shorter (reproductive) life spans, while the opposite is true for juvenile mortality (Stearns 1992). Low survivorships of incipient colonies would thus select for long adult colony life and, by Hochberg et al.'s (1992) argument, for younger age at maturity or shorter semelparous life spans.

Mortality is an important component of parasite virulence. Hence, we would expect that social insects with many or virulent parasites of the mature colonies should have shorter adult colony life spans and late maturity, and those with many effective parasites at the juvenile stage should have longer colony life spans and early maturity. In social insects, we have seen that large colony size correlates with more parasites (chap. 4, fig. 4.10). Adult colonies of small-sized species may be able to "hide" from parasites and are thus expected to develop slowly and to live long. Many ponerine ants may qualify in this category. On the other hand, long-lasting perennial and iteroparous life cycles are often associated with large colony sizes, such as in honeybees and stingless bees, and vice versa in the bumblebees or sweat bees. This would thus not correspond to the expectation formulated above. However, honeybees and stingless bees reproduce by fission, which produces a different scenario altogether. Also, phylogenetic effects are clearly present and need to be taken into account in all of these analyses. For the time being, all of this must remain speculative. It seems nevertheless evident that avoiding parasites must be a tremendously difficult task for any long-lived social insect colony. In contrast, temporarily social groups, such as bumblebees or sweat bees, shut down their operations (colonies) on a regular basis and hence lose a few parasites every now and then (e.g., Packer 1986). The same may happen as a consequence of swarming and colony fission (Royce et al. 1991), as was already discussed in chapter 4.

Schmid-Hempel and Durrer (1991) investigated how parasite frequency (parasitoid flies in this case) relates to the timing of reproductive events in populations of bumblebee hosts. In fact, when a number of populations were compared, interesting correlations emerged so that heavy parasite pressure was associated with earlier reproduction in *Bombus lucorum* and later reproduction in the *B. terrestris.* These differences could be explained with the effects of higher mortality during different stages of the host life cycle (early or late). For example, if not associated with castration, parasites acting late could lead delayed reproduction, as the colony is buffered (see chap. 4) and may gain more from continuing to grow than from reproducing now. This should be enhanced by the problem of synchronizing reproduction with other colonies in the population. The same logic would suggest that reproduction should be accelerated when parasites act early in the life cycle. However, explicit laboratory experiments did not reveal a direct effect of mortality on reproductive timing (Müller and Schmid-Hempel 1992b) nor of environmental quality on (immune) defense (Schmid-Hempel and Schmid-Hempel 1998). This issue thus remains clouded but clearly merits further study.

As was shown in chapter 3, few parasites are exclusively specialized on sexuals of social insects. However, the timing of reproductive events should be important in relation to parasitism, too. Massive swarming, such as found in desert seed-harvester ants or termites, are known to attract predators. Among parasites, a number of dipteran flies qualify as associates of mating periods. For example, *Menozziola* (Phoridae) is found in the abdomen of *Camponotus* queens (Brown et al. 1991), and *Pseudaceton* spp. are known to hover over ant nests during mating flights (Wojcik 1990; Pesquero et al. 1993). Also from this point of view, the timing of reproduction should be important to reduce parasite impact. Nevertheless, I must conclude by saying that the extent to which social insects have adapted their life history and reproductive behavior to parasites is—unfortunately—not known, nor has it really been formulated as a research question. The prospects are brighter in the context of social parasites, but they have not been the focus here.

This chapter has illuminated a number of approaches that can lead to a better understanding of the evolutionary ecology of resistance in social insects. To date, none of these ideas can be tested rigorously, since the necessary data have not been collected. As for any model, the considerations presented here will become replaced by better ones later. However, if it stimulates research on the comparative analysis of resistance strategies in social insects, physiological or life history variants, it will have served its purpose.

7.5 Summary

An important key to host-parasite interaction is variation in host resistance and parasite virulence. Resistance in social insects can involve a number of behavioral and physiological defenses. The former can be illustrated by classical studies in the behavioral genetics of American foulbrood in the honeybee. The latter comprise hemoral and cellular components of the immune system that are known to be mounted against all major parasite groups that attack social insects. Although some of the underlying processes and proteins are similar to those of vertebrates, an immunological memory is virtually absent in insects. Hence, long-lasting protection against infections does not occur. Typically also, resistance levels show the genotypic variation of quantitative traits, and thus simple gene-for-gene interactions are not known. Heritability has been exploited by breeders to improve lineages of, for example, honeybees. Also, a number of nongenetic factors contribute to host resistance. These include stress induced by adverse conditions, age, nutrition, or interaction among parasites. Defensive secretions have been identified from a number of glands, mostly in ants and bees. Some of the components have probably evolved to combat soil pathogens (e.g., myrmicacin in ants) or to weed out unwanted fungi in leaf-cutting ants and termites. Little is known about the fitness costs of defense, but costs of encapsulation against parasitoids and resistance against foulbrood bacteria have been demonstrated for both.

Parasite virulence can be considered an adaptive trait that increases fitness of the parasite through its interaction with transmission and/or within-host multiplication. A large body of theory on this subject has accumulated in recent years. Quite generally, these theories suggest that parasite virulence should be most strongly affected by the rates of horizontal and vertical transmission, by population structure, and by the presence of multiple infections. A strongly partitioned population structure in the form of colonies, coupling of horizontal and vertical transmission, and multiple infections is typical for many social insect parasites. This should select for generally low levels of virulence in host species with large-sized colonies, and for species where horizontal transmission is common. It should select for high levels where multiple infections are common. A survey of the known cases remains ambiguous; some of the expected correlates can be confirmed, but many exceptions still exist. In addition, the relevance of genetic variation within and among colonies for parasite virulence, and indeed host specificity, remains to be explored. Nevertheless, explicit expectations can now be formulated for most situations and will hopefully be tested in the near future. Several expectations can also be formulated for the level of resistance that social insects should adopt. For example, good habitats of high resource availability should select for the production of many but poorly defended workers, while in habitats with predictable parasitism, defense should be stepped up. Finally, parasitism may be linked to variation in life history traits, as repeatedly alluded to in other chapters. In particular, parasites should pose a problem for long-lived iteroparous species, while temporarily social species, those with annual cycles and semelparous ones, can shut down their operations and thus lose some parasites every now and then. Although there are many reasons to expect that the scenarios discussed in this chapter are relevant, corresponding empirical data or tests are sadly lacking.

8 Social Evolution

It is almost impossible to write a book on social insects and not mention the evolution of sociality. This issue has been somewhat overemphasized, to the disadvantage of ecological issues. Hence, it should suffice to briefly treat two aspects that are particularly relevant to our topic: how parasites may be involved in the evolution of social behaviors, and how kin recognition relates to parasitism.

8.1 Parasites and the Evolution of Sociality

Parasites and Origins of Sociality

Sociality is a complex phenomenon (table 1.1) and has arisen independently many times in different phylogenetic lineages. One scenario by Alexander (1974) and Michener and Brothers (1974) suggested that parents may be able to manipulate the behavior of their offspring and force them to stay home and help; this eventually leads to sociality. A second scenario by Lin and Michener (1974) and Itô (1993a), for wasps in particular, proposed that aggregations of nesting females coupled with mutualistic interactions will lead toward sociality. These scenarios also characterize two different routes to sociality, the subsocial and parasocial route, respectively (see table 1.1). The third major idea is the benefit of helping relatives, i.e., kin selection (Hamilton 1964). One must keep in mind, however, the distinction between the causes for the evolution of sociality and the actual pathways taken. Additional factors include nest site limitation and small chances to raise a brood independently (Brockmann 1984; Seger 1993). From the taxonomic distribution of eusociality and the habitat characteristics, it is evident that protected nest sites and the defense against enemies are important ingredients in the evolution toward sociality (Seger and Moran 1996). The well-known concept of inclusive fitness (Hamilton 1964) is a powerful tool to explain the evolution toward sociality from first principles, although some authors express doubts about its value in explaining eusociality in Hymenoptera (e.g., Bourke and Franks 1995). It predicts that a rare allele coding for an altruistic behavior (i.e., a social behavior) will spread in a population if $rb - c > 0$, where $r =$ the degree of relatedness of recipient to donor, $b =$ the benefit to the recipient, and $c =$ the cost of the act to the donor. This condition

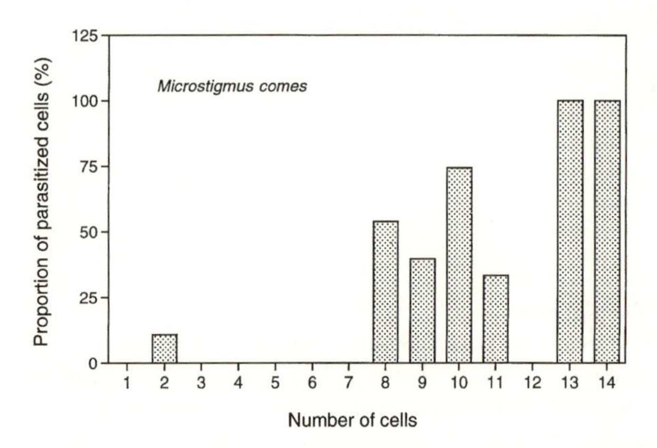

Figure 8.1 Percentage of nest cells of the social sphecid wasp *Microstigmus comes* parasitized by the braconid parasite *Heterospilus microstigmi. M. comes* is a social wasp with 3–8 females, of which one is the laying reproductive. (Reprinted from Matthews 1991. Used by permission of Cornell University Press.)

has been intensively discussed, criticized, and extended, but is found to be robust against additional assumptions and refinements (e.g., Charlesworth 1978; Taylor 1988). However, an important element is that altruistic behavior is not indiscriminately directed toward both sexes alike (Grafen 1986). Seger (1983) pointed out that certain partially bivoltine life cycles—as found in halictid bees, for example—select for different sex ratio biases in successive, overlapping generations. This can favor strategies of staying with the mother and help to raise closely related sisters. In addition, the existence of unmated females in the population does promote evolution toward eusociality due to the resulting shifts in the sex ratio produced by different females (Godfray and Grafen 1988).

In the present context, Hamilton's rule is useful because it focuses attention on two main aspects: on one hand, the propagation of copies of genes identical by descent (reflected in the value of r); and on the other, the fact that benefits and costs are embedded in an ecological context, affecting the terms c and b. The first part helps to explain why true eusociality evolved so many times in the Hymenoptera (perhaps more often than in any other lineage), i.e., because of the unusually close genetic relatedness among sisters. The second part explains why social behavior may be favored in some environments but not in others. In particular, the value of communal nest foundation or cooperative breeding is affected by the presence of predators and nest parasites (Lin 1964; Lin and Michener 1974; Strassmann 1981b). For example, figure 8.1 shows that larger nests of social wasps suffer more from nest parasites than small nests. Hence, defense against parasites is paramount for any cooperatively breeding group. In fact, the brood is quite vulnerable if not guarded. For example, in solitary and presocial wasps, many parasitoids are serious threats to the brood. This threat may be important in the evolution toward sociality (Cowan 1991). Of course, predation is also a major mortality factor for these wasps where defense by large nests is also more effective (Matthews 1991). Gadagkar (1991) gives detailed accounts of the importance of predators for the nesting behavior of social wasps in the genera *Belonogaster, Mischocyttarus, Parapolybia,* and *Ropalidia.*

Nevertheless, parasites have played an important role in the discussion. For example, West-Eberhard (1975) suggested that individuals with reduced fecundity profit more from adopting a helper role than by breeding independently (the "subfertility-hypothesis"). The same suggestion was later extended by Craig (1983) by assuming that an altruistic gene is expressed only in subfertile females. Hence, the cost of altruism would be small and the gene more likely to spread. Also, Godfray and Grafen (1988) implicitly assume some sort of subfertility in the sense that unmated females cannot produce both sexes of offspring. In all of these analyses, subfertility has been shown to facilitate the evolution of social behavior. Obviously, an important cause of subfertility is parasitic infection. For example, Shykoff and Schmid-Hempel (1992) found that a protozoan infection reduced the development of worker ovaries and thus made bee colonies more "social." Similarly, infections such as those by *Wolbachia* and other rickettsias (see chaps. 2 and 3) that cause cytoplasmic incompatibility, or sex ratio distortions, produce a phenotype of unmatedness and cause subfertility. Hence, subfertile females may rather help their sisters than breed on their own. Such infectious agents appear to be widespread, probably also in extant social insects (Werren et al. 1995; appendix 2).

Similarly, parasites have been reported to affect host behavior so that proximity to conspecifics is promoted. For example, ungulates infected by the Rinderpest virus suffer from increased thirst. Infected animals therefore seek to aggregate at water holes more often. This increases contact with other ungulates, i.e., with potential hosts (Plowright 1982). Also, many behavioral changes have been found in social insects (table 3.3). It is therefore quite possible that parasitic infections make the hosts more "social" or more prone to stay near nests or in dense aggregations.

Group living and association with closely related individuals is typical for sociality in almost all animals. As discussed earlier (chap. 4), larger groups often have more parasites (cf. fig. 4.10). In addition, parasites are more likely to be transmitted among closely related individuals (chap. 5). When comparing the two major alternative routes to sociality, subsocial and parasocial, there are interesting differences from the parasitological point of view. With subsociality, the offspring stays at home. Thus, a genetically rather homogeneous group is generated where parasites can be transmitted vertically from parent to offspring. If parasites can affect this, they should be suspected to manipulate their host's behavior in favor of staying with relatives, hence, in favor of this route to eusociality. With parasociality, basically unrelated (or less related) individuals gather together and can bring their parasites with them. In addition, with subsociality parasites should generally evolve toward lower levels of virulence because of the relative importance of vertical transmission. With parasociality, higher levels are expected as well as a more diverse parasite fauna to begin with. Obviously, it is not necessary that these characteristics persist over the subsequent evolutionary steps that lead to the full-fledged social species we can observe today. Nevertheless, if they are host-parasite associations that go back in deep time, then such associations may well be affected by how sociality evolved in the first place. With the progress in the analysis of phylogenetic relationships and with an increasing knowledge of host-parasite associations and their history, such scenarios are likely to become testable in the future.

Some Conflicts Reconsidered

Empirical data suggest that in the queen-worker conflict over sex allocation to offspring (Trivers and Hare 1976; Nonacs 1986; Boomsma 1991; Crozier and Pamilo 1996), the workers are typically in control. With a decrease in relatedness asymmetry (the ratio of relatedness to females : relatedness to males within the colony, weighted as life-for-life values) among workers in the colony (e.g., when the colony is more polyandrous), more resources are invested into the production of sons relative to daughters (e.g., Sundström 1994; Sundström et al. 1996). This explanation is based on kin-selected benefits accrued from producing more closely related individuals. But similar predictions would result if one assumes that parasites directly affect sex and offspring production.

Recall that a parasite needs to be transmitted to new hosts. This can occur by way of horizontal transmission, i.e., by the spread of infective stages to other colonies in a local population. In social insect species with overlapping generations, this process should take care of the parasite's problem: as it is transmitted among colonies, the parasite maintains itself within successive generations of the host. A somewhat different picture emerges when annual species of social insects are considered. In this case, the parasite can only persist in the population if it is transmitted to the next generation of hosts across the diapause. This could be achieved by the formation of durable stages that are able to infect the next generation. An example are the spores of *Paenibacillus larvae* which stay infective for years (Haseman 1961), although their host, the honeybee, is perennial but migrates to new nest sites. Such a pathway should be particularly important where nest sites are themselves limited, a constraint that is associated with polygyny (e.g., Herbers 1986).

Alternatively, the parasite could infect those individuals that survive during the diapause. In temperate climates and for hymenopteran social insects, this is typically the queen. It may sometimes also include males in species where both sexes overwinter as adults (e.g., in halictid bees). In the situation where the parasite can only survive in the queen, there should be an obvious advantage for the parasite to bias the sex ratio in favor of the sex that it uses as a vehicle. So, a shift toward females is predicted for many such situations. For example, in the case of the trypanosome *Crithidia bombi,* which infects bumblebees, workers are less likely to lay eggs and produce sons and this effect should, in principle, bias the colony-level sex ratio toward females (Shykoff and Schmid-Hempel 1992). In addition, a sex ratio shift should be relatively more favored in monogynous, monandrous colonies, because the produced daughters are genetically more similar and hence more susceptible to the parasite already established in the colony. Parasites in polygynous, polyandrous colonies would find themselves in a more difficult situation, since infection levels in such variable colonies may be lower to begin with. We may suspect that the parasite effect is correspondingly lower and that the sex ratio bias toward males preferred by the workers is more likely to persist. Therefore, the parasite-induced sex ratio bias would qualitatively be the same as with the split-sex-ratio hypothesis (Boomsma and Grafen 1991). Admittedly, with the current knowledge, the split-sex-ratio hypothesis is considerably more likely. Nevertheless, the patterns should be examined more carefully with the parasite perspective in mind as well, particularly in annual bees and wasps that construct their own nests (and are thus less likely to contract spores from the nest cavity) and have a strict diapause. Moreover, as parasite and worker interests in this respect are to a large degree the same, each party is probably even more likely to succeed, although for different reasons and with different benefit-to-cost ratios.

Sherman and Shellman-Reeve (1994) proposed that sex ratios produced by colonies may be affected by parasites if dispersal between sexes is different. The authors considered their idea an alternative to the split-sex ratio hypothesis of Boomsma and Grafen (1991), i.e., as a strategy for the benefit of the host (as Boomsma and Grafen 1991 discuss, this does not hold steady under closer scrutiny). But it may actually also be relevant in the context of how dispersal could be selected for by parasites in general. For example, males of ants typically disperse farther than females. Hence, locally successful parasites should prefer females. Obviously, this could only work if both sexes would in principle allow dispersal for the parasite, i.e., when the parasite can get from a male to a founding queen, otherwise males have no value, and no meaningful variation in sex ratio can be expected. There may actually be some cases where this requirement is met. Among these are sexually transmitted diseases, mites that can transfer to the female during copulation of the host, or the termites where both sexes are long-lived and functionally integrated into the colony.

In many cases, unmated workers in hymenopteran societies do reproduce, leading to male offspring (Bourke 1988; Crozier and Pamilo 1996, their table 4.3). But a fundamental conflict in insect societies is between the workers and the queen(s) over opportunities to reproduce. Sociality can thus been seen as having evolved and being maintained by carefully balanced "stay" and "peace" incentives (e.g., Keller and Reeve 1994b). With the current perspective, it is tempting to attribute a role to parasites. For example, subfertility caused by parasitic infection, as in the case of the *Crithidia* mentioned above, obviously changes the balance of such stay and peace incentives. In addition, only queens found new colonies, ignoring the cases of budding and fission. Thus any parasite causing its host worker to reproduce could only gain by being transmitted vertically to male offspring and then during mating to a founding queen. But sexually transmitted diseases appear to be rare in social insects (chap. 3). As far as worker reproduction in eusocial societies goes, parasites should therefore have quite similar interests to those of the queen. Given the fact that research into the causes of the evolution and maintenance of sociality is very active, progress in this particular area is thus likely to be fast.

8.2 Kin Recognition and Parasites

Kin recognition is important in the genetic theory of social evolution (Hamilton 1964). But not only hymenopteran societies are characterized by intrinsically high degrees of relatedness. In the termites as well, where sex is not determined by the haplo-diploid system, close genetic relatedness can result from cyclical inbreeding or genetic linkage (sex-linked translocations) within the genome (e.g., Bartz 1979; Luykx 1985). Theory predicts that altruistic behavior should favor individuals that share genes due to a common descent, i.e., relatives. With the introduction of kin selection theory, therefore, the problem of kin recognition arises too. As the discussion of genetics in chapter 5 suggested, kin recognition should also be important with respect to parasitism. Here, it is necessary first to discuss the current knowledge on kin selection before turning to parasitism.

The core function of kin recognition is a differential treatment of conspecifics based on genetic relatedness. While mating may occur whenever individuals in the right condition and of the opposite sex meet, kin discrimination among conspecifics in the above sense is intimately associated with social behavior. In fact, almost all social interactions such as food sharing, care for dependent young, dominance behavior, group defense, and so forth include a component of discrimination, although it may not always be based on genetic relatedness alone. In our discussion it is furthermore important to keep apart kin "recognition," i.e., the ability to recognize kin, and "discrimination," i.e., the visibly expressed differential treatment of individuals (Waldman 1987). The obvious reason is that discrimination may not be expressed in every situation, although the capacity to recognize different levels of relatedness could exist.

Research over the past decades has generated evidence for the existence of kin recognition in a wide variety of taxa, ranging from hydrozoa to mammals (table 8.1). In its most basic form, mate selection in nonsocial species has been considered a problem of kin recognition. The evidence is usually based on the observation that pairings are not random with respect to relatedness (Spiess 1987). Such nonrandom matings are common in many species and reveal the tendency to prefer or avoid a mate that is genetically close (Pusey and Wolf

Table 8.1

Examples of Kin Recognition (Definitions Used in a Broad Sense)

Species	*Finding*	*Type**	*References*
INVERTEBRATES			
Myxomycetes (slime molds)	Cell fusion according to genotype ("kin").	C	Carlile 1973
Sponges, sea anemones, corals	Recognition of foreign vs. self tissue ("kin").	C	Lubbock 1980; Neigel and Avise 1983
Cryptocercus punctulatus (cockroach)	Family recognition.	C	Seelinger and Seelinger 1983
Hemilepistus reaumuri (desert wood louse)	Family recognition.	G	Linsenmair 1985
SOCIAL INSECTS			
Lasioglossum zephyrum	Guard bees recognize sibships differing in degree of relatedness.	C	Greenberg 1979
Apis mellifera	Full sisters are fed more.	C	Breed et al. 1985
Polistes fuscatus	Cofounding queens recognize their sibling sisters.	C	Pfennig et al. 1983
Catagylphis cursor	Related brood is tended more.	C	Lenoir 1984
Coptotermes acinaciformis	Nest vs. non-nest mates.	G	Howick and Creffield 1980
VERTEBRATES (CHORDATES)			
Botryllus schlosseri (Ascidian)	Incompatibility among clones, based on MHC-like system.	C, I?	Scofield et al. 1982
Rana cascadae (Cascade frog), *Bufo americanus* (American toad)	Tadpoles recognize their full vs. half-sibs.	C	Waldman 1981; Blaustein and O'Hara 1982
Riparia riparia (Bank swallow)	Parents recognize chick by calls.	I	Beecher et al. 1981a
Spermophilus parryii, Spermophilus beldingi (ground squirrels)	Sibling recognition by familiarity.	C	Holmes and Sherman 1982
Cercopithecus aethiops (vervet monkey)	Recognition by vocalizations.	I	Cheney and Seyfarth 1980

*Types of recognition: G = recognition of own vs. alien group member; C = recognition of closely vs. distantly related class; I = recognition of individuals.

1996). More sophisticated cases of kin-directed helping behavior are known from higher taxa (table 8.1). Almost all authors thus consider kin recognition as an adaptive trait and a mechanism by which inclusive fitness is increased or an optimal level of outbreeding is ensured.

The actual mechanisms by which kin recognition is achieved varies among taxa, and may reach from simple familiarity to the detection of specific odor signals that have a heritable basis (e.g., Fletcher and Michener 1987). For example, bank swallows use spatial association of nests to infer relatedness (Beecher et al. 1981a,b). There is little evidence for this mechanism in social insects, although in the wasp *Polistes fuscatus* spatial cues are indeed used (Klahn 1979). With allelic recognition, it is assumed that genes code for cues that signal relatedness and, at the same time, for the mechanism needed to recognize these cues. In its clearest form, it has been suggested that a "green beard" may indicate whether its bearer possesses some particular alleles (Dawkins 1982). The evolution of an allelic recognition system has to overcome a positive frequency-dependent selection regime that selects against rare bearers of the trait (Crozier 1987; see also Mintzer 1989). As Grosberg and Quinn (1988) and Grafen (1990) have shown, the occasional occurrence of nondiscriminating mutants is crucial for the spread of such allelic systems.

A third alternative is to learn the identity of individuals in the group to which one is associated (Waldman 1987). This is a variant of the idea of phenotype matching. In this model, recognition is according to perceived cues. These are compared with a template against which individuals match others, and discrimination would be based on the degree of overlap. Two basic scenarios are possible (Crozier and Dix 1979): genotype matching could be based on an individualistic model, where recognition is associated with a set of common alleles or other characteristics linked to particular individuals. Alternatively, it could be based on a *gestalt* model, in which recognition would be associated with a mixture of cues, originating from a set of individuals. In both cases, phenotype matching leaves open the question from where cues originate, e.g., whether they are environmental or genetic in origin.

Kin Recognition in Social Insects

Kin recognition in social insects has been studied almost exclusively in the social Hymenoptera. Taxonomically, the ability to discriminate sibs from nonsibs is known for a number of primitively social species, for instance halictid bees (Kukuk et al. 1977; Greenberg 1979; Smith and Ayasse 1987) and wasps (Gamboa et al. 1986). In the sweat bee *Lasioglossum zephyrum,* discrimination by nest guards against intruders follows genetic relatedness (Greenberg 1979). In the wasp *Polistes,* nestmates may overwinter together, but discrimination is not based on this characteristic alone (Post and Jeanne 1982). Kin discrimination has also been reported from other subsocial arthropods, such as the sphecid wasps *Cerceris* (McCorquodale 1989) and *Sphecius* (Pfennig and Reeve 1989). In these communally nesting species, individuals tend to be more aggressive toward non-neighbors. This suggests that this capacity was already present in the ancestors of the extant social Hymenoptera.

It is important to distinguish "between-colony recognition" (nestmate recognition) from "within-colony recognition" (recognition of closely vs. distantly related kin). The latter is the major controversial issue in kin recognition of social insects. In fact, discrimination against individuals from other colonies or against individuals with low relatedness has been found in many contexts in the highly advanced social insects, e.g., in vespine wasps (*Dolichovespula:* Ryan et al. 1985; Michener and Smith 1987). This is less obvious for the termites, where little is known. In the genera *Reticulitermes, Nasutitermes,* and *Coptotermes* (Breed and Bennett 1987), at least, non-nestmates are discriminated against. In ants, it is typically found that nestmates are treated differently from non-nestmates (e.g., Hölldobler and Wilson 1990). Again the honeybee has served as the model organism for social insects. These studies have encompassed contexts such as colony defense, swarming, and brood care. In general, discrimination in social insects seems to be associated with monogyny but less so with polygyny, polydomy, and with patchy or ephemeral habitats (Hölldobler and Wilson 1990; Brian 1983; Frumhoff and Ward 1992). Breed and Bennett (1987) point out that polygyny is associated with polydomy as well as with patchy or ephemeral habitats (Frumhoff and Ward 1992). Hence, variation in the ability to discriminate among nestmates within a colony may also be dependent on environmentally determined selective pressures on queen numbers rather than polygyny itself.

In terms of the actual mechanisms involved, the available evidence suggests that phenotype matching, chemical cues, and learning shortly after eclosion from the pupa (such metamorphosis is characteristic for the holometabolic Hymenoptera but not for the hemimetabolic termites) are all important. Learning of nestmate identities has been documented in the honeybee, where it occurs shortly after eclosion (Masson and Arnold 1984). Genetically determined markers are suspected in a number of cases. For example, a genetic marker could be displayed as the recognition signal, while the template on the receiving end is based on learning. Such systems have been found in bees (Greenberg 1979; Buckle and Greenberg 1981; Breed 1983) and in many species of ants (Haskins and Haskins 1983; Jutsum et al. 1979; Mintzer and Vinson 1985; Stuart 1987a; Wallis 1962). The genetic effects appear generally to override the environmental cues (Wallis 1962; Jutsum et al. 1979). However, additional environmental effects on the markers, e.g., diet, have been found in almost every case listed above and are known in many social insects (Breed and Bennett 1987), including the honeybee (Kalmus and Ribbands 1952; Boch and Morse 1981), some wasps (e.g., *Polistes:* Gamboa et al. 1986), and in many ant species (Jutsum et al. 1979; Mabelis 1979; Haskins and Haskins 1983; Stuart 1987a), particularly the fire ant *Solenopsis* (Obin 1986; Obin and Vander Meer 1988, 1989). Sole reliance on external (i.e., nongenetic) cues is likely to carry a higher risk of false-positive identification among neighboring colonies, which should be discriminated against otherwise.

Whereas the foregoing examples have referred to discrimination among workers, discriminators originating from the queen can be involved, too. This has been found in the honeybee (Breed 1981; Boch and Morse 1982), may occur in wasps (*Dolichovespula:* Ryan et al. 1985), and perhaps in some ants of the genera *Camponotus, Leptothorax,* and *Myrmica* (Watkins and Cole 1966; Jouvenaz et al. 1974; Brian 1986; Carlin 1986; Carlin and Hölldobler 1987, 1988; Provost 1985, 1987). However, no queen contribution to discrimination could be established in *Pseudomyrmex* (Mintzer 1982), *Solenopsis* (Obin and Vander Meer 1988, 1989), and some ants of the genus *Leptothorax* (Stuart 1987a,b,c). Queen discriminators are therefore not the general rule. Discrimination based on the queen alone would be compatible with the individualistic model discussed above. In contrast, a Gestalt effect would rely on the blending of the cues, e.g., odors from all members of the colony, as in the ant *Leptothorax curvispinosus* (Stuart 1987a,b,c) and *Camponotus* (Carlin and Hölldobler 1986, 1987).

It is also important to consider in which contexts these discriminations occur, since selection can only act through the consequences of the differential treatment of individuals. For example, many of the tests for recognition abilities have been carried out in the context of nest defense. In the honeybee, workers may also be able to use genetically based differences in the context of swarming (Getz et al. 1982), aggressive interactions (Getz et al. 1986; Getz and Smith 1983; Breed et al. 1985), oviposition by workers in queenless colonies (Evers and Seeley 1986), during exchange of food, grooming (Frumhoff and Schneider 1987), or during queen rearing (Page and Erickson 1984; Visscher 1986; Noonan 1986). However, many of these studies have been criticized on various grounds and thus should be accepted with caution. Antennating in *Camponotus* ants also seems to vary according to kinship (Carlin et al. 1987). In some species it is known that individual ants may follow individually different odor trails (Maschwitz et al. 1986).

Nevertheless, discrimination among brood by the workers is generally less prominent than among adults (Breed and Bennett 1987). Brood discrimination is known from ants (*Iridomyrmex:* Hölldobler and Carlin 1985) and honeybees (e.g.,Visscher 1986). Any bias in this respect should have straightforward fitness consequences in favor of the discriminating genetic lines. But strangely enough, a foreign but conspecific brood seems to be recognized in many ants (Hölldobler and Wilson 1990, their table 5.2), yet is typically tolerated. In the case of *Formica,* discrimination against brood is associated with geographic distance (Rosengren and Chérix 1981). This was also found in *Leptothorax* (Stuart 1987c), as an additional element of the *gestalt* effect. Discrimination gradually disappears when the ants are kept in the laboratory under identical conditions for some time. This is in fact rather typical under many other circumstances. Also, queen acceptance into colonies of *Myrmica rubra* also depends on distance, so that acceptance is more likely for queens coming from nearby nests (Evesham and Cammaerts 1984). It is possible that queen odor is an important component in small but not in large colonies, or that it is important in "core areas" around the queen but not at the nest periphery. In very large colonies formed by some species of ants (e.g., *Pheidole, Formica, Wassmannia*), the cues may be so variable that the system breaks down entirely. This allows the formation of supercolonies that extend over large areas and comprise a very large number of workers.

Only in a few cases is it known what the precise nature of the cues involved in discrimination might be. Evidence points to cuticular hydrocarbons as the source of nestmate identification (e.g., Hölldobler and Wilson 1990). This fact is exploited, for instance, by a myrmecophilous beetle that gains entrance to ant nests by mimicking the hydrocarbons of its hosts (Vander Meer and Wojcik 1982). Glandular sources of chemical signatures have been identified in the mandibular glands of the carpenter ant *Camponotus rufipes* (Jaffe and Puche 1984; Bonavita-Cougourdan et al. 1987). Genetic components seem involved in these kinds of signatures (e.g., Haskins and Haskins 1979; Mintzer and Vinson 1985). In the well-studied *Camponotus*, a hierarchical system of recognition cues has been unraveled, were cues derived from fertile queens are most important, followed by worker discriminators and environmental odors, depending, among other things, on colony size (Carlin and Hölldobler 1986, 1987; Carlin et al. 1987).

It is therefore quite clear that members of social insect colonies can discriminate against non-kin, and, in particular, against non-nestmates in a variety of contexts. All tests have been based on behavioral interactions, so nothing is known about other discrimination processes. The nature of the signatures and cues that allow discrimination is not well known but they generally seem to be odor cues. These cues are likely to be based on genotype and to be learned by the "receiving system" as a template used in the recognition. It is also possible that queens dominate the cues in small colonies but not in large ones. On the other hand, the ability to discriminate kin can have serious drawbacks, for it allows nepotism within polyandrous and polygynous colonies. This may lead to a loss of overall colony efficiency and reproductive output (Hamilton 1972, 1987; Schmid-Hempel 1990). Hence, kin recognition is both a bane and a boon.

Parasites and Kin Recognition

In social insects, the paramount problem must be to avoid or eliminate an infection for the colony as a whole. This is also true for polyandrous or polygynous colonies. Few if any parasites will only be infectious or virulent for just one particular worker patriline or matriline within a colony (see appendix 2 for the variety of hosts used by many parasites). Social groups thus have a problem; once a parasite breaches the group barrier it may rapidly infect its members. In addition, many dangerous parasites are spread by the proximity of individuals or by ingestion of contaminated food in the colony (chap. 3). Therefore, in terms of parasite avoidance, the recognition of close kin vs. more distantly related individuals within a colony is not nearly as effective as discrimination against non-colony members. Infectious parasites should thus select for discrimination between colonies rather than for discrimination within a colony. Different genetic lines within the colony should even be tolerated under certain ecological conditions because this decreases the likelihood that a parasite can spread within the group, as with the arguments in chapter 5.

As briefly outlined above, many studies have indeed failed to show discrimination within the colony in several important contexts (e.g., Carlin et al. 1993). This failure has, for example, been attributed to the difficulty of distinguishing many different lines at the same time (in honeybees: Hogendoorn and Velthuis 1988, but see Crosland 1990). In contrast to within-colony kin discrimination, discrimination against non-colony members is the rule and typically overrides the former (see Michener and Smith 1987; Breed and Bennett 1987; Breed et al. 1988; Carlin et al. 1993; Peeters 1993). This is in line with the parasite hypothesis. We should furthermore expect that exclusion of non-colony members is particularly stringent when parasitism is a major problem. This could apply to cases where the cost of being infected is high, as for example for young colonies where novel infections are potentially more devastating. Hence, small colonies should be more prone to exclude non-nestmates. At least qualitatively, such a difference between large and small colonies is observed in ants (see Hölldobler and Wilson 1990). But parasites also pose a problem where the force of infection is high, such as under high population densities (i.e., many colonies per square meter). Similarly, the incidence of novel infections should be higher in large as compared to small colonies because more foragers circulate in the area and can acquire parasites. This allometry of discrimination in relation to colony size and parasitism certainly deserves some attention. More discrimination between colonies with increasing geographical distance is also known (e.g., queen acceptance in *Myrmica:* Evesham and Cammaerts 1984). This would be expected if novel infections acquired by carriers from distant sites are more problematic. Quite typically, though, it seems that local parasites are potentially more virulent (e.g., Ebert 1994b), although important counterarguments exist (see Morand et al. 1996 for a discussion).

Benest (1976) showed that workers of bumblebees and honeybees at artificial food sources were more aggressive to non-sisters than to sisters. On the one hand, this is an instance of simple resource competition; but on the other, displacement of others from food sources could be favored by selection if alien workers were likely to be carriers of parasites (Durrer and Schmid-Hempel

1994). Also, territoriality or defense of food resources by social insects may have evolved in the context of containing infection risks. This reasoning has been used by Freeland (1976, 1979) to discuss the evolution of many characteristics of primate social organization, e.g., xenophobic behavior toward foreigners, choice of sleeping sites and feeding areas, territories, and so on. These are exactly the same problems that also social insects face.

Distinguishing kin from non-kin is also similar in kind to the problem of distinguishing self from non-self in immune defense (Coombe et al. 1984). This capacity of organisms is central in evading parasitism and to make a living in the face of a variety of parasitic organisms. As mentioned in chapter 7, such immune functions in insects are phylogenetically ancient (Beck and Habicht 1996). The involved biochemical pathways are not well understood, however. Therefore, one can speculate whether the pathways involved in immune defense overlap with those involved in producing recognition markers, similar to the role of certain key substances in vertebrates (e.g., Folstad and Karter 1992). If so, then kin recognition would directly be under selection by defense against parasites.

It would leave a wrong impression to conclude that all the characteristics of social insects that have fascinated us for so long are due to coevolution with parasites. Parasitism is certainly one factor among others. However, it is true that compared to the potential importance of parasites, our knowledge is rather limited. It remains to be seen which characteristics of social insects are tied to parasitism. This will inevitably not only deepen our understanding of this fascinating group of organisms but also provide insight into the intricacies of the most successful lifestyles in our living world—sociality and parasitism.

8.3. Summary

Parasites may be prominently involved in a number of ways in all major hypotheses about the evolution of social behavior and eusociality. Communal nesting may, for example, be selected for by defense against nest parasites. Similarly, subfertility predisposes females to join others, either as hopeful reproductives or via the population-wide selection on sex ratio of offspring. In many instances, parasitic infection does in fact lead to subfertility and thus may have been instrumental in promoting social evolution. Sex ratio shifts may also be directly selected for by parasites because of the problem of further transmission via one particular sex, e.g., the hibernating queens in annual bees and wasps. Parasite-induced behavioral changes also sometimes lead to more frequent or closer contacts in groups of individuals, which is a prerequisite for social interactions. Because host-parasite interactions are intimately linked with genotypes, the genetical theory of eusociality suggests that parasitism should also have a role to play, although clear concepts and evidence for this problem are still lacking. Parasites could also stabilize the maintenance of sociality by causing subfertility in workers and therefore affect the balance of stay and peace incentives in favor of cooperation.

Social behavior is to a large degree tied to the problem of interacting with relatives. This raises the issue of kin recognition and discrimination. Kin recognition in social insects prominently involves genetic cues, learning, and odors. Discrimination has been shown in a variety of contexts. However, discrimination within a colony among different matri- or patrilines has remained a controversial issue. On the other hand, nestmate recognition, i.e., discrimination between colonies, is almost universally developed. This difference makes sense in the light of parasitism, as it should be more important to prevent that novel parasites are brought into the colony in the first place. Because kin discrimination is similar in kind to the basic immunological problem of distinguishing self from non-self, intriguing questions arise that may be analyzed in future studies.

Appendix 1
Glossary

Absconding. The more or less regular occurrence of a colony leaving the nest site, especially in bees.

Age-dependent polyethism. The orderly change of tasks that workers of a colony perform as they age.

Allele. An alternative gene that can occupy the same locus.

Allozyme, isozyme. An alternative enzyme variant. Detectable by electrophoretic methods.

Antigen. A foreign, non-self object that is typically able to induce the immune defense.

Breeding system. Characterizes the number of males a queen has mated with and is using sperm from, as well as the number of cooperatively breeding females and their genetic relationship with one another.

Budding. Colony multiplication by the departure of a (relatively small) part of the colony of workers with accompanying queen(s).

Caste. A set of individuals that perform a subset of tasks in the colony. Characterized by age class and/or morphology.

Cellular defense. A major component of immune defense insects. Based on functionally different hemocytes (blood cells).

Division of labor. See Polyethism.

Drifting. The process whereby workers of a colony go to and enter foreign colonies rather than their own.

Drone. A male sexual form.

Effective population size. The hypothetical number of individuals that in genetic terms behave like an ideal population of fully outbreeding individuals of the same size in equilibrium. The effective population size is affected by the level of inbreeding, sex ratio of breeding individuals, fluctuation in population size, and a nonrandom distribution of offspring number over breeders.

Encapsulation. A process of cellular defense that leads to the aggregation of hemocytes around an antigen and eventual full enclosure. A major defensive reaction in insects.

Endemic. A state where the parasite is more or less permanently maintained in the host population. Characterized by a host-parasite dynamic that has reached an equilibrium.

Epidemic. A state where the parasite is able to invade a host population and typically to increase in frequency (outbreak of a disease). Characterized by a host-parasite dynamic that is not in equilibrium. More precisely, in animals this is called "epizootic."

Epidemiology. The study of epidemic and endemic host-parasite dynamics, of the origin and spread of diseases. In medicine, also the study of statistical patterns that indicate causal factors for the occurrence of a disease.

Ergonomics. The science of analyzing the efficiency of social organization in animals by means of economic principles.

Female. In social insects, the sexually competent female forms, e.g., queens, gynes.

Fission. Colony multiplication by splitting of the colony into one or more parts, each accompanied by at least one queen.

Force of infection. The per capita chance of a susceptible host to become infected per unit time.

Gynes. A sexually competent female, whether acting as a reproductive or not. Here, often used to designate the females before they become a queen.

Haplometrosis. The founding of a colony by a single queen.

Hemocoel. The body cavity in insects. Contains the freely circulating hemolymph.

Hemocyte. A blood cell. Hemocytes vary in morphology and function.

Hemolymph. The "blood" of insects. Contains humoral components (e.g., proteins) and cellular components (the hemocytes).

Homeostasis. The maintenance of a steady state of physiological or sociological parameters upon external disturbance.

Horizontal transmission. The transfer of a parasite to individuals in the population other than one's own offspring. In social insects, mostly reserved for transmission between different colonies.

Humoral defense. A major component of immune defense insects. Based on proteins, etc., dissolved in the hemolymph.

Immunity. The result of biochemical processes in the immune system to defend a host against a parasite.

Immunocompetence. A loosely defined term that indicates the potential of hosts to defend themselves against parasites.

Infectiousness. The ability of a parasite to invade a host and to establish itself. (infectivity).

Intensity of infection. The degree to which a host is infected. Often measured as the number of cells or correlates thereof (parasite titer), or number of parasite individuals per host (e.g., worm burden).

Iteroparous. Characterizing a life history where reproduction occurs repeatedly, e.g., over several years.

Karyotype. The genotype of an individual characterized by the number and types of chromosomes.

Kin recognition. The recognition of others according to the genetic relatedness by descent relative to the acting individual.

Latency. The time period between infection of the host and the appearance of symptoms (typically in medicine) or transmissible stages (in ecology).

Macroparasite. A "large" parasite that can be individually addressed. Typically, the analysis requires that individual parasites are followed with their growth or reproduction.

Mating system. Characterized by the mating pattern in terms of number of mates and selection among potential mates.

Microparasite. A "small" parasite that occurs in large numbers so that the analysis is simplified by considering the hosts rather than the parasites themselves, e.g., in terms of infected, susceptible hosts, etc.

Microsatellite. A short piece of repetitive DNA, mostly as repeats in duplets or triplets, that varies in length. Similar to allozymes used to distinguish alternative alleles in population genetic studies.

Monandry. A single father is present because the queen typically has mated just once. Functional monandry occurs if only the sperm of one male are used.

Monogyny. The presence of one functional queen in the colony.

Net reproductive rate (R_0). The per capita rate of increase of a population. In epidemiology, used as a fitness measure for the parasite.

Parasite load. A loosely defined term that characterizes the number and/or diversity of parasites in the hosts. Here, used as the number of parasite species present in the host.

Parasite richness. Here, defined as the number of parasite species (parasite load) per host species. Standardized parasite richness is corrected for the sampling effort for a given host species. A parasitoid typically kills its host toward the end of its development.

Per os (*by os*). Infection via the mouth, e.g., by ingestion of spores.

Pleometrosis. The founding of a colony by several queens in social Hymenoptera.

Polyandry. The presence of several fathers.

Polycalic. A polydomous colony that has tight interactions. Often same as polydomous.

Polydomy. The fact that a colony occupies more than one nest.

Polyethism. The division of labor among members of the colony.

Polygynandry. A breeding system whereby queens mate with several males and males in turn mate with several females.

Polygyny. The presence of more than one functional queen in the colony.

Polymorphism. In social insects, the presence of more than one worker morph in the colony.

Queen. The functional female in the colony.

Red queen. A hypothesis that explains the constant coevolutionary race between two parties (e.g., host and parasite) due to regularly changing selection pressure caused by the other party (e.g., negative frequency-dependent selection).

Refractoriness. Same as resistance.

Resiliency. In social insects, the ability of the colony to resist external disturbance and to return to the previous state. A process by which homeostasis is achieved.

Resistance. A loose term describing how easily a host becomes infected when exposed to a parasite.

Semelparous. Characterizing a life history where reproduction occurs only once.

Sexuals. In social insects, the sexually competent forms, i.e., males (drones, kings) and females (queens, gynes).

Social parasite. A parasite that exploits the breeding effort of its social host. In social insects, typically a form closely related to the host (Emery's rule).

Susceptible. Describing a host that can potentially become infected.

Tangled bank. A hypothesis explaining the advantage of variability among offspring by the need to avoid competition among them. In host-parasite terms, this should help to reduce the risk of losing offspring to parasites.

Taxon (taxa). A named group of organisms, e.g., species, genus, order.

Transmission. The transfer of a parasite from one host to the next.

Vector. A carrier of the parasite to new hosts. Typically, specialized parasite stages inhabit a vector.

Vertical transmission. The transfer of a parasite to offspring of its current host. In social insects, this is the transmission to offspring sexuals. Sometimes, within-colony transmission is considered a case of vertical transmission.

Virulence. A general term describing the adverse effect of a parasite on its host.

Appendix 2
The Parasites of Social Insects

This appendix lists the reported associations of parasites with social insects. The sections are ordered according to parasite group and social insect taxa. Many uncertainties exist about the exact relationship between host and parasite. This synopsis adopts a more conservative approach to classify parasites. Mites are somewhat exceptional, since their relationships to social insects are manifold and range from probably mutualistic to severely pathogenic. Appendixes 2.7 and 2.8 mention these relationships explicitly, at least where known. Social parasites are not a topic of this book and are not listed here.

The taxonomy of many parasites and host groups is not resolved. This appendix is not intended to provide a valid systematic account. Ambiguities and synonyms have been resolved and taken into account wherever possible, but otherwise the original author's usage is followed. In cases of uncertainty, the reader is referred to these sources. For the same reason, host and parasite species are listed alphabetically and not according to systematic position.

Appendix 2.1 Viruses

Viruses	Host	Remarks	Refs.
	BEES:		
APV(acute paralysis)	*Apis mellifera*, *Bombus* spp.	Varroa mites can "activate" or release virus in *Apis*. In *Bombus*, experimental infection.	21,95
BQCV(black queen cell), BVX, BVY(bee virus X,Y), CBPV (chronic bee paralysis), CBPV (CPBV-associate), CWV(cloudy wing), Deformed wing V, Egypt bee V and JEBV, FV(filamentous), SPV(slow paralysis), Arkansas bee V	*Apis mellifera*	BVX associated with *Malphigamoeba mellificae*, BVY with *Nosema apis*, CBPV with *Acarapis woodi*. Known as "Waldtrachtkrankheit" (syndrome 1) or "hairless black syndrome," "Schwarzsucht" (type 2). Deformed wing, SPV activated by *Varroa*.	16,18,38, 232,323
Entomopox-like virus	*Bombus* spp.	Uncertain in nature.	61
IV (Apis) iridescent virus	*Apis cerana*	Clustering disease.	20
Kashmir bee virus	*Apis cerana*, *A.mellifera* Stingless bees?	Multiplies quickly and kills host within 3 days when injected.	
SBV(sacbrood virus) TSBV(Thai sacbrood virus)	*Apis mellifera*, *A.cerana*	Strains are different in the two *Apis* species.	1,14,15, 18
	ANTS:		
Nonoccluded virus(NOB), virus-like particles	*Solenopsis* sp. *Formica lugubris* ?	Associated with microsporidian (*Thelohania*) infections. Also occur together with rickettsias and mycoplasms.	10,149, 302
	WASPS:		
CrPV(cricket paralysis virus)	*Vespula germanica*		117
CPV(cytoplasmic polyhedrosis), EPV(entomopox), NPV(nuclear polyhedrosis)	*Polistes hebraceus*	In fecal pellets of larvae (meconia).	224
Kashmir bee virus	*Vespula germanica*		117
	TERMITES:		
Paralysis-like virus	*Coptotermes*, *Nasutitermes*, *Porotermes*	Up to 60% of experimentally infected termite workers die within 4-5 days.	114

NOTE: Bees after ref.(17), additional references given.

Appendix 2.2 Bacteria and Mollicutes

Bacteria and Mollicutes	Host	Remarks	Refs.
	BEES:[a]		
Spiroplasma[b] *apis, S.melliferum*	*Apis mellifera, Bombus* spp.	Also in solitary bees. Found on flowers. In *Bombus* in hemolymph.	59,60,62, 227
Aerobacter cloaca	*Apis mellifera, Bombus*	In ovaries of queens. Causes B-melanosis.	106,298, 304
Bacillus alvei, B.laterosporus	*Apis mellifera*	Some are secondary invaders with *P.larvae* after years of endemic foulbrood.	13
Bacillus pulvifaciens	*Apis mellifera*	Causes "powdery scale" of larvae. Perhaps a saprophyte that occasionally infects larvae.	130
Bacterium eurydice	*Apis mellifera*	Secondary invader with *M.pluton.*	13
Hafnia alvei (= *Bacillus paratyphi alvei*)	*Apis mellifera*	Associated with infection by *Varroa* mites. Causes septicemia and death when in hemolymph.	
Melissococcus (*Streptococcus*) *pluton*	*Apis mellifera*	Causes European foulbrood. More benign than American foulbrood.	
Paenibacillus (*Bacillus*) *larvae*	*Apis mellifera*	Causes American foulbrood. Kills larvae after cocoon is spun.	
Pseudomonas aeriginosa, P.apiseptica	*Apis mellifera*	In hemolymph of moribund bees near hives. Also in soil.	53,186
Pseudomonas fluorescens, Yersinia pseudotuberculosis	*Apis mellifera*	From hemolymph of bees with "Schwarzsucht." Associated with chronic bee paralysis virus. In soil and water.	134,186
Serratia marcescens *Streptococcus fascialis*	*Apis mellifera*	A secondary invader with *P.larvae.*	
Nonidentified bacterium (gram-positive)	*Bombus melanopygus*	Dead larvae characteristically hard.	200
	ANTS:		
Pseudomonas Nonidentified bacterium	*Solenopsis invicta, S.richteri*	Found in dead ants. Shown to be highly toxic to fire ants.	152,191
	WASPS:		
Serratia marcescens	*Vespula germanica*	In surveys in New Zealand.	117
	TERMITES:		
Bacillus thuringiensis	*Bifiditermes beesoni*	Also pathogenic to other termite species. Transmitted by trophallaxis within colony.	168,169, 192

NOTES: (a) For bees based on ref.(17). (b) *Spiroplasma* is a member of the mollicutes.

Appendix 2.3 Fungi

Fungi	Host	Remarks	Refs.
	BEES:		
Acrostalagmus sp.	*Bombus* spp.	Diseased queens with short hibernation.	298
Ascosphaera alvei, A.apis, A.flavus, A.fumigatus	*Apis mellifera*	*A.apis* causes chalkbrood disease; *A.flavus, A.fumigatus* causes stonebrood of larvae.	17,124, 177,188
Aspergillus candidus, A.niger	*Bombus* spp., *Apis mellifera*	*A.niger* probably opportunistic infections.	17,198, 263
Beauveria bassiana, B.tenella	*Bombus* spp., *Apis mellifera*	From worker pupae in *Apis*.	188,198, 263
Candida pulcherrima, Candida sp. Various yeasts	*Apis mellifera, Bombus* spp.	Appears as a consequence of stress. Diseased queens with short hibernation.	116, 298
Cephalosporium	*Apis mellifera, Bombus* spp.	Causes typical discolorations. Serious effects in *Bombus*.	28,263
Chactophoma sp. *Cladosporium cladosporoides*	*Apis mellifera*	Causes typical discolorations. Also in combs.	28,263
Hirsutella sp. *Metarhizium aniospliae*	*Bombus* spp.	Mycel extends beyond host body.	198
Paecilomyces farinosus	Apidae, *Bombus* spp.	Pathogenic in *Bombus*.	188,198
Penicillium funiculosum, P.cyclopium	*Apis mellifera*	From all stages, workers, drones.	263
Phoma sp., *Rhodotorula glutinis, Rh. mucilaginosa* (=*rubra*)	*Apis mellifera*	Causes typical discolorations. In drone larvae.	263
Torulopsis sp.	*Apis mellifera*	Pathogenic yeast. In sick bees.	17
Verticilium lecanii	*Bombus* spp.		198
Nonidentified fungus	*Apis mellifera*	In epithelium of ovaries, poison sac, and rectum of queens.	17,106
	ANTS:		
Aegeritella lenkoi, A.rousillonensis, A.superficialis, A.tuberculata	*Camponotus, Cataglyphis, Formica* spp., *Lasius*	*A.rousillonensis* grows on mouth and body surface of workers and queens.	22,242
Alternaria tenuis	*Formica rufa*	Pathogenicity unclear.	7,211
Aspergillus flavus	*Atta texana*		191,305
Beauveria bassiana (=*B.densa*), *B.tenella*	*Atta, Lasius, Myrmecia, Solenopsis* Formicidae	May be an important pathogen for fire ants in South America.	7,44,166, 188, 237

Fungi (cont.)	Host	Remarks	Refs.
Conidiobolus (*Entomophthora*)	*Solenopsis invicta*		237
Cordyceps sp.	*Camponotus*, *Cephalotes*, *Dolichoderus*, *Mesoponera*, *Monacis*, *Pachycondyla*, *Paraponera*	For all *Cordyceps* and anamorphs behavioral changes in host known to occur. Obligately parasitic on ants. Regulators for host populations in tropics?	7,92,237, 246,282
Cordyceps australis, *C.bicephala*, *C.cucumispora*, *C.formicarum*, *C.kniphofioides* (Anamorph: *Hirsutella stilbelliformis*), *C. lloydii*, *C.necator*, *C.proliferans*, *C.pseudolloydii*, *C.subdiscoidea*, *C.unilateralis*	Formicidae Ponerinae *Cephalotes*, *Camponotus*, *Dolichoderus*, *Echinopla*, *Megaponera*, *Monacis*, *Pachycondyla*, *Paraponera*, *Paltothyreus*, *Phasmomyrmex*, *Polyrhachis*	Behavioral changes. Epizootics observed on *Camponotus*, *Phasmomyrmex*, *Polyrhachis*. Experimental infections lead to host death.	7,91,92, 132,282, 320
Desmidiospora myrmecophila	*Camponotus* spp.	May be specific to *Camponotus* queens.	92,132
Entomophthora (*Conidiobolus*) *coronata*	*Acromyrmex octospinosus*, *Camponotus* spp.	In *Acromyrmex*, fungus acts with toxic secretions.	166,194
Erynia formicae, *E.* (=*Entomophthora*) *myrmecophaga*	*Formica* spp., *Serviformica*, *Lasius*	Epizootics observed. Behavioral change.	91
Hirsutella acerosa, *H.formicarum*, *H.sporodochialis*, *H.stilbelliformis var. gnamptogenys* *Hymenostilbi australienses* (=*Stilbum formicarium*)	*Camponotus* spp., *Polyrhachis* spp. Formicidae		92
Laboulbenia formicarum	Many ant species		132
Metarhizium anisopliae	*Atta* spp., *Solenopsis* spp.	In *Atta* "queen disease," but behavioral change likely.	7,305
Myrmecomyces annelisae	*Solenopsis* spp.	In hemolymph.	40,42,154
Myrmicinosporidium durum (a fungus?, close to Entomophthorales)	*Chalepoxenus*, *Leptothorax* spp., *Myrmecia*?, *Pheidole*, *Plagiolepis*, *Pogonomyrmex*, *Solenopsis*	Chronic infection in hemocoel. Infected ants live longer and are darker in color. Described as "Näpfchenkrankheit" in Europe.	132,284

Fungi (cont.)	Host	Remarks	Refs.
Pandora (=*Erynia* =*Entomophthora*) *fomicae*	*Formica pratensis*		194,237
Paecilomyces lilacinus (Teleomorph: *Cordyceps, Torrubiella*) *Penicillium glaucum*	*Myrmecia nigriscapa* *Myrmica* sp.		237
Polycephalomyces sp. *Pseudogibellula formicarum* (Teleomorph: *Torrubiella*)	Formicidae *Oecophylla, Paltothyreus*		283
Rickia wasmanni	*Myrmica laevinodis*	Host-specific. Ectoparasitic.	132
Sporothrix insectorum (Teleomorph: ?*Cordyceps*)	*Paltothyreus tarsatus*		282
Stilbella (*Stilbum*) *burmensis, S.buquetii* var. *formicarum, S.dolichoderinarum*	*Camponotus, Oecophylla, Phasmomyrmex, Polyrhachis* spp., *Paltothyreus, Technoymrmex* Formicidae	Epizootic in *Oecophylla*. Fungal stemmata and position of fruit bodies characteristic for different parts of ant body.	7,92,282
Tarichium sp.	*Tetramorium caespitum*	Perhaps an *Erynia* species.	91,211
Tilachlidium liberianum (Teleomorph: *Torrubiella*) *Torrubiella carnata*	*Paltothyreus tarsatus*		282
Nonindentified fungi, yeasts	*Mesoponera, Polyrhachis* spp., *Solenopsis*		147,150
	WASPS:		
Beauveria bassiana *B.tenella* (=*brongniartii*) *Cordyceps sphecocephala* *Hirsutella sausseri, Hirsutella* sp.	*Vespula* spp., *Vespa, Polistes*	Most also occur in nonsocial insects.	110,117 182,188
Hymenostilbe sphecophila *Paecilomyces farinosus*	*Vespula* spp.		110,188
	TERMITES:		
Amphoromorpha entomophila (?)	*Acanthotermes, Reticulitermes*		33
Antennopsis gallica, A.grassei, A.gayi	*Kalotermes, Reticulitermes* spp., *Glyptotermes, Neotermes* spp.	A mild ectoparasite. Host becomes sluggish.	33,316
Beauveria brongniartii	*Odontotermes wallonensis*	On newly dealated reproductives in founding chamber.	266

Fungi (cont.)	Host	Remarks	Refs.
Cordycepioides bisporus, C.octosporus	*Macrotermes* spp., *Tenvirostitermes*	A virulent ectoparasite.	33,282
Coreomycetopsis oedipus *Dimeromyces isopterus, D.majewskii*	*Nasutitermes* (*Eutermes*), *Reticulitermes* spp., *Alyscotermes*	A mild ectoparasite.	33
Hormiscioideus filamentosus	*Amitermes neotenicus*	Only known from type locality.	33
Laboulbenia hageni, L.felicis-caprae, L. geminata, L.ghanaensis, L. brignolii *Laboulbeniopsis termitarius*	*Macrotermes* spp., *Anacanthotermes, Odontotermes, Amitermes, Macrotermes, Ahmaditermes, Coptotermes, Leptomyxotermes, Nasutitermes, Reticulitermes* spp.	A mild ectoparasite.	33,315
Mattirolella crustosa, M.silvestrii	*Nasutitermes* spp., *Reticulitermes, Rhinotermes*	Typically not many individuals infected, mostly soldiers.	33,315
Metarhizium anisopliae	*Cryptotermes brevis*	Very virulent.	161
Termitaria coronata, T.macrospora, T.hombicarpa, T.longiphialidis, T.snyderi, T.thaxteri, Termitaria spp. *Termitaropsis*	*Allodontotermes, Amitermes, Coptotermes* spp., *Cornitermes, Mastotermes, Microhodotermes, Microtermes, Nasutitermes* (*Eutermes*) spp., *Neotermes, Odontotermes, Porotermes* spp., *Reticulitermes* spp., *Rhinotermes*	Mild ectoparasites.	33,217, 315

Appendix 2.4 Protozoa

Protozoa	Host	Remarks	Refs.
MICROSPORIDIA:			
	BEES:		
Nosema apis, N. cerana	*Apis mellifera, A.cerana*	Association with BQCV, BVY viruses, and with *Malpighamoeba*. Queens become sterilized. Colony growth reduced, lower honey yield.	12,19,30, 103,183, 184, 311, 323,329, 336
Nosema bombi	*Bombus* spp., *Apis florea*?	Can cross-infect among *Bombus* species. Workers die soon. Colonies develop badly.	93,99,179, 200, 296, 300
	ANTS:		
Burenella dimorpha *Nosema*? *Thelohania solenopsae* *Vairimorpha invictae* Four nonidentified microsporidia	*Solenopsis geminata, S.invicta, S.quinquecuspis, S.richteri, S.saevissima*-complex	Many dimorphic with different spores. *Burenella* well investigated. Only one type of spore infective. *Thelohania* perhaps a set of different species.	7,8,40,41, 42,149, 150,153, 155,340, 245
	WASPS:		300
Nosema bombi	*Vespa germanica*		
	TERMITES:		
Duboscqia coptotermi, D.legeri	*Coptotermes, Reticulitermes* spp., *Termes*	In midgut, fat body. Generally light infections. Infected cells hypertrophied.	160,180, 181,300
Gurleya spraguei	*Macrotermes estherae*	Max. infection at end of winter.	159
Microsporidium termitis	*Reticulitermes flavipes*	Peculiar trophozoites divide into sporonts. Described as *Nosema* by Kudo.	180,181, 300
Microsporidium sp.	*Microtermes championi*	In workers.	192
*Plistophora ganaptii, P.weiseri, P.*sp.	*Odontotermes, Coptotermes, Reticulitermes*	In epithelial cells of foregut or Malpighian tubules.	159,300
Stempellia odontotermi	*Odontotermes* sp.	Epithelial cells of foregut of workers.	159
FLAGELLATES:			
	BEES:		82,296,298
Crithidia bombi	*Bombus* spp., *Psithyrus vestalis*	Highly infective. In *Psithyrus*: known from males only.	
Crithidia mellificae *Leptomonas apis*	*Apis mellifera*	Common. No harmful effects known.	17,187,327
	WASPS:		
Crithidia sp.	*Vespa squamosa*	Different from *C.mellificae.*	218,319

Protozoa (cont.)	Host	Remarks	Refs.
GREGARINIDA:			
	BEES:		
Apicystis (=*Mattesia*) *bombi*	*Bombus* spp., *Psithyrus*	Also found in queens.	200
Leidyana sp.	*Apis mellifera*	Attached to epithelium of midgut.	17
Neogregarina sp.n. Nonidentified gregarines	*Apis mellifera*, *Bombus* spp.	*Apis*: Worldwide. Morphologically different from other microsporidia described in *Bombus*.	17,94,129, 189,225, 235,306
	ANTS:		
A gregarine sp.	*Myrmecia* spp.	Color aberrations.	70,148
Mattesia geminata Neogregarinidae sp.	*Solenopsis geminata*, *S.invicta*, *S.richteri*	Unusual, because infects hypodermis.	149,150, 151
	TERMITES:		
Nonidentified gregarine	*Coptotermes acinaciformis*		136
AMOEBA:			
	BEES:		
Malpighamoeba mellificae	*Apis mellifera*	Associated with Bee Virus X and *Nosema apis*. Few effects.	17,19,121, 323

Appendix 2.5 Nematodes

Nematodes	Host	Remarks	Refs.
RHABDITIDAE:[a]			
	ANTS:		
Caenorhabditis dolichura	*Acanthomyops, Camponotus, Formica* spp., *Lasius* spp., *Tetramorium*	In pharyngeal glands of head capsule. Originally described as *Pelodera janeti* by Janet 1893.	231
Diploscapter lycostoma (=*D.coronata*)	*Camponotus* spp., *Formica* spp., *Iridomyrmex* spp., *Lasius* spp., *Myrmica*	Nematode resides in pharyngeal glands.	212
Pristionchus lheritieri	*Lasius* spp.	In head glands.	
Rhabditis janeti (*Protorhabditis*)	*Camponotus, Formica, Lasius*	In post-pharyngeal glands.	243,256
Various rhabditid nematodes	*Acantomyops, Camponotus, Formica* spp., *Iridomyrmex, Lasius* spp., *Myrmica, Tetramorium*	Can be common.	231,243
	TERMITES:		
Rhabditis sp.	*Neotermes, Reticulitermes, Capritermes,Copto-termes,Macrotermes, Microtermes, Rhinotermes,Termes*		249
Rhabditis janeti (=*Caenorhabditis dolichura*) *Rhabpanus ossiculum* *Termirhabditis fastidiosus*	*Reticulitermes flavipes*	Found in head of termite.	231
MERMITHIDAE[b]			
	BEES:		
Agamomermis sp. *Mermis* (*Hexamermis*) *albicans, M. nigrescens, M.subnigriscens*	*Apis mellifera*	In workers, queens, and drones. Perhaps workers take up eggs with water from foliage. Occasionally high infection levels of nonidentified nematodes.	17
	ANTS:		
Allomermis lasiusi, A.myrmecophila	*Lasius* spp.	Infection associated with change in morphology.	69,243
Hexamermis sp.	*Pheidole, Polyrhachis*		243,321
Mermis racovitzai	*Camponotus, Formica*		
Mermis spp.	*Aphaenogaster, Colobopsis, Lasius* spp., *Pheidole, Solenopsis, Tetramorium*	Partly aberrant color morphs (yellow ants). Sometimes morphological and behavioral changes.	177,182, 205,244, 334

Nematodes (cont.)	Host	Remarks	Refs.
Pheromermis villosa	*Lasius* spp.	Morphological changes. Mostly near wetland habitats.	158
Various mermithid nematodes	*Aphaenogaster, Camponotus, Dinoponera, Ectatomma, Epimyrma, Formica, Lasius, Leptothorax, Myrmecia, Myrmica, Neoponera, Odontomachus, Pachycondyla, Paraponera, Pheidole, Plagiolepis, Solenopsis, Tetramorium*	Mostly records of morphologically abberant ("mermithized") ants. In *Formica, Lasius, M.scabrinodis, Solenopsis, Leptothorax* found in spring colonies. In *Solenopsis*, parasitized sexuals have no mating flights.	69,177, 219,243
Nonidentified nematodes, mermithids?	*Solenopsis* spp.		149,156, 223
	WASPS:		
Agamomermis (*Mermis*) *pachysoma*	*Vespula* spp.	Only report for wasps, except *Sphaerularia* (Cobbold 1888).	32
Pheromermis pachysoma	*Vespula pennsylvanica*	Complex life cycle with paratenic host (an aquatic insect).	206,252, 257
	TERMITES:		
Mermis sp.	*Cornitermes, Thoracotermes*	One specimen found each.	278
APHELENCHOIDEA:			
	BEES:		
Huntaphelenchoides sp.	*Halictus farinosus*		115
CHORDIDAE:			
	WASPS:		
Nonidentified sp.	*Paravespula germanica*	Infection by contaminated food or materials.	113
DIPLOGASTERIDAE:			
	BEES:		
Acrostichus sp.	*Halictus farinosus*	Occurs in reproductive tract. Perhaps sexually transmitted disease.	115
	ANTS:		
Eudiplogaster histophorus	*Lasius* spp.	In head glands.	325
	TERMITES:		
Diplogaster aerivora	*Leucotermes lucifugus*	In head around mouth parts; becomes sluggish.	221
Isakis migrans	*Reticulitermes* spp.		
Mikoletzkya aerivora	*Reticulitermes* spp.	Occurs in head glands. Host becomes sluggish.	25,221

Nematodes (cont.)	Host	Remarks	Refs.
Pseudodiplogasteroides sp.	*Mastotermes darwiniensis*	From head of termite.	325
SPIRURIDEA:			
	TERMITES:		
Harteria gallinarum	*Hodotermes pretoriensis*		
TETRADONEMATIDAE:			
	ANTS:		
Tetradonema solenopsis	*Solenopsis invicta*		156
TYLENCHIDAE:			
	BEES:		
Sphaerularia bombi	*Bombus* spp., *Psithyrus* spp.	In overwintering queens. Behavioral change: queen seeks for overwintering instead of nest sites.	5,72, 105,122, 196,220, 260,258, 303
	WASPS:		
Sphaerularia bombi	*Vespula* spp.		
NONIDENTIFIED NEMATODES:			
	ANTS:		
	Acromyrmex, *Solenopsis*, *Teleutomyrmex*	Established as pathogenic. Some encysted in thorax.	191,243, 330
	TERMITES:		
	Microcerotermes championi	Nematodes infect per os, but later enter the body cavity and head.	170

NOTES: (a) Based on refs.(29), (256), and (325). (b) The following groups based on refs.(29) and (256). Additional references given.

Appendix 2.6 Helminths

Helminths[a]	Host·	Remarks·	Refs.·
CESTODA:			
Anomotaenia brevis	*Leptothorax* spp.	Infected as larvae. Aberrant colors. Final hosts are woodpeckers.	254,255,309
Chaenotania craterformis, *Ch.musculosa*, *Ch.unicornata*? *Anomotaenia borealis*?	*Crematogaster*, *Harpagoxenus*, *Leptothorax* spp.	Aberrant colors (yellow ants), sluggish behavior. Final hosts are woodpeckers.	54,90,243, 253,309
Cotugnia digonopora	*Monomorium* spp.		243
Moniezia expensa	*Solenopsis invicta*		104
Raillietina circumvalata, *R.echinobothrida*	*Monomorium*, *Pheidole*, *Pheidologeton*, *Prenolepis*, *Tetramorium*	*R.echinobothrida*: final hosts are poultry.	54,118,135, 157,229,234, 243,310
Raillietina fedjushini, *R.friedbergeri*, *R.georgiensis*, *R.kashiwarensis*, *R.loeweni*, *R.tetragona*, *R.urogalli*	*Brachyponera*, *Crematogaster* spp., *Dorymyrmex*, *Euponera*, *Formica*, *Iridomyrmex*, *Monomorium* spp., *Myrmecocystus*, *Myrmica* spp.,*Pheidole* spp., *Prenolepis*, *Solenopsis* spp., *Tetramorium* spp.	Final hosts are turkeys, pheasants, peacocks. Infections decrease when final host excluded from range. For *R.loeweni*: final host is a rabbit. According to Passera 1975, *R.tetragona* found in 32 ant species, with 1-20 cysticercoids per ant.	27,57,90, 118,135,229, 234,243,269, 310
Nonidentified cestodes (probably *Anomotaenia*, *Choanotaenia*)	*Euponera* spp., *Leptothorax* spp., *Monomorium*, *Myrmica*, *Pheidole*,*Pheidologeton*, *Solenopsis*, *Tetramorium*	Host specificity seems not very strict. Can produce aberrant color morphs. Final host probably birds.	54,107,146, 309
TREMATODES:			
Brachylecithum mosquensis	*Camponotus* spp.	Typically 1-2 metacercaria found in host brain. Infected ants show sluggish behavior, prefer open spaces. Have two other hosts: land snail and a bird.	55,243
Dicrocoelium dendriticum, *D.hospes*, *D.lanceolatum*	*Cataglyphis*, *Camponotus*, *Formica* spp., *Proformica*	In gaster. Typically 1-2 metacercaria found in host brain. Behavioral change: infected ants climb and cling to grasses. Final host an ungulate.	39,131,195, 234,243,288, 310
Eurytrema pancreaticum	*Technomyrmex detorquens*		310

NOTE: (a) The only known examples are in ants.

Appendix 2.7 Mites of Social Bees

Mites [a]: Number of Species	Type[b]	Host Group. Remarks·	Refs.[c]
ASTIGMATA:			
Acarus: 5 spp.; *Acotyledon*:1 sp.; *Aeroglyphus*: 1 sp.	C?, C	*Apis*, *A.mellifera*	58
Anoetus: 10 spp.	S, P	Halictinae, *Apis*	
Caloglyphus: 1 sp.; *Carpoglyphus*: 2 spp.	C	Apinae	58
Cerophagopsis: 1 sp.	S?	Meliponinae	
Cerophagus: 1 sp.	?	Bombinae	
Chaetodactylus: 1 sp.; *Cosmoglyphus*:1 sp.; *Dermacarus*: 1 sp.; *Dermatophagoides*: 2 spp.; *Forcellinia*: 1 sp.	?, S	*Apis*. *F.galleriella* normally a myrmecophile mite.	37
Gaudiella: 2 spp.	S?	Meliponinae	
Glycyphagus: 2 spp.	?	*Apis*	58
Halictacarus: 1 sp.; *Histiostoma*: 1 sp.; *Konoglyphus*: 1 sp.	P, S?	Halictinae, Bombinae	
Kuzinia (formerly =*Tyrophagus*): 6 spp.	S, C, H	*Apis*, Bombinae	58,119, 298
Lasioacarus: 1 sp.	C	*Apis mellifera*	58
Meliponocoptes: 3 spp.; *Meliponoecius*: 1 sp.; *Meliponopus*: 1 sp.; *Myianoetus*: 1 sp.	S?	Meliponinae *Apis*	
Partamonacoptes:1 sp.; *Platyglyphus*: 1sp.	S?	Meliponinae	
Rhizoglyphus: 1 sp.; *Sancassania*: 2 spp.	? S, C	*Apis*, Halictinae. *Sancassiana* consumes provisions and dead brood in halictid bees.	
Saproglyphus: 1 sp.	?	*Apis*	
Schulzea: 2 spp.	S?	Halictinae	
Schwiebea: 1 sp.; *Suidasia*: 1 sp.; *Thyreophagus*: 1 sp.; *Trigonacoptes*: 1 sp.	?	*Apis*, Meliponinae. *Trigonacoptes* in *Trigona*.	
Tyrolichus: 1 sp.; *Tyrophagus*: 3 spp.	?, S	*Apis*	
PROSTIGMATA:			
Acarapis: 3 spp.	P	Apinae. A pest in honeybees.	89,240
Allothrombium: 1 sp.; *Anystis*: 1 sp.; *Cheyletus*: 2 spp.; *Eucheyletia*: 1 sp.	?, R	*Apis*	
Ereynetes: 1 sp.; *Eriophyes*: 1 sp.	?	Meliponinae, *Apis*	
Imparipes: 13 spp.; *Leptus*: 1 sp.	S, P	Halictinae, Apinae, Bombinae. *I.apicola* feeds on fungus.	

Mites [a]: Number of Species (cont.)	Type[b]	Host Group. Remarks·	Refs.[c]
Locustacarus (=*Bombacarus*): 1 sp.	P	Bombinae. Puncture trachea and suck hemolymph.	6,72,138, 197,298, 301
Melissotydens: 1 sp.	S?	Meliponinae	
Neotarsonemoides: several spp.	H	Social bees	
Neotydeolus: 1 sp.	?	Meliponinae, *Scaptotrigona*. Mutualistic?, removes fungi. In genital chamber of males. Perhaps sexually transmitted.	274
Parapygmephorus(=*Sicilipes*): 5 spp.	S, P?	Halictinae. Perhaps egg parasites.	274
Parascutacarus: 1 sp.	S	Bombinae	
Proctotydeus: 3 spp.	S, P?	Meliponinae	
Pseudacarapis: 1 sp.	?, H	Apinae	
Pyemotes: 2 spp.; *Pygmephorus*: 1 sp.	?, P	*Apis*, *Bombus*. Polyphagous ectoparasite.	71
Scutacarus: 4 spp.; *Siteroptes*: 1 sp.	S, P?	Halictinae, Apinae, Bombinae. Also on *Parasitellus* mites.	4,27,119, 290
Tarsonemus: 3 spp.; *Tetranychus*: 1 sp.	S?, H	Apinae, *Apis*	
Trochometridium: 1 sp.	P, S, R?	Halictinae. A larval parasite. Can transmit fungal spores.	
Trombidium: 1 sp.; *Tydeus*: 1 sp.	?	*Apis*	
MESOSTIGMATA:			
Afrocypholaelaps: 1 sp.	C, H	Meliponinae, Apinae	294
Alliphis: 1 sp.; *Allodinychus*: 2 spp.; *Androlaelaps*: 2 spp.	?	*Apis*	31
Antennophours: ? spp.; *Bisternalis*: 5 spp.	C	Meliponinae	176
Blattisocius: 3 spp.	?, R	*Apis*	
Calaenostanus: 1 sp.; *Edbarellus*: 1 sp.; *Eumellitiphis*: 3 spp.	C, H P?, R?	Meliponinae, Apinae	
Euvarroa: 1 sp.	P	Apinae	31,226
Garmaniella: 2 spp.	?	Bombinae	293
Grafia: 3 spp.	P, R?	Meliponinae	
Haemogamasus: 1 sp.	?	*Apis*	

Mites [a]: Number of Species (cont.)	Type[b]	Host Group. Remarks·	Refs.[c]
Holostaspella: 1 sp.; *Hunteria*: 1 sp.	P?, R?	Meliponinae	
Hypoaspis s.l.: ca. 24 spp. ("*Pneumolaelaps*": 8 spp.) ("*Holostaspis*": 2 spp.)	P, C, R	*Apis*, Meliponinae, Bombinae. May feed on injured bees.	67,137, 275,293
"*Iphis*": 1 sp.	?	*Apis*. Nonidentified species.	
Laelapidae: 5 genera	?	Meliponinae	
Laelaps: 1 sp.	?	*Apis*	
Laelaspoides: 1 sp.	C	Halictinae. A pollen eater.	
Lasioseius: 3 spp.	?, R	*Apis*	56
Macrocheles: 3 spp.; *Melichares*: 1 sp.	P, R, S	Apinae, *Apis*, Bombinae	271
Meliponaspis: 2 spp.; *Meliponapachys*: 1 sp.; *Mellitiphis*: 1 sp.; *Mellitiphisoides*: 1 sp.; *Neocypholaelaps*: 7 spp.; *Neohypoaspis*: 1 sp.; *Oplitis*: 1 sp.	P?, R?	Meliponinae	176
Parasitellus (formerly = *Parasitus*): 16 spp.; *Parasitus*: 2 spp.	C ?, P	Bombinae	271,291
Pergamasus: 2 spp.	?	*Apis*	
Proctolaelaps: 2 spp.	C, R	Bombinae, *Apis*	193,239
Stevellus: 1 sp.; *Trigonholaspis*: 2 spp.; *Triplogynium*: 1 sp.	P?, R?	Meliponinae. *Trigonholaspis* parasitizes pupae.	
Tropilaelaps: 2 spp.	P	Apinae. *T.clareae* causes morphological aberrations. In *A.mellifera* harmful.	2,31,50,51
Urodiscella: 1 sp.; *Urobovella*: several spp.; *Uroplitana*: 1 sp.; *Urozercon*: 1 sp.	P, R? S?	Meliponinae, Bombinae	
Varroa: 2 spp.	P	Apinae, *Bombus*. *Apis cerana* is orginial host.	167,314
Veigaia: 1 sp.	?	*Apis*	
Zontia: 1 sp.	P?, R?	Meliponinae	
ORIBATIDIA (CRYPTOSTIGMATA): *Pediculochelus*: 1 sp.; *Scheloribates*: 1 sp.	?	*Apis*	

NOTES: (a) The taxonomy of mites is in disarray. Names of genera and species keep changing.
(b) Purely predatory mites (if known) are not listed. Where known, mode of life is indicated as: scavenger/saprophagous (S), cleptoparasitic (C), parasitic (P), phoretic (H), predatory (R; only where in combination with other reported lifestyles); (?) is unknown relationship.
(c) Based on refs.(75), (84-87), additional references given.

Appendix 2.8 Mites of Ants, Wasps, and Termites

Mites[a] (Group, Genus)	Host	Remarks	Refs.[b]
ASTIGMATA:			
	ANTS:		
Acaridae, Anoetidae	*Eciton*	Commensals? Phoretic.	
Cosmoglyphus, Dorylacarus, Forcellinia, Froriepia, Garsaultia, Histiostoma, Lasioacarus, Lemaniella, Mycetoglyphus, Ocellacarus, Rettacarus, Sancassiana, Tyrophagus	*Lasius*, various other ants	Associates. *Forcellinia* abundant nest inhabitant. Consumes dead ants and other remains.	233
	WASPS:		
Acarus sp.	*Dolichovespula*	Feeds on larval hemolymph.	299
Glycyphagus, Medeus, Myianeotus, Neottiglyphus, Sphexicozela, Tortonia	*Vespula, Polistes* Social wasps	Saprophagous, commensals? In nests.	233,324
Tyroglyphus fucorum (= *Kuzinia laevis*)			
	TERMITES:		
Anoetus myrmicarum	*Reticulitermes lucifugus*	High mortality in lab colonies.	176
Cosmoglyphus, Dorylacarus, Forcellinia, Froriepia, Histiostoma, Lasioacarus, Lemaniella, Ocellacarus, Rettacarus, Sancassiana, Tyrophagus	Termites, *Velocitermes*	Probably commensals. Most genera also found in ants.	233
MESOSTIGMATA:			
	ANTS:		
Aenicteques chapmani	*Aenictus margini*	Cleptoparasitic.	
Amblyseius, Androlaelaps *Anoplocelaeno sulcata*	*Atta sexdens, Solenopsis* spp.	Predatory, also saprophagous.	88,127
Antennophorus grandis, A.uhlmanni, A.wasmanni	*Acanthomyops, Lasius* spp.	Cleptoparasitic? *A.uhlmanni* solicits food.	68,101, 176,338
Calotrachytes eidmanni *Ceratozercon similis* *Coprholaspis:* 5 spp.	*Atta sexdens*	Possibly saprophagous.	88
Cillibano comato	*Lasius* spp.	Sucks hemolymph.	338
Circocyllibanidae, Coxequesomidae	*Eciton, Labidus*	Phoretic; also on larvae.	176
Cosmolaelaps Gamasidae	Several ant genera	Saprophagous. Free-living predators otherwise.	176
Digamasellus prominulus *Dinychus fustipilis*	*Atta sexdens*	From refuse chambers.	88
Garmania , Holostaspis, Hoploseius, Hypoaspis	*Atta, Cyphomyrmex, Formica, Lasius, Myrmecia, Myrmica, Tetramorium*	Saprophagous. Commensals. Some may feed on other mites.	88,176

Mites[a] (Group, Genus) (cont.)	Host·	Remarks·	Refs.[b]
Laelapidae, *Laelaps*	*Solenopsis* spp., *Formica*		127
Laelaspis brevichelis, L.dubitatus, L.equitans	*Aphaenogaster, Crematogaster, Tetramorium*	Phoretic on adults.	176
Lasioseius sp.	*Atta sexdens*	Saprophagous?	88
Macrocheles dibamos, M.rettenmeyeri	*Eciton* spp.	Feeds on body fluid.	176,270
Messocarus mirandus *Myrmozercon* (= *Myrmonyssus*)	Various ants	Some cleptoparasitic.	
Neoberlesia mexicana	*Pheidole acolhua*	Found in nests.	24
Oolaelaps oophilus	*Formica* spp.	Feed on salivary secretions deposited on eggs.	176,339
Oplitis exopodi, O.virgilinus	*Atta, Solenopsis* spp.	Many phoretic.	128
Planodiscidae	*Eciton*	Phoretic on legs.	
Rhodocarus angustiscutatus	*Atta sexdens*	Saprophagous?	88
Sphaerolaelaps holothyroides	*Solenopis, Lasius*	Cleptoparasitic, saprophagous.	176
Trachyuropoda,Trachytidae *Trhypochthonius*	*Atta, Formica, Solenopsis* spp.		23,127
Dinychus, Ipiduropoda, Oodinychus, Phaulodinchyus, Prodynchius, Trematurella, Urobovella, Urodiscella, Uroseius (Uropodidae, Trachyuropididae)	Various ants	Commonly found in nests. Some may be parasitic. Phoretic on emerging queens.	176
	WASPS:		
Parasitus vesparum	*Vespula vulgaris* Various social wasps	May feed on nest associates.	299,324
	TERMITES:		
Eickwortius termes	*Macrotermes michaelseni*		344
Histiostoma formosana *Laelaptonyssus* spp. *Termitacarius cuneiformis*	*Amitermes, Coptotermes,* on various termites	Phoretic; heads of workers and soldiers.	176
Macrochelidae, Uropodidae, Trachyuropoda, Urobovella, Urodiscella, Urozercon	*Odontotermes, Glyptotermes,* other genera	Phoretic; heads of workers.	176

Mites[a] (Group, Genus) (cont.)	Host·	Remarks·	Refs.[b]
PROSTIGMATA:			
	ANTS:		
Acinogaster, A.marianae	*Eciton, Labidus, Neivamyrmex, Nomamyrmex*	Phoretic.	71
Bakerdania, Forania (Erythraeidae)	On various ants		
Ereynetidae, *Ereynetes, Glyphidomastax, Imparipes*	*Atta, Eciton,* army ants	Parasitic, commensal, saprophagous, predatory.	88
Lophodispus	*Lasius niger*	Phoretic on heads.	
Myrmecodispus, Parapygmephorus, Perperipes	*Eciton, Neivamyrmex, Nomamyrmex*	Larval or egg parasites?	71
Pseudopygmephurs	*Lasius flavus*		71
Pyemotes ventricosus	*Solenopsis*, other ants	Can destroy colonies.	
Scutacaridae *Scutacarus minutus, S.attae*	*Eciton, Atta sexdens*	Phoretic on legs of adults. Saprophagous.	88
Tarsenomidae sp. *Tarsocheylus atomarius*	*Atta, Solenopsis*	Feeds on excretions of ant's anus.	88,317
Thaumatopelvis sp.	Various ants		
	WASPS:		
Pyemotes sp.	*Polistes* spp.	Parasitic ?	230
Scutacarus acarorum	Social wasps		324
	TERMITES:		71
Bakerdania, Parapygmephorus, Premicrodispus: 5 spp. *Pyemotes ventricosus*	Various termites, *Reticulitermes*	Phoretic? *Pyemotes* ectoparasitic on immatures.	
An unnamed trombidid mite	*Nasutitermes*	Larva is parasitic.	
ORIBATIDIA (CRYPTOSTIGMATA):			
	ANTS:		
Galumna acutifrons	*Pheidole vasliti acolhua*	Found in nests. Unknown habits.	24
Hypochthoniidae	*Eciton*	Possible commensals.	
Oppia: 3 spp., *Oribata novus*	*Atta sexdens*	Saprophagous?	88
	WASPS:		
Corynetes caerulus	*Vespa crabro*	Phoretic (?) on larvae.	176

NOTES: (a) The taxonomy of mites is in disarray. Names of genera or species change. (b) Records for all gropus after ref.(86), for ants after ref.(132). Additional references given. Parasitic status of most species unkown. Phoretic mites have also been listed. Ref.(88) lists additional species found in debris chambers of leaf-cutter ants. Purely predatory taxa, where known, are not listed.

Appendix 2.9 Hymenoptera

Hymenoptera·	Host·	Stage[a]	Remarks·	Refs.·
EUCHARITIDAE: [b]				
Austeucharis, Epimetagea (*Austeucharis*) sp.	*Myrmecia* spp.	L, P	Overwinters in cocoons.	312
Chalcura sp., *Ch.deprivata*	*Formica, Odontomaches*	P	Only small pupae infected.	66,109
Eucharis ascendens, E.bedeli, E.myrmeciae, E.punctata, E.scutellaris	*Cataglyphis, Formica, Messor, Myrmecia, Myrmecocystus*	L, P	From cocoons.	63,277
Eucharomorpha, Isomerala. Kapala floridana,K.terminalis Obeza floridana	*Pheidole, Ectatomma Pogonomyrmex, Odontomachus, Camponotus, Pachycondyla*	L, P	Females disperse only a few meters away from host nest.	65, 66,74, 247
Orasema aenea, O.assectator, O.coloradensis, O. costaricensis, O.crassa, O.minutissima, O.rapo, O.robertsoni, O.sixaolae, O.tolteca, O.viridis, O.wheeleri	*Formica* spp., *Pheidole* spp., *Solenopsis* spp. *Wasmannia*	L, P	Leafhoppers or thrips carriers of planidia. Ectoparasitic on pupae. Parasites mimic host chemically.	65,73,109, 145,247,322, 339
Pheidoloxeton wheeleri Pseudochalcura gibbosa Pseudometagea schwarzii Psilogaster antennatus, P.fasciventris	*Pheidole* spp., *Camponotus* spp., *Lasius* spp., *Coelogyne*	L, P, W?		11,109,125, 126,210,247, 332,333
Rhipipallus affinis Tricoryna	*Odontomachus, Myrmecia, Chalcoponera, Ectatomma*	P	Superparasitism found.	144,332
Schizaspidia convergens, S.doddi, S.polyrhachicida, S.tenuicornis Stilbula cynipiformis, S.(Schizaspidia) polyrhachicida, S.tenuicornis	*Calomyrmex, Camponotus* spp., *Odontomachus, Polyrhachis*	P	Females or males hover over nest entrance.	63,66,241, 277,332
BETHYLIDAE:				
	ANTS:			
Pseudisobrachium terresi	*Aphaenogaster, Solenopsis*	L, P	Larvae probably endoparasitic.	132,176
BRACONIDAE:				
	BEES:			
Blacus hastatus, B.masoni Syntretus splendidus	*Bombus* sp., *Psithyrus vestalis*	W, S	Some are parasites of nest associates.	4,6,119,200, 261,285,326
Syntretomorpha szaboi	*Apis cerana*	W		326
	ANTS:			
Elasmosoma beroliense, E.petulans, E.vigilans Hybrizon buccatum, H.cemieri Neoneurus mantis	*Formica* sp., other genera	W	Females oviposit into gaster.	132,176,213, 228,295

Hymenoptera· (cont.)	Host·	Stage[a]	Remarks·	Refs.·
	WASPS (SPHECIDAE):			
Heterospilus microstigmi	*Microstigmus* spp.	L, P		216,272
	TERMITES:			
Termitobracon emersoni *Ypsistocercus manni*, *Y.vestigialis*	*Nasutitermes* spp.		Immature stages unknown.	176
CHALCIDIDAE:				
	WASPS:			
Brachymeria discreta *Dibrachys cavus*, *D.vesparum* *Dimmockia incongrua*	*Polistes* spp., *Dolichovespula* *Vespula maculata*	L, P		113,230,247, 268,341
Elasmus biroi, *E.japonicus*, *E.lamborni*, *E.polistis*, *E.schmitti*	*Polistes* spp.	L, P	Some external parasites on pupae.	52,140,209
Nasonia vitripennis *Trichokaleva microstigmi*	*Dolichovespula*, *Microstigmus* (Sphecidae)	L, P		113,216
CHRYSIDIDAE:				
Chrysis ignitia	*Vespa rufa*	L		299
DIAPRIIDAE:				
	ANTS:			
Auxopaedeutes lyriformis, *A.sodalis* *Bruesopria americana*, *B.seevesi* *Gymnopria lucens* *Hemilexis jessei* *Lepidopria pedestris* *Solenopsia imitatrix*	*Formica*, *Paratrechina*, *Solenopsis*, *Acromyrmex*	L, P	*Hemilexis* parasitic? *Solenopsia* migrates with host colony (*S.fugax*).	193,210,215, 337
Mimopria ecitophila *Szelenyiopria* (*Doliopria*) *reichenspergeri*	*Eciton* spp., *Neivamyrnex legionis*	A	On soldiers?	193,215
Plagiopria passeria	*Plagiolepis pygmaea*	P	Queen pupae.	185
Ashmeadopria, *Basalys*, *Bruchopria*, *Geodiapria*, *Loxotropa*, *Mimopria*, *Myrmecopria*, *Phaenopria*, *Tetramopria*, *Trichopria*	*Ecitonini*, *Plagiolepis* *Solenopsis* (*Diplorhoptrum*), *Tetramorium*	L, P	Larvae of these species probably endoparasitic in host brood.	133
ENCYRTIDAE:				
	WASPS:			
Signiphora polistomyiella	*Polistes* sp.			272
EULOPHIDAE:				
	BEES:			
Brachicoma devia, *B.sarcophagina*, *B. setosa*	*Bombus* spp.	L, P		203,289
Mellitobia acasta, *M.chalybii*, *M. europaea* *Pediobius williamsoni*	*Bombus* spp., *Psithyrus*	L, P		6,83,199, 203,247

Hymenoptera (cont.)	Host	Stage[a]	Remarks	Refs.
	WASPS:			
Elasmus biroi, E. japonicus, E. polistis, E. schmitti (= *E.invreae*)	*Polistes* spp.			207,341
Melittobia australica, M.chalybii *Pediobius ropalidiae* *Tetrastichus nidulans*	*Vespula, Polistes* spp. *Belonogaster* spp.	P	Can kill colony. Hyperparasites of tachinid flies in nest.	165,201,207, 230,341
	ANTS:			
Alucha floridensis *Paracrias*?	*Camponotus, Crematogaster*	L, P		74,332
	WASPS:			
*Melittobia australica, M.acasta?, M.chalybii, M.*sp. *Tetrastichus* *Pediobius ropalidia* *Syntomosphyrum*	*Polistes, Vespula* spp., other Vespidae *Belonogaster, Parischnogaster*	B, L, P	*M.australica* only found in incipient colonies.	201,230,247, 268,273,299, 318
ICHNEUMONIDAE:				
	BEES:			335
Venturia canescens	*Bombus* spp.			
	WASPS:			
Arthula flavofasciata, A.formosana, A. sp.	*Polistes* spp., *Ropalidia* spp.	B, L	Many cells infected.	141,207,341
Christelia arvalis *Ephialtus extensor* *Latibulus argiolus, L.siculus* *Mesostenus gladiator*	*Polistes* spp., *Sulcopolistes*	B, L, P		207,208,268, 341
Camptotypus (*Hemipimpla*) *apicalis* *Hemipimpla pulchripennis*	*Belonogaster* spp., *Ropalidia*	L, P	*Hemipimpla* a parasite of Lepidoptera.	43,165,207
Pachysomoides iheringi, P.flavescens, P.(*Sphecophaga*) *fulvus* (= *P.fulvipes, P.arvalis*), *P.stupidus, P.vespicola*	*Polistes* spp.	B, L, P	*P.fulvus = P. fulvescens* of Nelson 1968	176,207,230, 265,331,341
Polistophaga spp. *Sphecophaga vesparum burra, S.* (*Anomalon*) *v. vesparum*	*Dolichovespula* spp., *Polistes* spp., *Vespula* spp., *Paravespula* spp.	L, P		80,112,207, 230,265,299, 341
Toechorychus abactus, T.albimaculatus, T.cassunungae	*Mischocyttarus* spp., *Polistes canadensis*	B		207,341
MUTILIDAE:				
	BEES:			
Mutilla europaea (*Smicromyrme* ?)	*Bombus* spp., Apidae	L, P	Also feeds on the larval provisions.	6,111

Hymenoptera· (cont.)	Host·	Stage[a]	Remarks·	Refs.·
	ANTS:			
Ponerotilla clarki, P.crinata, P.lissantyx	*Brachyponera lutea*		Likely an endoparasite of larval workers.	45
	WASPS:			
Dasymutilla castor (=*D.mediatoria*) *Pycnotilla barvara brutia* *Tropidotilla littoralis*	*Polistes* spp.	B		207,230,341
PTEROMALIDAE:				
	ANTS:			
Pheidoloxenus wheeleri	*Pheidole instabilis*	L, W?		210,247
	WASPS:			
Nonidentified pteromalids	*Dolichovespula saxonica*	L, P		112
THYSANIDAE:				
Signiphora polistimyiella	*Polistes* sp.			230,272
TIPHIIDAE:				
Myrmosa melanocephala	*Dolichovespula sylvestris*	W?	Only known from one colony.	299
TORYMIDAE:				
	BEES:			
Monodontomerus montivagus	*Bombus morrisoni*			247
	WASPS:			
Monodontomerus minor, M.sp.	*Polistes, Mischocyttarus*	L, P	Rare.	190,230,341
TRIGONALIDAE:				
Bakeronymus typicus seidakka	*Parapolybia*	L		343
Barcogonalos canadensis *Poecilogonalos* spp. *Pseudogonalos hahni*	*Vespa* spp.	L	*Poecilogonalos* first endo- then ectoparasitic.	142,342
Nomadina cisandina *Pseudonomadina biceps*	*Stelopolybia, Polybia* spp. *Ropalidia* (*Icarelia*)	L, P	Resembles host in appearance.	142,207,342
Seminota depressa, S. sp. (nr. *depressa*), *S.marginata, S. mejicana*	*Parachartergus*	L, P		142,143,207, 341

NOTES: (a) Stages of host attacked are E: egg, L: larvae, P: pupa, W: adult worker, S: sexual. May only apply to some cases in the group. Species listed here are generally regarded as true parasites, but exact relationship often not precisely known. (b) This group of parasitic wasps (Eucharitidae) is specialized on ants; after ref.(144).

Appendix 2.10 Diptera

Diptera[a]	Host·	Stage[b]	Remarks·	Refs.·
PHORIDAE:[c]				
	BEES:[d]			
Borophaga incrassata	*Apis mellifera*	L	Lays eggs on larvae.	17
Apocephalus borealis *Megaselia* sp.	*Bombus* spp., *Apis cerana*	A	Feeds on thorax muscles.	
Melaloncha: 5 spp.	*Trigona* spp., *Apis, Bombus*	A	In thorax or abdomen.	274
Pseudohypocera kerteszi, P. spp.	*Apis, Melipona* spp., *Cephalotrigona, Oxytrigona, Tetragona, Trigona*	L, A?	Attack larvae of entire colonies. Can be very destructive. (D.Roubik, pers. comm.)	262
Nonidentified phorids	*Lestrimelitta*		Many phorids arrive when nest damaged.	274
	ANTS:			
Acanthophorides: 9 spp.	*Eciton, Labidus*	A		
Aenigmatis dorni, A.franzi, A.lubbocki	*Formica* spp., *Lasius, Myrmica*	P	*Aenigmatis* females are apterous.	
Apocephalus: 39 spp.	*Acromyrmex, Atta, Azteca,Camponotus, Eciton, Formica, Iridomyrmex, Labidus,Nomamyrmex, Paraponera, Pheidole, Solenopsis*	A, (L)	Typically female attacks host ant from air outside nest and oviposits on head and thorax, or inside gaster. Often attacks majors.	35,47, 97,98, 328,338
Apodicrania termilophila	*Solenopsis* spp.	L	Parasite tended by the ants.	338,339
Ceratoconus ecitophilus, C.setipennis	*Eciton, Iridomyrmex* (*Linepithema*)	A?		262
Apterophora attophila, A.caliginosa *Auxanommatidia myrmecophila* *Borgmeierphora kempfi, B.multisetosa* *Cataclinusa* sp.	*Atta, Camponotus, Eciton* spp., *Iridomyrmex* (*Linepithema*), *Pachycondyla, Solenopsis*	A(?)	*Cataclinusa*: female lays egg on adult ant.	132,262
Cremersia: 10 spp.	*Eciton, Labidus, Neivamyrmex, Nomamyrmex*	A		
Dacnophora legionis, D.setithorax	*Labidus, Neivamyrmex*	A		
Diocophora: 5 spp.	*Camponotus, Eciton, Nomamyrmex*	A	Some attack head.	
Iridophora clarki *Macrocerides* sp. *Megaselia kondongi, M.miki, M.persecutrix, M.sembeli*	*Camponotus* spp., *Crematogaster, Iridomyrmex, Pheidologeton*	A	Female attacks foraging workers.	34,36, 76,79

Diptera[a] (cont.)	Host·	Stage[b]	Remarks·	Refs.·
Menozziola schmitzi (=*M.camponoti*) *Microselia deemingi, M.rivierae, M.southwoodi, M.texana*	*Camponotus* spp., *Crematogaster, Paratrechina, Pheidole, Solenopsis*	A	*Menozziola* in gaster of *C.herculaneus* queens.	49
Myrmosicarius: 11 spp.	*Acromyrmex, Atta, Eciton, Labidus, Nomamyrmex, Solenopsis*	A		328
Neodohrniphora: 5 spp.	*Acromyrmex, Atta*	A	Typically lays egg into head capsule. Often majors attacked.	88,98, 239,262
Plastophora *Plastophorides bequaerti* *Pradea iniqua* *Procliniella hostilis*	*Solenopsis, Pheidole, Labidus, Acromyrmex*	A	Female lays egg on workers.	340
Pseudaceton: 31 spp.	*Camponotus, Crematogaster, Dorymyrmex, Formica, Iridomyrmex, Lasius, Mecrocerides, Myrmica, Neivamyrmex, Paratrechina, Solenopsis, Tapinoma*	A	Females typically hover over foraging columns or mating flights, oviposit into thorax or gaster. Larvae may migrate to head. Many species synonyms exist.	77,96, 250,251, 259,339
Puliciphora boltoni *Rhyncophoromyia conica, R.maculineura, R.gymnopleura* *Stenoneurellis laticeps*	*Acropyga, Camponotus* spp., *Dolichoderus* *Acromyrmex*	A	In gaster. Attacks injured workers.	48,262
Styletta camponoti, S.crocea *Thalloptera comes* *Trucidophora camponoti*	*Camponotus* spp., *Eciton*	?A	Female may enter colony. *Trucidophora* prefers queens in all species.	35,46
	WASPS:			
Apocephalus borealis *Gymnoptera vitripennis*	*Vespula* spp., *Vespa*	A	A factor for colony failure.	299
Megaselia sp.	*Mischocyttarus labiatus*	L?		190,341
Triphleba lugubris	*Dolichovespula, Vespa, Vespula* spp.	A		
	TERMITES:			
Apterophtora attophila, A.caliginosa	*Nasutitermes* sp.	A?		262
Dicranopteron setipennis *Diplonevra mortimeri, D.watsoni*	*Odontotermes, Bulbitermes, Nasutitermes* spp.	N, A	In young nymphs. Eggs in abdomen of adults.	

Diptera[a] (cont.)	Host·	Stage[b]	Remarks·	Refs.·
Misotermes exenterans *Palpiclavina kistneri* *Puliciphora collinsi*, *P.decachete* *Thaumatoxena andreinii*, *Th.wasmanni*	*Macrotermes*, *Odontotermes*	A, N	*P.kistneri* lays eggs on nymphs. Often larvae in head of soldiers.	
Nonidentified phorid	*Acanthotermes*	A	In head.	
CONOPIDAE:[e]				
	BEES:			
Conops: 6 spp.	*Bombus* spp.			
Dalmannia marginata. *D. pacifica*, *D. signata*	*Halictus*, *Lasioglossum* spp.			
Myopa buccata	*Bombus* spp. *Apis mellifera*			
Physocephala: 13 spp.	*Bombus* spp., *Apis mellifera*, *Halictus*		Hyperparasitized by pteromalid wasps (*Habrocytus*).	204,222, 287
Sicus ferruguineus	*Bombus* spp.		Prevalences 20-30%.	287
Thecophora modesta, *Th.occidensis* *Zodion cinereum*, *Z.fulvifrons*	*Dialictus*, *Halictus* spp., *Lasioglossum* *Apis mellifera*, *Bombus* spp		*Th.* mostly found in primitively social halictid bees.	
	WASPS:			
Conops flavipes, *C.scutellatus*	*Vespula* spp., *Vespa* spp.		Said to wait at nest entrance and to pounce on wasps.	64,299
Leopoldius coronatus, *L.diadematum* *Myopa buccata* *Physocephala rufipes*	*Vespa*, *Vespula* spp.			
BRAULIDAE:[f]				
	BEES:			
Braula coeca	*Apis mellifera*	B	In brood, later in adults.	
Megabraula sp.	*Apis* spp.	L	Similar to *Braula*.	
DROSOPHILIDAE:				
Drosophila bucksii	*Apis mellifera*	A	In thorax.	
EPHYDRIDAE:				
	ANTS:			
Rhynchopsilopa nitidissima	*Crematogaster* spp.	A	Predatory?	102
HELOSCIOMYZIDAE:				
Heliosciomyza subalpina	*Chelaner*, *Prolasius*	L	Predatory?	26

Diptera[a] (cont.)	Host	Stage[b]	Remarks	Refs.
SARCOPHAGIDAE:				
	BEES:			
Boettcharia litorosa	*Bombus* spp.	A		279
Brachicoma devia, *B. sarcophagina*	*Bombus* spp.	B	Can be very destructive.	6,72
Helicobia morionella	*Bombus sonorus*			279
Sarcophaga sarracenioides, *S.surrubea* *Senotainia tricuspis*	*Bombus* spp., *Apis mellifera*	A	*Senotainia*: larva deposited externally, burrows into host.	202
	WASPS:			
Macronychia conica	*Polistes gallicus*			207
Sarcophaga bullata (?), *S. polistensis*	*Polistes* spp.	L	At base of brood cells. Destroys larvae.	207,230
SYRPHIDAE:				
Volucella inanis, *V.pellucens*, *V.zonaria*	*Paravespula* spp.	L, P	Larva actively seeks out host.	276
TACHINIDAE:				
	BEES:			
Rondanioestrus apivorus	*Apis mellifera*	W, D?	Attacks workers in flight and deposits larva on bee.	297
	ANTS:			
Strongylogaster (*Tamiclea*) *globula*	*Lasius*	A	Endoparasites of young queens, later cared for by workers.	120
	WASPS:			
Anacamptomyia: 5 spp. (*Roubaudia*)	*Belonogaster*, *Icaria*, *Polistes*, *Ropalidida*	P		43,108, 165,207
Envespivora decipiens *Koralliomyia portentosa*	*Polistes*, *Ropalidia*, *Liostenogaster*, *Parischnogaster*	?, L		108,207
Ophiron sp. *Polybiophila fitzgeraldi*, *P.* sp. *Telothyriosoma polybia*	*Polybia* spp., *Mischocyttarus*			207

NOTES: (a) For large groups, only number of identified species given. (b) The stages attacked are adults (A), workers (W), drones (D), larvae (L), pupa (P), brood (pupae, larvae, or eggs), nymphs (N, in termites). (c) Based on ref.(78); additional references given. Phorid flies where parasitic status is unclear are not listed. (d) According to ref.(78), the identifications in honeybees before ref.(238) are not reliable. (e) Based on ref.(286); additional references given. Stage attacked is the adult in all known cases. (f) In the following groups: for *Apis* based on ref.(17), for *Bombus* based on ref.(200); additional references given.

Appendix 2.11 Other Arthropods

Arthropods[a]	Host·	Remarks·	Refs·
STREPSIPTERA:[b]			
	BEES:		
Halictoxenos: 6 spp.	*Halictus*, *Lasioglossum* (*Evylaeus*) spp.	Female parasite. Male is unknown. Perhaps identical to ref.(280).	248
	ANTS:		
Caenocholax brasiliensis	*Pheidole* spp.	Only a few workers infected in nest.	236,313
Myrmecolacidae spp.	*Camponotus*, *Eciton*, *Pheidole*, *Pseudomyrmex*, *Solenopsis*	A heteroxenic taxon, i.e., males and females use different hosts.	338
Myrmecolax sp.	*Camponotus*, *Pseudomyrmex*, *Solenopsis*	Regularly found.	236
Stichotrema dallatorr-eanum, *S.vilhena*, *S.wigodzinsky*	*Camponotus*, *Pseudomyrmex*, *Solenopsis*	Male on queen. Females parasitic on Orthoptera and mantids.	147,162, 313
Caenocholax fenyesi	*Solenopsis invicta*	Endoparasite of larvae.	163,164
	WASPS:		
Belonogastechtrus zavattarii	*Belonogaster lateritius elegans*		
Clypoxenus americanus	*Clypeopolybia duckei*		
Megacanthopus flavitarsis	*Mischocyttarus flavitarsis*		
Vespaxenos crabronis, *V.moutoni* (*buyssoni*)	*Paravespula*, *Vespa* spp.		299
Xenos: 26 spp.	*Polistes*, *Provespa.*, *Vespa*, *Vespula*	Female in workers and males. Male parasite unknown. As adult female in queens.	81,171, 173,175, 307
Unspecified Xeninae	*Mischocyttarus flavitarsis*	Stylopized queens overwinter but do not found colonies.	190
Unspecified spp.	*Polybia* (*Parapolybia*), *Ropalidia*	Found in adults.	108,173, 178
Non-identified strepsiptera	*Belonogaster*, *Mischocytt-arus*, *Nectarina*, *Polistes*, *Polybia*, *Ropalidia*, *Vespa*	Parasite prefers dorsal position in abdomen, with multiple infections.	190
COLEOPTERA:[c]			
	BEES:		
Melöe cicatricosus, *M.laevis*, *M.variegatus*, *M.* spp.	*Apis mellifera*	Triungulins burrow into joints of abdomen and extract hemolymph. Common locally.	17,264

Arthropods[a] (cont.)	Host	Remarks	Refs.
Opimelöe sp.	*Bombus* spp.	A flightless beetle. Probably a parasite.	274
Rhipiphorus sp.	*Apis florea*	Larva endoparasitic in bee larva.	3
	WASPS:		
Metoecus paradoxus, *M.vespae*	*Dolichovespula* spp., *Paravespula* spp., *Vespula* spp.	Initially develop as endoparasites of wasp larvae, later become ectoparasitic.	9,64,32, 100,112, 123,299
LEPIDOPTERA:[c]			
	BEES:		
Achroia grisella *Ephestia cautella*	*Apis mellifera*, *Bombus*?	Attacks brood comb.	6,17
Aphomia sociella	*Bombus* spp.	Usually destroys entire colonies. Attacks brood. Host nest located by smell.	6,289
Ephestia kühniella	*Bombus fervidus*	Feeds mainly on provisions, perhaps on brood.	6
Galleria melonella	*Apis mellifera*, *Bombus*?	Can kill small colonies. A serious threat.	17
Vitula edmandsii	*Bombus* spp.	The North American equivalent to *Aphomia*.	6
	ANTS:		
Maculinea teleius	*Myrmica laevinodis*	Attacks ant larvae. Finds host by following ant trails.	292
Tarucus waterstradti vileja	*Crematogaster*	Said to damage "habitat" of ants.	214
	WASPS:		
Antipolistes anthracella *Antispila* sp.	*Polistes* spp.	Seen emerging from nest cells.	268
Aphomia sociella	*Dolichovespula* (*Vespula*)		299
Chalcoela iphitalis	*Mischocyttarus* spp., *Polistes* spp.	Attacks larvae. Common cause of colony mortality.	165,190, 207,230, 268,308
Dicymolomia pegasalis	*Polistes* spp.	Attacks larvae. Nests at sheltered places have fewer parasites. May have eliminated wasps from islands.	207,230, 267,268
Epithectis sphecophila *Taeniodictys servicella* *Tinea fuscipunctella*	*Polistes* spp.	Found emerging from nest.	268

NOTES: (a) In large groups, only number of identified species given. (b) Strepsiptera after refs.(172),(174), and (281); additional references given. Ref.(207) also lists a number of Neuroptera as parasites of social wasps (e.g. in *Polybia* spp.), but status is unconfirmed. (c) These groups also contain many species with unclear or commensalistic associatons which are not listed here; see ref.(176) for details.

REFERENCES FOR APPENDIX 2

(1) Abrol, D.P. and Bhat, A.A. 1990. Journal of Animal Morphology and Physiology 37: 1-2; (2) Abrol, D.P. and Putatunda, B.N. 1995. Current Science 68: 90; (3) Ahmadi, A.A. 1988. Bee World 69: 159-161; (4) Alford, D.V. 1968. Transactions of the Royal Entomological Society 120: 375-393; (5) Alford, D.V. 1969. Journal of Apicultural Research 8: 49-54; (6) Alford, D.V. 1975. Bumblebees. Davis-Poynter, London; (7) Allen, G.E. and Buren, W.F. 1974. Journal of the New York Entomological Society 82: 125-130; (8) Allen, G.E. and Silveira-Guido, A. 1974. Florida Entomologist 57: 327-329; (9) Askew, R.R. 1971. Parasitic Insects. Heinemann, London; (10) Avery, S.W. et al. 1977. Florida Entomologist 60: 17-20; (11) Ayre, G.L. 1962. Canadian Journal of Zoology 40: 157-164; (12) Bailey, L. 1958. Parasitology 48: 493-506; (13) Bailey, L. 1963. Journal of Insect Pathology 5: 198-205; (14) Bailey, L. 1967. Annals of Applied Biology 60: 43-38; (15) Bailey, L. 1969. Annals of Applied Biology 63: 483-491; (16) Bailey, L. 1981. Honey Bee Pathology. Academic Press, London; (17) Bailey, L. and Ball, B.V. 1991. Honey Bee Pathology. Academic Press, London; (18) Bailey, L. et al. 1981. Annals of Applied Biology 97: 109-118; (19) Bailey, L. et al. 1983. Annals of Applied Biology 103: 13-20; (20) Bailey, L. et al. 1976. Journal of General Virology 31: 459-461; (21) Bailey, L. and Gibbs, A.J. 1964. Journal of Insect Pathology 6: 395-407; (22) Balazy, S. et al. 1986. Cryptogamie, Mycologie 7: 37-45; (23) Balazy, S. and Wisniewski, J. 1982. Bulletin de L'Academie Polonaise des Sciences , Sciences Biologiques 30: 81-84; (24) Banks, N. 1915. Psyche 22: 60-61; (25) Banks, N. and Snyder, T.E. 1920. U.S. National Museum Bulletin 108: 1-228; (26) Barnes, J.K. 1980. New Zealand Journal of Zoology 7: 221-229; (27) Bartel, M.H. 1965. Journal of Parasitology 51: 800-806; (28) Batra, S.W. 1980. Journal of the Kansas Entomological Society 53: 112-114; (29) Bedding, R.A. 1985. In W.R. Nickle, eds. Plant and Insect Nematodes, pp.755-95. Marcel Dekker, New York; (30) Beutler, R. and Opfiger, E. 1949. Zeitschrift für Vergleichende Physiologie 32: 383-421; (31) Bhasker, S. et al. 1989. Progress in Acarology 2: 287-289; (32) Blackith, R.E. and Stevenson, J.H. 1958. Insectes Sociaux 5: 347-352; (33) Blackwell, M. and Rossi, W. 1986. Mycotaxon 25: 581-601; (34) Borgmeier, T. 1968. Studia Entomologica 11: 1-367; (35) Borgmeier, T. 1971. Studia Entomologica 14: 1-172; (36) Borgmeier, T. 1971. Studia Entomologica 14: 177-224; (37) Bowman, C.E. and Ferguson, C.A. 1985. Bee World 66: 51-53; (38) Bradbear, N. 1988. Bee World 69: 15-39; (39) Brangham, A.N. 1959. Entomologist's Gazette 10: 111-131; (40) Briano, J. et al. 1995. Florida Entomologist 78: 531-537; (41) Briano, J. et al. 1995. Journal of Economic Entomology 88: 1233-1237; (42) Briano, J.A. et al. 1995. Environmental Entomology 24: 1328-1332; (43) Brooks, R.W. and Wahl, D.B. 1987. Journal of the New York Entomological Society 95: 547-552; (44) Broome, J.R. et al. 1976. Journal of Invertebrate Pathology 28: 87-91; (45) Brothers, D.J. 1994. Journal of the Australian Entomological Society 33: 143-152; (46) Brown, B.V. 1988. The Canadian Entomologist 120: 307-322; (47) Brown, B.V. and Feener, D.H. 1991. Biotropica 23: 182-187; (48) Brown, B.V. and Feener, D.H. 1993. Journal of Natural History 27: 429-434; (49) Brown, B.V. et al. 1991. Entomologica Scandinavica 22: 241-250; (50) Burgett, D.M. et al. 1990. Canadian Journal of Zoology 68: 1423-1427; (51) Burgett, M. et al. 1983. Bee World 64: 25-28; (52) Burks, B.D. 1971. Journal of the Washington Academy of Sciences 61: 194-196; (53) Burnside, C.E. 1928. Journal of Economic Entomology 21: 379-386; (54) Buschinger, A. 1973. Zoologischer Anzeiger 191: 369-380; (55) Carney, W.P. 1969. American Midlands Naturalist 82: 605-611; (56) Casanueva, M.E. 1992. Boletin da Sociedad de Biologia 63: 51-53; (57) Case, A.A. and Ackert, J.E. 1940. Transcations of the Kansas Academy of Sciences 43: 393-396; (58) Chmielewski, W. 1991. In F. Dusbabek and V. Bukva, eds. Modern Acarology, pp.615-619. SPB Academic Publishing, The Hague; (59) Clark. 1982. Science 217: 57-59; (60) Clark, T.B. 1977. Journal of Invertbrate Pathology 29: 112-113; (61) Clark, T.B. 1982. Journal of Invertebrate Pathology 39: 119-122; (62) Clark, T.B. et al. 1985. International Journal of Systematic Bacteriology 35: 296-308; (63) Clausen, C.P. 1923. Annals of the Entomological Society of America 16: 195-217; (64) Clausen, C.P. 1940. Entomophagous Insects. McGraw-Hill Book Company, New York; (65) Clausen, C.P. 1940. Journal of the Washington Academy of Sciences 30: 504-516; (66) Clausen, C.P. 1941. Psyche 48: 57-69; (67) Costa, M. 1966. Journal of Zoology 148: 191-200; (68) Crawley, W.C. 1911. Entomologist's Record and Journal of Variation 23: 22-23; (69) Crawley, W.C. and Baylis, H.A. 1921. Journal of the Royal Microscopic Society London (1921): 353-364; (70) Crosland, M.W.J. 1988. Annals of the Entomological Society of America 81: 481-484; (71) Cross, E.A. 1965. University of Kansas Science Bulletin 45: 29-275; (72) Cumber, R.A. 1949. Transactions of the Royal Entomological Society, London A 24: 119-127; (73) Das, G.M. 1963. Bulletin of Entomological Research 54: 373-378; (74) Davis, L.R., Jr. and Jouvenaz, D.P. 1990. Florida Entomologist 73: 335-337; (75) De Jong, D. et al. 1982. Annual Review of Entomology 27: 229-252; (76) Disney, R.H.L. 1986. Journal of Natural History 20: 777-787; (77) Disney, R.H.L. 1991. Sociobiology 18: 283-298; (78) Disney, R.H.L. 1994. Scuttle Flies: The Phoridae. Chapman & Hall, London; (79) Disney, R.H.L. and Shaw, M.R. 1994.

Entomologist's Monthly Magazine 130: 1564-67; (80) Donovan, B.J. and Read, P.E.C. 1987. New Zealand Journal of Zoology 14: 329-336; (81) Dunkle, S.W. 1979. Psyche 86: 327-336; (82) Durrer, S. and Schmid-Hempel, P. 1995. Ecography 18: 114-122; (83) Edwards, C.J. and Pengelly, D.H. 1966. Proceedings of the Entomological Society of Ontario 96: 98-99; (84) Eickwort, G.C. 1979. In J. Rodriguez, eds. Recent Advances in Acarology, Vol. I, pp.575-581. Academic Press, New York; (85) Eickwort, G.C. 1988. In G.R. Needham, eds. Africanized Honey Bees and Bee Mites, pp.327-338. Ellis Horwood, Chichester; (86) Eickwort, G.C. 1990. Annual Review of Entomology 35: 469-488; (87) Eickwort, G.C. 1994. In M.A. Houck, ed. Mites, pp.218-251. Chapman and Hall, New York; (88) Eidmann, H. 1937. Zeitschrift für Morphologie und Ökologie der Tiere 32: 391-462; (89) Eischen, F.A. et al. 1989. Apidologie 20: 1-8; (90) Espadaler Gelabert, X. and Riasol Boixant, J.M. 1983. Revista Ibérica de Parasitologica 43: 219-227; (91) Evans, H.C. 1989. In N. Wilding et al., eds. Insect-Fungus Interactions, pp.205-238. Academic Press, London; (92) Evans, H.C. and Samson, R.A. 1982. Transactions of the British Mycological Society 79: 431-453; (93) Fantham, H.B. and Porter, A. 1914. Annals of Tropical Medicine and Parasitology 8: 623-638; (94) Fantham, H.B. et al. 1941. Parasitology 38: 186-208; (95) Faucon, J.P. et al. 1992. Apidologie 23: 139-146; (96) Feener, D.H. 1987. Annals of the Entomological Society of America 80: 148-151; (97) Feener, D.H. 1988. Behavioural Ecology and Sociobiology 22: 421-427; (98) Feener, D.H. and Brown, B.V. 1993. Journal of Insect Behavior 6: 675-688; (99) Fisher, R.M. 1989. Journal of the Kansas Entomological Society 62: 581-589; (100) Fox-Wilson, G. 1946. Proceedings of the Royal Entomological Society London 21: 17-27; (101) Franks, N.R. et al. 1991. Journal of Zoology 225: 59-70; (102) Freidberg, A. and Mathis, W.N. 1985. Entomophaga 30: 13-21; (103) Fries, I. et al. 1996. European Journal of Protistology 32: 356-365; (104) Fritz, G.N. 1985. Proceedings of the Helminthological Society Washington 52: 51-53; (105) Fye, R.E. 1966. The Canadian Entomologist 98: 88-89; (106) Fyg, W. 1964. Annual Review of Entomology 9: 207-224; (107) Gabrion, C. et al. 1976. Annales de parasitologie 51: 407-420; (108) Gadagkar, R. 1991. In K.G. Ross and R.W. Matthews, eds. The Social Biology of Wasps, pp.149-190. Comstock Publishing Associates, Ithaca, New York; (109) Gahan, A.B. 1940. Proceedings of the US National Museum 88: 425-458; (110) Gambino, P. and Thomas, G.M. 1988. Pan-Pacific Entomologist 64: 107-113; (111) Gauld, I. and Bolton, B. 1988. The Hymenoptera. Oxford University Press, Oxford; (112) Gauss, R. 1968. Zeitschrift für Angewandte Entomologie 61: 453-545; (113) Gauss, R. 1970. Zeitschrift für Angewandte Entomologie 65: 239-244; (114) Gibbs, A.J. et al. 1970. Virology 40: 1063-1065; (115) Giblin, R.M. et al. 1981. Nematologica 27: 20-27; (116) Gilliam, M. 1973. Annals of the Entomological Society of America 66: 1176; (117) Glare, T.R. et al. 1993. New Zealand Journal of Zoology 20: 95-120; (118) Gogoi, A.R. and Chaudhuri, R.P. 1982. Indian Journal of Animal Science 52: 246-253; (119) Goldblatt, J.W. 1984. Environmental Entomology 13: 1661-1665; (120) Gösswald, K. 1950. Verhandlungen der Deutschen Zoologischen Gesellschaft, Mainz 1949: 256-264; (121) Hassanein, M.H. 1952. Bee World 33: 109-112; (122) Hattingen, R. 1956. Zentralblatt für Bakteriologie, Parasitenkunde, Infektionskrankheiten und Hygiene Abt. II 109: 236-249; (123) Hattori, T. and Yamane, S. 1975. New Entomologist 24 (May): 1-7; (124) Heath, L.A.F. 1985. Bee World 66: 9-15; (125) Heraty, J.M. 1985. Proceedings of the Entomological Society of Ontario 116: 61-103; (126) Heraty, J.M. 1986. Systematic Entomology 11: 183-212; (127) Hermann, H.R. et al. 1970. Proceedings of the Louisiana Academy of Sciences 33: 13-18; (128) Hirschmann, W. 1972. Acarologie 17: 28-29; (129) Hitchcock, J.D. 1948. Journal of Economic Entomology 41: 854-858; (130) Hitchcock, J.D. et al. 1979. Journal of the Kansas Entomological Society 52: 238-246; (131) Hohorst, W. and Graefe, G. 1961. Naturwissenschaften 48: 229-230; (132) Hölldobler, B. and Wilson, E.O. 1990. The Ants. Springer, Berlin; (133) Hölldobler, K. 1928. Biologisches Zentralblatt 48: 129-142; (134) Horn, H. and Eberspächer, J. 1976. Apidologie 17: 307-324; (135) Horsfall, M.W. 1938. Journal of Parasitology 24: 409-421; (136) Huger, A.M. and Lenz, M. 1976. Zeitschrift für Angewandte Entomologie 81: 252-258; (137) Hunter, P.E. and Husband, R.W. 1973. Florida Entomologist 56: 77-91; (138) Husband, R.W. 1967. Proceedings of the 2nd International Congress of Acarology, pp. 287-288; (139) Husband, R.W. 1968. Papers of the Michigan Academy of Science, Arts and Letters 53: 109-112; (140) Ikeda, E. and Sayama, K. 1994. Japanese Journal of Entomology 62: 265-266; (141) Itô, Y. 1983. Journal of Ethology 1: 1-14; (142) Jeanne, R.L. 1970. Psyche 77: 54-69; (143) Jeanne, R.L. 1979. Behavioural Ecology and Sociobiology 4: 293-310; (144) Johnson, D.W. 1988. Florida Entomologist 71: 528-537; (145) Johnson, J.B. et al. 1986. Proceedings of the Entomological Society of Washington 88: 542-549; (146) Jones, M.F. and Horsfall, M.W. 1936. Science 83: 303-304; (147) Jouvenaz, D.P. 1983. Florida Entomologist 66: 111-121; (148) Jouvenaz, D.P. 1984. In T.C. Cheng, ed. Comparative Pathobiology, pp.195-203. Plenum Press, New York; (149) Jouvenaz, D.P. 1986. In C.S. Lofgren and R.K. Vander Meer, eds. Fire Ants and Leaf-Cutting Ants, Biology and Management, pp.327-38. Westview Press, Boulder, Colorado; (150) Jouvenaz, D.P. et al. 1977. Florida Entomologist 60: 275-279; (151) Jouvenaz, D.P. and Anthony, D.W. 1979. Journal of Protozoology 26: 354-356; (152) Jouvenaz, D.P. et al. 1980. Florida Entomologist 63:

345-346; (153) Jouvenaz, D.P. and Ellis, E.A. 1986. Journal of Protozoology 33: 457-461; (154) Jouvenaz, D.P. and Kimbrough, J.W. 1991. Mycological Research 95: 1395-1401; (155) Jouvenaz, D.P. et al. 1981. Journal of Invertebrate Pathology 37: 265-268; (156) Jouvenaz, D.P. and Wojcik, D.P. 1990. Florida Entomologist 73: 674-675; (157) Joyeux, C. and Baer, J.-G. 1936. In Faune de France, Federation Francaise des Sociétés de Sciences Naturelles. Office Central de Faunistique, Paris; (158) Kaiser, H. 1986. Zoologischer Anzeiger 217: 156-177; (159) Kalavati, C. 1976. Acta Protozoologica 15: 411-422; (160) Kalavati, C. and Narasimhamurti, C.C. 1976. Journal of Parasitology 62: 323-324; (161) Kaschef, A. and Abou-Zeid. 1987. In J. Eder and H. Rembold, eds. Chemistry and Biology of Social Insects, pp.638. Verlag J.Peperny, München; (162) Kathiritamby, J. 1991. In Insects of Australia, pp.684-695. Melbourne University Press, Melbourne; (163) Kathirithamby, J. and Hamilton, W.D. 1995. Nature 374: 769-770; (164) Kathirithamby, J. and Johnston, J.S. 1992. Annals of the Entomological Society of America 85: 293-297; (165) Keeping, M.G. and Crewe, R.M. 1983. Journal of the Entomological Society of South Africa 46: 309-323; (166) Kermarrec, A. and Mauleon, H. 1975. Annales de Parasitologie 50: 351-360; (167) Kevan, P.G. et al. 1990. Bee World 71: 119-121; (168) Khan, K.I. et al. 1977. Pakistan Journal of Scientific Research 29: 12-13; (169) Khan, K.I. et al. 1981. Material und Organismen 16: 189-197; (170) Khan, K.I. et al. 1994. Pakistan Journal of Zoology 26: 109-111; (171) Kifune, T. 1986. Kontyû 54: 84-88; (172) Kifune, T. 1991. Japanese Journal of Entomology 59: 367-374; (173) Kifune, T. and Maeta, Y. 1985. Kontyû 53: 426-435; (174) Kifune, T. et al. 1994. Esakai 34: 209-214; (175) Kifune, T. and Yamane, S. 1991. Japanese Journal of Entomology 59: 104; (176) Kistner, D.H. 1982. In H.R. Hermann, ed. Social Insects. Vol.III., pp.1-244. Academic Press, London; (177) Kloft, W. 1950. Zoologisches Jahrbuch. Abteilung für Systematik, Ökologie und Geographie der Tiere 78: 526-30; (178) Krombein, K.V. 1967. Trap-nesting Wasps and Bees: Life histories, Nests and Associates. Smithsonian Institution, Washington D.C.; (179) Kudo, R. 1924. Illinois Biological Monographs 9: 9-60, 90-101, 103-105; (180) Kudo, R.R. 1941. Journal of Morphology 71: 307-333; (181) Kudo, R.R. 1943. Journal of Morphology 72: 263-277; (182) Kutter, H. 1958. Mitteilungen der Schweizerischen Entomologischen Gesellschaft 31: 313-316; (183) L'Arrivée, J.C.M. 1965. American Bee Journal 105: 246-248; (184) L'Arrivée, J.C.M. 1965. Journal of Invertebrate Pathology 7: 408-413; (185) Lachaud, J.P. and Passera, L. 1982. Insectes Sociaux 29: 561-567; (186) Landerkin, G.B. and Katznelson, H. 1959. Canadian Journal of Microbiology 5: 169-172; (187) Langridge, D.F. and McGhee, R.B. 1967. Journal of Protozoology 14: 485-487; (188) Leatherdale, D. 1970. Entomophaga 15: 419-435; (189) Lipa, J.J. and Triggiani, O. 1992. Apidologie 23: 533-536; (190) Litte, M. 1979. Zeitschrift für Tierpsychologie 50: 282-312; (191) Lofgren, C.S. et al. 1975. Annual Review of Entomology 20: 1-30; (192) Logan, J.W.M. et al. 1990. Bulletin of Entomological Research 80: 309-330; (193) Loiacono, M.S. 1985 (1987). Revista de la Societa Entomologica Argentina 44: 129-36; (194) Loos-Frank, B. and Zimmermann, G. 1976. Zeitschrift für Parasitenkunde 49: 281-289; (195) Lucius, R. et al. 1980. Zeitschrift für Parasitenkunde 63: 271-275; (196) Lundberg, H. and Svensson, B.G. 1975. Norwegian Journal of Entomology 22: 129-134; (197) MacFarlane, R.P. 1975. New Zealand Entomologist 6: 79; (198) MacFarlane, R.P. 1976. Mycopathologia 59: 41-42; (199) MacFarlane, R.P. and Donovan, B.J. 1989. Proceedings of the 42nd New Zealand Weed and Pest Control Conference 274-277; (200) MacFarlane, R.P. et al. 1995. Bee World 76: 130-148; (201) MacFarlane, R.P. and Palma, R.L. 1987. New Zealand Journal of Zoology 14: 423-425; (202) MacFarlane, R.P. and Pengelly, D.H. 1974. Proceedings of the Entomological Society of Ontario 105: 55-59; (203) MacFarlane, R.P. and Pengelly, D.H. 1977. Proceedings of the Entomological Society of Ontario 108: 31-35; (204) Maeta, Y. and MacFarlane, R.P. 1993. Japanese Journal of Entomology 61: 493-509; (205) Maeyama, T. et al. 1994. Sociobiology 24: 115-119; (206) Magis, M.N. 1989. Bulletin et Annales de la Société Royale Belge d'Entomologie 125: 296; (207) Makino, S. 1985. Sphecos 10: 19; (208) Makino, S. 1989. Researches on Population Ecology 31: 1-10; (209) Makino, S. and Sayama, K. 1994. Japanese Journal of Entomology 62: 377-383; (210) Mann, W.M. 1914. Psyche 21: 171-184; (211) Marikovsky, P.I. 1962. Insectes Sociaux 9: 173-179; (212) Markin, G.P. and McCoy, C.W. 1968. Annals of the Entomological Society of America 61: 505-509; (213) Marsh, P.M. et al. 1987. Memoirs of the Entomological Society of Washington 13: 1-98; (214) Maschwitz, U. et al. 1985. Nachrichten des Entomologischen Vereins Apollo, N.S. 6: 181-200; (215) Masner, L. 1959. Insectes Sociaux 6: 361-367; (216) Matthews, R.W. 1991. In K.G. Ross and R.W. Matthews, eds. The Social Biology of Wasps, pp.570-602. Comstock Publishing, Ithaca, New York; (217) McCoy, C.W. et al. 1988. In C.M. Ignoffo, ed. CRC Handbook of Natural Pesticides. Vol. V Microbial Insecticides. Part A Entomogenous Protozoa and Fungi, pp.237-236. CRC Press, Boca Raton, Florida; (218) McGhee, R.B. and Cosgrove, W.B. 1980. Microbiological Review 44: 140-173; (219) McInnes, D.A. and Tschinkel, W.R. 1996. Annals of the Entomological Society of America 89: 231-237; (220) Medler, J.T. 1962. The Canadian Entomologist 94: 825-833; (221) Merrill, J.H. and Ford, A.L. 1916. Journal of Agricultural Research 6: 115-127; (222) Mihajlovic, L. et al. 1989. Journal of the Kansas Entomological Society 62: 418-420; (223) Mitchell, G.B. and Jouvenaz, D.P.

1985. Florida Entomologist 68: 492-493; (224) Morel, G. and Fouillaud, M. 1992. Journal of Invertebrate Pathology 60: 210-212; (225) Morgenthaler, O. 1926. Schweizerische Bienenzeitung 49: 176-224; (226) Mossadegh, M.S. 1991. Apidologie 22: 127-134; (227) Mouches, C. et al. 1984. Annales de microbiologie 135: 151-155; (228) Muesebeck, C.F.W. 1941. Bulletin of the Brooklyn Entomological Society 36: 200-201; (229) Muir, D.A. 1954. Nature 173: 688-689; (230) Nelson, J.M. 1968. Annals of the Entomological Society of America 61: 1528-1539; (231) Nickle, W.R. and Ayre, G.L. 1966. Proceedings of the Entomological Society of Ontario 96: 96-98; (232) Nixon, M. 1982. Bee World 63: 23-42; (233) O'Connor, B.M. 1994. In M.A. Houck, ed. Mites, pp.136-159. Chapman & Hall, New York; (234) O'Rourke, F.J. 1956. Insectes Sociaux 3: 107-118; (235) Oertel, E. 1965. American Bee Journal 105: 10-11; (236) Ogloblin, A.A. 1939. Verhandlungen des VII Internationalen Kongress für Entomologie 2: 1277-1285; (237) Oi, D.H. and Pereira, R.M. 1993. Florida Entomologist 76: 63-73; (238) Örösi-Pal, Z. 1938. Bee World 19: 64-68; (239) Orr, M.R. 1992. Behavioural Ecology and Sociobiology 30: 395-402; (240) Otis, G.W. and Scott-Dupree, C.D. 1992. Journal of Economic Entomology 85: 40-46; (241) Parker, H.L. 1937. Proceedings of the Entomological Society of Washington 39: 1-3; (242) Pascovici, V.D. 1983. Rev Padurilor Ind Lemnului Celul Hirtie Cel Hirtie 98: 148-149; (243) Passera, L. 1975. Année Biologique 14: 228-259; (244) Passera, L. 1976. Insectes Sociaux 23: 559-575; (245) Patterson, R.S. and Briano, J. 1990. In G.K. Veeresh et al., eds. Social Insects and the Environment, pp.625-626. Oxford & IBH Publishing, New Delhi; (246) Paulian, R. 1949. Une Naturaliste en Côte d'Ivoire. Editions Stock, Paris; (247) Peck, O. 1963. The Canadian Entomologist, Supplement 30: 1-162; (248) Pekkarinen, A. 1971. Acta Entomologica Fennica 28: 40-41; (249) Pemberton, C.E. 1928. Proceedings of the Hawaiian Entomological Society 7: 148-150; (250) Pesquero, M.A. et al. 1993. Florida Entomologist 76: 179-181; (251) Pesquero, M.A. et al. 1995. Journal of Applied Entomology 119: 677-678; (252) Petersen, J.J. 1985. Advances in Parasitology 24: 307-344; (253) Plateaux, L. 1972. Annales des Sciences Naturelles, ser. 2 14: 203-220; (254) Plateaux, L. 1990. Actes Colloques Insectes Sociaux 6: 43-50; (255) Plateaux, L. and Péru, L. 1987. In J. Eder and H. Rembold, eds. Chemistry and Biology of Social Insects, pp.46-47. Peperny Verlag, München; (256) Poinar, G.O. 1975. Entomogenous Nematodes - A Manual and Host List of Insect-Nematode Associations. E.J.Brill, Leiden; (257) Poinar, G.O. et al. 1976. Nematologica 22: 360-370; (258) Poinar, G.O. and Van der Laan, P.A. 1972. Nematologica 18: 239-252; (259) Porter, S.D. et al. 1995. Environmental Entomology 24: 474-479; (260) Pouvreau, A. 1964. Bulletin de la Société Zoologique de France 89: 717-719; (261) Pouvreau, A. 1974. Apidologie 5: 39-62; (262) Prado, A.P.d. 1980. Studia Entomologica 19: 561-609; (263) Prest, D.B. et al. 1974. Journal of Invertebrate Pathology 24: 253-255; (264) Quintero M., M.T. and Canales, I. 1987. Veterinaria México 18: 135-138; (265) Rabb, R.L. 1960. Annals of the Entomological Society of America 53: 111-121; (266) Rajagopal, D. 1984. Journal of Soil Biology and Ecology 4: 102-107; (267) Rau, P. 1930. The Canadian Entomologist 62: 119-120; (268) Rau, P. 1941. Annals of the Entomological Society of America 34: 355-366; (269) Reid, W.M. and Nugara, D. 1961. Journal of Parasitology 47: 885-889; (270) Rettenmeyer, C.W. 1962. Journal of the Kansas Entomological Society 35: 358-360; (271) Richards, L.A. and Richards, K.W. 1976. University of Kansas Science Bulletin. 51: 1-18; (272) Richards, O.W. 1935. Stylops 4: 131-133; (273) Richards, O.W. 1969. Memoria della Societa Entomologica Italiana 48: 79-93; (274) Roubik, D.W. 1989. Ecology and Natural History of Tropical Bees. Cambridge University Press, Cambridge; (275) Royce, L.A. and Krantz, G.W. 1989. Experimental and Applied Acarology 7: 161-165; (276) Rupp, L. 1987. In J. Eder and D.H. Rembold, eds. Chemistry and Biology of Social Insects, pp.642-643. Peperny Verlag, München; (277) Ruschka, F. 1924. Deutsche Entomologische Zeitschrift (1924): 82-96; (278) Ruttledge, W. 1925. Parasitology 17: 187-188; (279) Ryckman, R.E. 1953. Pan-Pacific Entomologist 29: 144-146; (280) Sakagami, S.F. and Hayashida, K. 1968. Journal of the Faculty of Science Hokkaido University, Series VI, Zoology 16: 413-513; (281) Salt, G. and Bequaert, J. 1929. Psyche 36: 249-282; (282) Samson, R.A. et al. 1988. Atlas of Entomopathogenic Fungi. Springer, Berlin; (283) Samson, R.A. et al. 1981. Proceedings of the Koninklijke Nederlandse Akademie van Wetenschappen, Ser.C 84: 289-301; (284) Sanchez-Pena, S.R. et al. 1993. Journal of Invertebrate Pathology 61: 90-96; (285) Schmid-Hempel, P. et al. 1990. Insectes Sociaux 37: 14-30; (286) Schmid-Hempel, R. 1994. Evolutionary Ecology of a Host-Parasite Interaction. Ph.D. Thesis, Universität Basel; (287) Schmid-Hempel, R. and Schmid-Hempel, P. 1996. Ecological Entomology 21: 63-70; (288) Schneider, G. and Hohorst, W. 1971. Naturwissenschaften 58: 327-328; (289) Schousboe, C. 1980. Entomologiske Meddelelser 48: 127-129; (290) Schousboe, C. 1986. Acarology 27: 151-158; (291) Schousboe, C. 1987. Acarology 28: 37-41; (292) Schroth, M. and Maschwitz, U. 1984. Entomologia Generalis 9: 225-230; (293) Schwarz, H.H. et al. 1996. Journal of the Kansas Entomological Society 69: 35-42; (294) Seeman, O.D. and Walter, D.E. 1995. Journal of the Australian Entomological Society 34: 45-50; (295) Shaw, S.R. 1993. Journal of Insect Behavior 6: 649-658; (296) Shykoff, J.A. and Schmid-Hempel, P. 1991. Apidologie 22: 117-125; (297) Skaife, S.H. 1921. South African Journal of Science 17: 196-200; (298) Skou, J.P. et al. 1963. Kongelige Veterinaerskole og Landbohoejskole Daenemark (1963): 27-41; (299) Spradbery, P.J. 1973. Wasps: An Account of the Biology and Natural History of Solitary and Social Wasps. University of Washington Press, Seattle, Washington; (300) Sprague, V. 1977. In L.A.J. Bulla and T.C. Cheng, eds. Comparative Pathobiology. Systematics of the Microsporidia, pp.31-334, 447-461. Plenum Press, New York; (301) Stammer, H.J. 1951. Zoologischer Anzeiger 146: 137-150; (302) Steiger, V. et al. 1969. Archiv für die gesamte Virusforschung 26: 271-282; (303) Stein, G. and Lohmar, E. 1972. Decheniana 124: 135-140; (304) Steinhaus, E.A. 1949. Principles of Insect Pathology. McGraw-Hill, New York; (305) Steinhaus, E.A. and Marsh, G.A. 1967. Journal of Invertebrate Pathology 9: 436-438; (306) Stejskal, M. 1955. Journal of Protozoology 2: 185-188; (307) Strambi, A. and Strambi, C. 1973. Archives d'anatomie microscopique 62: 39-54; (308) Strassmann, J.E. 1981. Behavioural Ecology and Sociobiology 8: 55-64; (309) Stuart, R.J. and Alloway, T.M. 1988. In J.C. Trager, ed. Advances in Myrmecology, pp.537-545. E.J.Brill, Leiden; (310) Svadzhyan, P.K. and Frolkova, L.V. 1966. Zoologiceskij Zurnal 45: 213-219; (311) Sylvester, H.A. and Rinderer, T.E. 1978. American Bee Journal Dec: 806-807; (312) Taylor, R.W. et al. 1993. Sociobiology 23: 109-114; (313) Teson, A. and de Remes Lenicov, A.M.M. 1979. Revista de la Sociedad Entomologica Argentina 38: 115-122; (314) Tewarson, N.C. et al. 1992. Apidologie 23: 161-171; (315) Thorne, B.L. and Kimbrough, J.W. 1982. Mycologia 74: 242-249; (316) Toumanoff, C. 1965. Annales de Parasitologie 40: 611-624; (317) Travis, B.V. 1941. Florida Entomologist 24: 15-22; (318) Turillazzi, S. 1991. In K.G. Ross and R.W. Matthews, eds. The Social Biology of Wasps, pp.74-98. Cornell University Press, Ithaca, New York; (319) Urdaneta-Morales, S. 1983. Revista Brasileira do Biologia 43: 409-412; (320) Van Pelt, A. 1958. Journal of the Tennessee Academy of Science 33: 120-122; (321) Vandel, A. 1934. Annales des Sciences Naturelles 17: 47-58; (322) Vander Meer, R.K. et al. 1989. Journal of Chemical Ecology 15: 2247-2261; (323) Varis, A.L. et al. 1992. Apidologie 23: 133-137; (324) Vitzthum, H.v. 1927. Zeitschrift der wissenschaftlichen Insekten-Biologie 22: 46-48; (325) Wahab, A. 1962. Zeitschrift für Morphologie und Ökologie der Tiere 52: 33-92; (326) Walker, A.K. et al. 1990. Bulletin of Entomological Research 80: 79-83; (327) Wallace, F.G. 1979. In W.H.R. Lumsden and D.A. Evans, eds. Biology of the Kinetoplastida, pp.213-240. Academic Press, London; (328) Waller, D.A. and Moser, J.C. 1990. In R.K. Vander Meer et al., eds. Applied Myrmecology, A World Perspective, pp.255-273. Westview Press, Boulder, CO. 741 p.; (329) Wang, D.I. and Moeller, F.E. 1969. Journal of Invertebrate Pathology 14: 135-142; (330) Weber, N.A. 1945. Review of Entomology 16: 1-88; (331) West-Eberhard, M.J. 1969. Miscellaneous Publications of the Museum of Zoology of the University of Michigan 140: 1-101; (332) Wheeler, G.C. and Wheeler, E.W. 1924. Psyche 31: 49-56; (333) Wheeler, G.C. and Wheeler, E.W. 1937. Annals of the Entomological Society of America 30: 163-175; (334) Wheeler, W.M. 1928. Journal of Experimental Zoology 50: 165-237; (335) Whitfield, J.B. and Cameron, S.A. 1993. Entomological News 104: 240-248; (336) Wilson, W.T. and Nunamaker, R.A. 1983. Bee World 64: 132-136; (337) Wing, M.W. 1951. Transactions of the Royal Entomological Society of London 102: 195-210; (338) Wojcik, D.P. 1989. Florida Entomologist 72: 43-51; (339) Wojcik, D.P. 1990. In R.K. Vander Meer et al., eds. Applied Myrmecology, A World Perspective, pp.329-344. Westview Press, Boulder, Colorado; (340) Wojcik, D.P. et al. 1987. In J. Eder and D.H. Rembold, eds. Chemistry and Biology of Social Insects, pp.627-628. Peperny Verlag, München; (341) Yamane, S. 1996. In S. Turilazzi and M.J. West-Eberhard, eds. Natural History and Evolution of Paper Wasps, pp.75-97. Oxford University Press, Oxford; (342) Yamane, S. and Kojima, J. 1982. Kontyû 50: 183-188; (343) Yamane, S. and Terayama, M. 1983. Memories of the Kagoshima University Research Center in the South Pacific 3: 169-173; (344) Zhang, Z.-Q. 1995. Systematic Entomology 20: 239-246.

References

Akre, R. D., and H. C. Reed. 1983. Evidence of a queen pheromone in *Vespula* (Hymenoptera: Vespidae). *The Canadian Entomologist* 115: 371–377.

Alexander, R. D. 1974. The evolution of social behaviour. *Annual Review of Ecology and Systematics* 15: 165–189.

Alexejenko, F. M., and A. M. Wowk. 1971. Alters- und Rassenwiderstandsfähigkeit der Bienen gegen Akariose und deren neue Bekämpfungsmittel. *Proceedings of the International Bee Breeder Congress* 23: 500–502.

Alford, D. V. 1968. The biology and immature stages of *Syntretus splendidus* (Marshall) (Hymenoptera: Braconidae, Euphorinae), a parasite of adult bumble bees. *Transactions of the Royal Entomological Society* 120: 375–393.

Alford, D. V. 1969. *Sphaerularia bombi* as a parasite of bumble bees in England. *Journal of Apicultural Research* 8: 49–54.

Alford, D. V. 1975. *Bumblebees.* Davis-Poynter, London.

Allen, G. E., and W. F. Buren. 1974. Microsporidian and fungal diseases of *Solenopsis invicta* Buren in Brazil. *Journal of the New York Entomological Society* 82: 125–130.

Allen, G. E., and C. M. Ignoffo. 1969. The nucleopolyhedrosis virus of *Heliothis:* Quantitative in vivo estimates of virulence. *Journal of Invertebrate Pathology* 13: 378–381.

Allen, G. E., and A. Silveira-Guido. 1974. Occurrence of microsporidia in *Solenopsis richteri* and *Solenopsis* sp. in Uruguay and Argentina. *Florida Entomologist* 57: 327–329.

Allen, M. F., B. V. Ball, R. F. White, and J. F. Antoniw. 1986. The detection of Acute Paralysis Virus in *Varroa jacobsoni* by the use of a simple indirect ELISA. *Journal of Apicultural Research* 25: 100–105.

Allison, A. C. 1982. Coevolution between hosts and infectious disease agents, and its effects on virulence. In R. M. Anderson and R. M. May, eds., *Population Biology of Infectious Diseases,* pp. 245–268. Springer-Verlag, Berlin.

Anderson, D. L., and H. Giacon. 1992. Reduced pollen collection by honey bee (Hymenoptera: Apidae) colonies infected with *Nosema apis* and sacbrood virus. *Journal of Economic Entomology* 85: 47–51.

Anderson, R. M. 1991. Populations and infectious diseases: Ecology or epidemiology? *Journal of Animal Ecology* 60: 1–50.

Anderson, R. M., and R. M. May. 1979. Population biology of infectious diseases. Part I. *Nature* 280: 361–367.

Anderson, R. M., and R. M. May. 1981. The population dynamics of microparasites and their invertebrate hosts. *Philosophical Transactions of the Royal Society London, B* 291: 451–524.

Anderson, R. M., and R. M. May. 1982. Coevolution of hosts and parasites. *Parasitology* 85: 411–426.

Anderson, R. M., and R. M. May. 1991. *Infectious Diseases of Humans.* Oxford University Press, Oxford.

Andersson, M. 1994. *Sexual Selection.* Princeton University Press, Princeton, N.J.

Antia, R., B. R. Levin, and R. M. May. 1994. Within-host population dynamics and the evolution and maintenance of microparasite virulence. *The American Naturalist* 144: 457–472.

Aoki, S. 1977. *Colophina clematis* (Homoptera: Pemphigidae), an aphid species with "soldiers." *Kontyû* 45: 276–282.

Apanius, V. 1991. Brood parasitism, immunity and reproduction in American Kestrels, *Falco spervidus.* Ph.D. diss., University of Pennsylvania, Philadelphia.

Armstrong, E. 1976. Transmission and infectivity studies on *Nosema kingi* in *Drosophila willistoni* and other drosophilids. *Zeitschrift für Parasitenkunde* 50: 161–165.

Aruga, H., and H. Watanabe. 1964. Resistance to per os infection with cytoplasmic-polyhedrosis virus in the silkworm, *Bombyx morio* (Linnaeus). *Journal of Insect Pathology* 6: 387–394.

Askew, R. R. 1971. *Parasitic Insects.* Heinemann, London.

Attygalle, A. B., B. Siegel, O. Vostrowsky, H. J. Bestmann, and U. Maschwitz. 1989. Chemical composition and function of the metapleural gland secretion of the ant, *Crematogaster deformis* Smith (Hymenoptera: Myrmicinae). *Journal of Chemical Ecology* 15: 317–329.

Avery, S. W., D. P. Jouvenaz, W. A. Banks, and D. W. Anthony. 1977. Virus-like particles in a fire ant, *Solenopsis* sp. (Hymenoptera: Formicidae) from Brazil. *Florida Entomologist* 60: 17–20.

Ayre, G. L. 1962. *Pseudometagea schwarzii* (Ashm.) (Eucharitidae: Hymenoptera), a parasite of *Lasius neoniger* Emery (Formicidae: Hymenoptera). *Canadian Journal of Zoology* 40: 157–164.

Bailey, L. 1955a. The epidemiology and control of *Nosema* disease of the honey bee. *Annals of Applied Biology* 43: 379–389.

Bailey, L. 1955b. The infection of the ventriculus of the adult honey bee by *Nosema apis* (Zander). *Parasitology* 45: 89–94.

Bailey, L. 1958a. The epidemiology of the infestation of the honey bee, *Apis mellifera* L., by the mite *Acarapis woodi* (Rennie), and the mortality of infested bees. *Parasitology* 48: 493–506.

Bailey, L. 1958b. Wild honeybees and disease. *Bee World* 39: 92–95.

Bailey, L. 1960. The epizootiology of European foulbrood of the larval honey bee, *Apis mellifera* L. *Journal of Insect Pathology* 2: 67–83.

Bailey, L. 1963. The pathenogenicity for honeybee larvae of microorganisms associated with European foulbrood. *Journal of Insect Pathology* 5: 198–205.

Bailey, L. 1965. Susceptibility of the honey bee, *Apis mellifera* Linnaeus, infected with *Acarapis woodi* (Rennie) to infection by airborne pathogens. *Journal of Invertebrate Pathology* 7: 141–143.

Bailey, L. 1967a. The effect of temperature on the pathogenicity of the fungus, *Ascosphaera apis,* for larvae of the honey bee *Apis mellifera.* In P. A. Van der Laan, ed., *Proceedings of the International Colloquium on Insect Pathology and Microbial Control,* pp. 162–167. North-Holland, Amsterdam.

Bailey, L. 1967b. The incidence of *Acarapis woodi* in North American strains of honey bees in Britain. *Journal of Apicultural Research* 6: 99–103.

Bailey, L. 1967c. The incidence of virus disease in the honey bee. *Annals of Applied Biology* 60: 43–38.

Bailey, L. 1968. The measurement of inter-relationships of infections by *Nosema apis* and *Malphigamoeba mellificae* of honey-bee populations. *Journal of Invertebrate Pathology* 12: 175–179.

Bailey, L. 1969. The multiplication and spread of sacbrood virus of bees. *Annals of Applied Biology* 63: 483–491.

Bailey, L. 1981. *Honey Bee Pathology.* Academic Press, London.

Bailey, L. 1982. Viruses of honeybees. *Bee World* 63: 165–173.

Bailey, L. 1985. *Acarapis woodi:* A modern appraisal. *Bee World* 66: 99–104.

Bailey, L., and B. V. Ball. 1978. *Apis* iridescent virus and "clustering disease" of *Apis cerana. Journal of Invertebrate Pathology* 31: 368–371.

Bailey, L., and B. V. Ball. 1991. *Honey Bee Pathology.* 2d ed. Academic Press, London.

Bailey, L., B. V. Ball, and J. N. Perry. 1981. The prevalence of viruses of honey bees in Britain. *Annals of Applied Biology* 97: 109–118.

Bailey, L., B. V. Ball, and J. N. Perry. 1983a. Association of viruses with two protozoal pathogens of the honey bee. *Annals of Applied Biology* 103: 13–20.

Bailey, L., B. V. Ball, and J. N. Perry. 1983b. Honeybee paralysis: Its natural spread and its diminished incidence in England and Wales. *Journal of Apicultural Research* 22: 191–195.

Bailey, L., and E. F. W. Fernando. 1972. Effects of sacbrood virus in adult honey bees. *Annals of Applied Biology* 72: 27–35.

Bailey, L., and A. J. Gibbs. 1964. Acute infection of bees with paralysis virus. *Journal of Insect Pathology* 6: 395–407.

Bailey, L., and R. D. Woods. 1977. Two more small RNA viruses from honey bees and further observations on sacbrood and acute bee-paralysis viruses. *Journal of General Virology* 37: 175–182.

Balazy, S., A. Lenoir, and J. Wisniewski. 1986. *Aegeritella roussillonensis* n. sp. (Hyphomycetales, Blastosporae) une espce nouvelle de champignon epizoique sur les fourmis *Cataglyphis cursor* (Fonscolombe) (Hymenoptera, Formicidae) en France. *Cryptogamie, Mycologie* 7: 37–45.

Balazy, S., and A. Sokolowski. 1977. Morphology and biology of *Entomophthora myrmecophaga. Transactions of the British Mycological Society* 68: 134–137.

Ball, B. V. 1989. *Varroa jacobsoni* as a virus vector. In R. Cavalloro, ed., *Proceedings of a Meeting of the EC-Experts' Group, Udine,* vol. EUR 11932 EN, pp. 241–244. Commission of the European Communities, Brussels, Belgium.

Ball, B. V., and M. F. Allen. 1988. The prevalence of pathogens in honey bee (*Apis mellifera*) colonies infested with the parasitic mite *Varroa jacobsoni. Annals of Applied Biology* 113: 237–244.

Ball, B. V., and L. Bailey. 1991. Viruses of the honey bee. In J. R. Adams and J. R. Bonami, eds., *Atlas of Invertebrate Viruses,* pp. 525–551. CRC Press, Boca Raton, Florida.

Bamrick, J. F. 1964. Resistance to American foulbrood in honeybees. V. Comparative pathogenesis in resistant and susceptible larvae. *Journal of Insect Pathology* 33: 284–304.

Bamrick, J. F., and W. C. Rothenbuhler. 1961. Resistance to American foulbrood in honey bees. IV. The relationship between larval age at inoculation and mortality in a resistant and in a susceptible line. *Journal of Insect Pathology* 3: 381–390.

Barlow, N. 1995. Critical evaluation of wildlife disease models. In B. T. Grenfell and A. P. Dobson, eds., *Ecology of Infectious Diseases in Natural Populations,* pp. 230–259. Cambridge University Press, Cambridge, U.K.

Baroni-Urbani, C., B. Bolton, and P. S. Ward. 1992. The internal phylogenies of ants (Hymenoptera: Formicidae). *Systematic Entomology* 17: 301–329.

Bartz, S. 1979. Evolution of eusociality in termites. *Proceedings of the National Academy of Sciences USA* 76: 5764–5768.

Batra, S. W. T. 1966. The life cycle and behaviour of the primitively social bee *Lasioglossum zephyrum* (Halictidae). *University of Kansas Science Bulletin* 46: 359–423.

Bauer, L. S., and G. L. Nordin. 1988. Nutritional physiology of the eastern spruce budworm, *Choristoneura fumiferana,* infected with *Nosema fumiferanae,* and interactions with dietary nitrogen. *Oecologia* 77: 44–50.

Beattie, A. J. 1986. *The Evolutionary Ecology of Ant-Plant Mutualism.* Cambridge University Press, Cambridge, U.K.

Beattie, A. J., C. L. Turnbull, T. Hough, S. Jobson, and R. B. Knox. 1985. The vulnerability of pollen and fungal spores to ant secretions: Evidence and some evolutionary implications. *American Journal of Botany* 72: 606–614.

Beattie, A. J., C. L. Turnbull, T. Hough, S. Jobson, and R. B. Knox. 1986. Antibiotic production: A possible function for the metapleural glands of ants (Hymenoptera: Formicidae). *Annals of the Entomological Society of America* 79: 448–450.

Beck, G., and G. S. Habicht. 1996. Immunity and the invertebrates. *Scientific American* (November) 42–46.

Bedding, R. A. 1985. Nematode parasites of Hymenoptera. In W. R. Nickle, ed., *Plant and Insect Nematodes,* pp. 755–795. Marcel Dekker, New York.

Beecher, M. D., I. M. Beecher, and S. Hahn. 1981a. Parent-offspring recognition in bank swallows (*Riparia riparia*): II. Development and acoustic basis. *Animal Behaviour* 29: 95–101.

Beecher, M. D., I. M. Beecher, and S. Lumpkin. 1981b. Parent-offspring recognition in the bank swallow (*Riparia riparia*): I. Natural history. *Animal Behaviour* 29: 86–94.

Beegle, C. C., and E. R. Oatman. 1974. Differential susceptibility of parasitized and nonparasitized larvae of *Trichoplusia ni* to a nuclear polyhedrosis virus. *Journal of Invertebrate Pathology* 24: 188–195.

Begon, M., C. B. Krimbas, and M. Loukas. 1980. The genetics of *Drosophila subobscura* populations. XV. Effective size of a natural population estimated by three independent methods. *Heredity* 45: 335–350.

Bell, G. 1982. *The Masterpiece of Nature.* University of California Press, Berkeley.

Benelli, E., and P. Schmid-Hempel. 1998. Immuno-competence of bumble bees in relation to environment and genotype. In prep.

Benest, G. 1976. Relations interspecifiques et intraspecifiques entre butineuses de *Bombus* sp. et d' *Apis mellifera* L. *Apidologie* 7: 113–127.

Benz, G. 1987. Environment. In J. R. Fuxa and Y. Tanada, eds., *Epizootology of Insect Diseases,* pp. 177–214. John Wiley and Sons, New York.

Bequaert, J. 1922. Ants and their diverse relations to the plant world. *Bulletin of the American Museum of Natural History* 45: 333–584.

Berkelhamer, R. C. 1983. Intraspecific genetic variation and haplodiploidy, eusociality, and polygyny in the hymenoptera. *Evolution* 37: 540–545.

Berkelhamer, R. C. 1984. An electrophoretic analysis of queen numbers in three species of dolichoderine ants. *Insectes Sociaux* 31: 132–141.

Beutler, R., and E. Opfiger. 1949. Pollenernährung und Nosemabefall der Honigbiene (*Apis mellifera*). *Zeitschrift für Vergleichende Physiologie* 32: 383–421.

Bhatkar, A. P. 1979a. Evidence of intercolonial food exchange in fire ants and other Myrmicinae using radioactive phosphorus. *Experientia* 35: 1172–1173.

Bhatkar, A. P. 1979b. Trophallactic appeasement in ants from distant colonies. *Folia Entomologica Mexicana* 41: 135–143.

Bhatkar, A. P. 1983. Interspecific trophallaxis in ants, its ecological and evolutionary significance. In P. Jaisson, ed., *Social Insects in the Tropics,* vol. 2, pp. 105–123. Presse de l'Université Paris-Nord, Paris.

Bhatkar, A. P., and W. J. Kloft. 1977. Evidence, using radioactive phosphorus, of interspecific food exchange in ants. *Nature* 265: 140–142.

Bienvenu, R. J., F. W. Atchinson, and E. A. Cross. 1968. Microbial inhibition by prepupae of the alkali bee, *Nomia melanderi. Journal of Invertebrate Pathology* 12: 278–282.

Birkhead, T. C., and G. A. Parker. 1996. Sperm competition and mating systems. In J. R. Krebs and N. B. Davies, eds., *Behavioural Ecology—An Evolutionary Approach,* 4th ed., pp. 121–145. Blackwells, Oxford.

Bitner, R. A., W. T. Wilson, and J. D. Hitchcock. 1972. Passage of *Bacillus* larvae spores from adult queen honeybees to attendant workers (*Apis mellifera*). *Annals of the Entomological Society of America* 65: 899–901.

Black, F. L. 1992. Why did they die? *Science* 258: 1739–1741.

Blaustein, A. R., and R. K. O'Hara. 1982. Kin recognition in *Rana cascadae* tadpoles: Maternal and paternal effects. *Animal Behaviour* 30: 1151–1157.

Blumberg, D. 1988. Encapsulation of eggs of the encrytid wasp *Metaphycus swirskii* by the hemispherical scale, *Saissetia coffeae:* Effects of host age and rearing temperature. *Entomologica Experimentalis et Applicata* 47: 95–99.

Blumberg, D., and P. DeBach. 1981. Effects of temperature and host age upon the encapsulation of *Metaphycus stanleyi* and *Metaphycus helvolus* eggs by brown soft scale *Coccus hesperidum. Journal of Invertebrate Pathology* 37: 73–79.

Boch, R., and R. A. Morse. 1981. Effects of artificial odors and pheromones on queen discrimination by honey bees (*Apis mellifera* L.). *Annals of the Entomological Society of America* 74: 66–67.

Boch, R., and R. A. Morse. 1982. Genetic factor in queen recognition odors of honey bees. *Annals of the Entomological Society of America* 76: 654–656.

Boecking, O., and W. Drescher. 1991. Response of *Apis mellifera* L. colonies infested with *Varroa jacobsoni* Oud. *Apidologie* 22: 237–241.

Boecking, O., and W. Drescher. 1992. The removal response of *Apis mellifera* L. colonies to brood in wax and plastic cells after artificial and natural infestation with *Varroa jacobsoni* Oud. and to freeze-killed brood. *Experimental and Applied Acaraology* 16: 321–329.

Bonavita-Cougourdan, A., J. L. Clement, and C. Lange. 1987. Nestmate recognition: The role of cuticular hydrocarbons in the ant *Camponotus vagus* Scop. *Journal of Entomological Science* 22: 1–10.

Bonhoeffer, S., R. E. Lenski, and D. Ebert. 1996. The curse of the pharaoh: The evolution of virulence in pathogens with long living propagules. *Proceedings of the Royal Society London, B* 263: 715–721.

Boomsma, J. J. 1991. Adaptive colony sex ratios in primitively eusocial bees. *Trends in Ecology and Evolution* 6: 92–95.

Boomsma, J. J., and A. Grafen. 1991. Colony-level sex ratio selection in the eusocial Hymenoptera. *Journal of Evolutionary Biology* 3: 383–407.

Boomsma, J. J., and F. L. W. Ratnieks. 1996. Paternity in eusocial hymenoptera. *Philosophical Transactions of the Royal Society London, B* 351: 947–975.

Boorstein, S. M., and P. W. Ewald. 1987. Costs and benefits of behavioral fever in *Melanoplus sanguinipes* infected by *Nosema acridophagus. Physiological Zoology* 60: 586–595.

Boots, M., and M. Begon. 1993. Trade-offs with resistance to granulosis virus in the Indian Meal Moth examined by laboratory evolution experiments. *Functional Ecology* 7: 528–534.

Bosch, R. v. 1964. Encapsulation of the eggs of *Bathyplectes curculionis* (Thomson) in larvae of *Hypera brunneipennis* (Boheman) and *Hypera postica* (Gyllenhal) (Coleoptera: Curculionidae). *Journal of Invertebrate Pathology* 6: 343–367.

Bot, A. N. M., and J. J. Boomsma. 1996. Variable metapleural gland size-allometries in *Acromyrmex* leafcutter ants (Hymenoptera: Formicidae). *Journal of the Kansas Entomological Society* 69 (Suppl.): 375–383.

Boulétreau, M. 1986. The genetic and coevolutionary interactions between parasitoids and their hosts. In J. Waage and D. Greathead, eds., *Insect Parasitoids,* pp. 169–200. Academic Press, London.

Bourke, A. F. G. 1988. Worker reproduction in the higher eusocial hymenoptera. *Quarterly Review of Biology* 63: 291–311.

Bourke, A. F. G., and N. R. Franks. 1991. Alternative adaptations, sympatric speciation and the evolution of parasitic, inquiline ants. *Biological Journal of the Linnean Society* 43: 157–178.

Bourke, A. F. G., and N. R. Franks. 1995. *Social Evolution in Ants.* Princeton University Press, Princeton, New Jersey.

Bowman, C. E., and C. A. Ferguson. 1985. *Forcellinia galleriella,* a mite new to British beehives. *Bee World* 66: 51–53.

Breed, M. D. 1981. Individual recognition and learning of queen odors by worker honeybees. *Proceedings of the National Academy of Sciences USA* 78: 2635–2637.

Breed, M. D. 1983. Nestmate recognition in honey bees. *Animal Behaviour* 31: 86–91.

Breed, M. D., and B. Bennett. 1987. Kin recognition in highly eusocial insects. In D. C. J. Fletcher and C. D. Michener, eds., *Kin Recognition in Animals,* pp. 243–285. John Wiley and Sons, Chichester, U.K.

Breed, M. D., L. Butler, and T. M. Stiller. 1985. Kin discrimination by worker honeybees in genetically mixed groups. *Proceedings of the National Academy of Sciences USA* 82: 3058–3061.

Breed, M. D., T. M. Stiller, and M. J. Moor. 1988. The ontogeny of kin discrimination cues in the honey bee, *Apis mellifera. Behavioural Genetics* 18: 439–448.

Bremerman, H. J., and J. Pickering. 1983. A game-theoretical model of parasite virulence. *Journal of theoretical Biology* 100: 411–426.

Bremerman, H. J., and H. R. Thieme. 1989. A competitive exclusion principle for pathogen virulence. *Journal of Mathematical Biology* 27: 179–190.

Brian, M. V. 1965. *Social Insect Populations.* Academic Press, London.

Brian, M. V. 1983. *Social Insects.* Chapman and Hall, London.

Brian, M. V. 1986. Bonding between workers and queens in the ant genus *Myrmica. Animal Behaviour* 34: 1135–1145.

Brian, M. V., and A. D. Brian. 1952. The wasp *Vespula sylvestris* (Scop.): Feeding, foraging, and colony development. *Transactions of the Royal Entomological Society London* 103: 1–26.

Briano, J., D. Jouvenaz, D. Wojcik, H. Cordo, and R. Patterson. 1995a. Protozoan and fungal diseases in *Solenopsis richteri* and *S. quinquecuspis* (Hymenoptera: Formicidae) in Buenos Aires province, Argentina. *Florida Entomologist* 78: 531–537.

Briano, J., R. Patterson, and H. Cordo. 1995b. Relationship between colony size of *Solenopsis richteri* (Hymenoptera: Formicidae) and infection with *Thelohania solenopsae* (Microsporida: Thelohaniidae) in Argentina. *Journal of Economic Entomology* 88: 1233–1237.

Briano, J. A., R. S. Patterson, and H. A. Cordo. 1995c. Colony movement of the black imported fire ant (Hymenoptera: Formicidae) in Argentina. *Environmental Entomology* 24: 1131–1134.

Briano, J. A., R. S. Patterson, and H. A. Cordo. 1995d. Long-term studies of the black imported fire ant (Hymenoptera: Formicidae) infected with a microsporidium. *Environmental Entomology* 24: 1328–1332.

Briese, D. T. 1982. Genetic basis for resistance to a granulosis virus in the potato moth *Phthorimaea operculella. Journal of Invertebrate Pathology* 39: 215–218.

Brockmann, H. J. 1984. The evolution of social behaviour in insects. In J. R. Krebs and N. B. Davies, eds., *Behavioural Ecology,* 2d ed., pp. 340–361. Blackwells, Oxford.

Brooks, D. R., and D. A. McLennan. 1993. *Parascript—Parasites and the Language of Evolution.* Smithsonian Institution Press, Washington, D.C.

Brooks, W. M., ed. 1988. *CRC Handbook of Natural Pesticides,* vol. 5: *Microbial Insecticides,* part A, *Entomogenous Protozoa and Fungi.* CRC Press, Boca Raton, Florida.

Broome, J. R., P. R. Sikorowski, and B. R. Norment. 1976. A mechanism of pathogenicity of *Beauveria bassiana* on larvae of the imported fire ant, *Solenopsis richteri. Journal of Invertebrate Pathology* 28: 87–91.

Brown, B. V., A. Francoeur, and R. L. Gibson. 1991. Review of the genus *Styletta* (Diptera: Phoridae) with a description of a new genus. *Entomologica Scandinavica* 22: 241–250.

Brown, C. R., and M. B. Brown. 1986. Ectoparasitism as a cost of coloniality in cliff swallows. *Ecology* 67: 1206–1218.

Brown, G. C. 1984. Stability in an insect-pathogen model incorporating age-dependent immunity and seasonal host reproduction. *Bulletin of Mathematical Biology* 46: 139–153.

Brown, J. L. 1987. *Helping and Communal Breeding in Birds: Ecology and Evolution.* Princeton University Press, Princeton, New Jersey.

Bruce, W. A., and G. L. LeCato. 1980. *Pyemotes tritici:* A potential new agent for biological control of the red imported fire ant, *Solenopsis invicta* (Acari: Pyemotidae). *International Journal of Acarology* 4: 271–274.

Brusca, R. C., and G. J. Brusca. 1990. *Invertebrates.* Sinauer Associates, Sunderland, Mass.

Buchli, H. H. R. 1958. L'origine des castes et les potentialités ontogéniques des termites européens du genre *Reticulitermes* Holmgren. *Annales des Sciences Naturelles* 11: 263–429.

Buckle, G. R., and L. Greenberg. 1981. Nestmate recognition in sweat bees (*Lasioglossum zephyrum*): Does an individual recognize its own odour or colony odours of its nestmates? *Animal Behaviour* 29: 802–809.

Bull, J. J. 1994. Virulence. *Evolution* 48: 1423–1437.

Bull, J. J., and I. J. Molineux. 1992. Molecular genetics of adaptation in an experimental model of cooperation. *Evolution* 45: 875–882.

Bull, J. J., I. J. Molineux, and W. R. Rice. 1991. Selection of benevolence in a host-parasite system. *Evolution* 45: 875–882.

Burda, H., and M. Kawalika. 1993. Evolution of eusociality in the Bathyergidae. The case of the giant mole rats (*Cryptomys mechowi*). *Naturwissenschaften* 80: 235–237.

Burgett, D. M., P. Akratanakul, and R. A. Morse. 1983. *Tropilaelaps clarae:* A parasite of honeybees in South-East Asia. *Bee World* 64: 25–28.

Burgett, D. M., P. A. Rossignol, and C. Kitprasert. 1990. A model of dispersion and regulation of brood mite parasitism (*Tropilaelaps clarae*) parasitism on the giant honey bee (*Apis dorsata*). *Canadian Journal of Zoology* 68: 1423–1427.

Burgett, D. M., L. A. Royce, and L. A. Ibay. 1989. Occurrence of the *Acarapis* species complex in a commercial honey bee apiary in the Pacific Northwest. *Experimental and Applied Acarology* 7: 251–255.

Burnside, C. E. 1928. A septicemic condition of adult bees. *Journal of Economic Entomology* 21: 379–386.

Burt, A., and G. Bell. 1987. Mammalian chiasma frequencies as a test of two theories of recombination. *Nature* 326: 803–805.

Buschinger, A. 1973. Ameisen des Tribus Leptothoracini (Hymenoptera: Formicidae) als Zwischenwirte von Cestoden. *Zoologischer Anzeiger* 191: 369–380.

Buschinger, A. 1986. Evolution of social parasitism in ants. *Trends in Ecology and Evolution* 1: 155–60.

Buschinger, A. 1989. Evolution, speciation, and inbreeding in the parasitic ant genus *Epimyrma* (Hymenoptera: Formicidae). *Journal of Evolutionary Biology* 2: 265–283.

Buschinger, A. 1990. Sympatric speciation and radiative evolution of socially parasitic ants—heretic hypotheses and their factual background. *Zeitschrift für Systematik und Evolutionsforschung* 28: 241–260.

Calabi, P., and S. D. Porter. 1989. Worker longevity in the fire ant *Solenopsis invicta:* Ergonomic considerations of correlations between temperature, size and metabolic rates. *Journal of Insect Physiology* 35: 643–649.

Calabi, P., and J. F. A. Traniello. 1989. Social organization in the ant *Pheidole dentata.* Physical and temporal caste ratios lack ecological correlates. *Behavioural Ecology and Sociobiology* 24: 69–78.

Calabi, P., J. F. A. Traniello, and M. H. Werner. 1983. Age polyethism: Its occurrence in the ant *Pheidole hortensis,* and some general considerations. *Psyche* 90: 395–412.

Calderone, N. W., and R. E. Page. 1991. Evolutionary genetics of division of labor in colonies of the honey bee (*Apis mellifera*). *American Naturalist* 138: 69–92.

Cale, G. H., and W. C. Rothenbuhler. 1975. Genetics and breeding of the honey bee. In Dadant and Sons, eds., *The Hive and the Honey Bee,* pp. 157–184. Dadant and Sons, Hamilton, Illinois.

Camazine, S. 1988. Factors affecting the severity of *Varroa jacobsoni* infestations on European and Africanized honeybees. In C. G. Needham et al., eds., *Africanized Honey Bees and Bee Mites,* pp. 444–451. Ellis Horwood, Chichester, U.K.

Camponovo, F., and G. Benz. 1984. Age-dependent tolerance to *Baculovirus* in last larval instars of the codling moth, *Cydia pomonella* L., induced either for pupation or for diapause. *Experientia* 40: 938–939.

Cane, J. H., S. Gerdin, and G. Wife. 1983. Mandibular gland secretions of solitary bees: Potential for nest cell disinfection. *Journal of the Kansas Entomological Society* 56: 199–204.

Canning, E. U. 1982. An evaluation of protozoal characteristics in relation to biological control of pests. *Parasitology* 84: 119–149.

Canning, E. U. 1990. Phylum Microsporidia. In L. Margulis, J. O. Corliss, M. Melkonian, and D. J. Chapman, eds., *Handbook of Protocista,* pp. 53–72. Jones and Bartlett, Boston.

Cantwell, G. E. 1970. Standard methods for counting *Nosema* spores. *American Bee Journal* 110: 222–223.

Carlile, M. J. 1973. Cell fusion and somatic incompatibility in myxomycetes. *Berichte der Deutschen Botanischen Gesellschaft* 86: 123–139.

Carlin, N. F. 1986. Recognition pheromones and the role of queens in ants. Ph.D. diss., Harvard University, Cambridge, Mass.

Carlin, N. F., and B. Hölldobler. 1986. The kin recognition system of carpenter ants (*Camponotus* spp.). I. Hierarchial cues in small colonies. *Behavioural Ecology and Sociobiology* 19: 123–134.

Carlin, N. F., and B. Hölldobler. 1987. The kin recognition system of carpenter ants (*Camponotus* spp.). II. Larger colonies. *Behavioural Ecology and Sociobiology* 20: 209–217.

Carlin, N. F., and B. Hölldobler. 1988. Influence of virgin queens on kin recognition in the carpenter ant *Camponotus floridanus* (Hymenoptera: Formicidae). *Insectes Sociaux* 35: 191–197.

Carlin, N. F., B. Hölldobler, and D. S. Gladstein. 1987. The kin recognition system of carpenter ants (*Camponotus* spp.). III. Within-colony discrimination. *Behavioural Ecology and Sociobiology* 20: 219–227.

Carlin, N. F., H. K. Reeve, and S. P. Cover. 1993. Kin discrimination and division of labour among matrilines in the polygynous carpenter ant, *Camponotus planatus.* In L. Keller, ed., *Queen Number and Sociality in Insects,* pp. 362–401. Oxford University Press, Oxford.

Carney, W. P. 1969. Behavioral and morphological changes in carpenter ants harbouring dicrocoeliid metacercaria. *American Midlands Naturalist* 82: 605–611.

Carpenter, J. M. 1991. Phylogenetic relationships and the origin of social behavior in the Vespidae. In K. G. Ross and R. W. Matthews, eds., *The Social Biology of Wasps,* pp. 7–32. Comstock Publishing Associates, Ithaca, New York.

Carruthers, R. I., and R. S. Soper. 1987. Fungal diseases. In J. R. Fuxa and Y. Tanada, eds., *Epizootiology of Insect Diseases,* pp. 357–416. John Wiley and Sons, New York.

Carton, Y., and M. Boulétreau. 1985. Encapsulation ability of *Drosophila melanogaster:* A genetic analysis. *Developmental and Comparative Immunology* 9: 21–219.

Carton, Y., and J. R. David. 1983. Reduction of fitness in *Drosophila* adults surviving parasitization by a cynipid wasp. *Experientia* 39: 231–233.

Carton, Y., F. Frey, and A. Nappi. 1992. Genetic determinism of the cellular immune reaction in *Drosophila melanogaster. Heredity* 69: 393–399.

Case, A. A., and J. E. Ackert. 1940. New intermediate hosts of fowl cestodes. *Transcations of the Kansas Academy of Sciences* 43: 393–396.

Charlesworth, B. 1978. Some models of the evolution of altruistic behaviour between siblings. *Journal of theoretical Biology* 72: 297–319.

Charnov, E. L. 1982. *The Theory of Sex Allocation.* Princeton University Press, Princeton, New Jersey.

Cheney, D. L., and R. M. Seyfarth. 1980. Vocal recognition in free-ranging velvet monkeys. *Animal Behaviour* 28: 362–367.

Cheng, T. C. 1986. *General Parasitology.* Academic Press, Orlando, Florida.

Chérix, D. 1980. Note préliminaire sur la structure, la phénologie et le régime alimentaire d'une super-colonie de *Formica lugubris* Zett. *Insectes Sociaux* 27: 226–236.

Cherrett, J. M., R. J. Powell, and D. J. Stradling. 1989. The mutualism between leaf-cutting ants and their fungus. In N. Wilding, N. M. Collins, P. M. Hammond, and J. F. Webber, eds., *Insect-Fungus Interactions,* pp. 93–120. Academic Press, London.

Chmielewski, W. 1971. The mites (Acarina) found on bumble bees (*Bombus* Latr.) and in their nests. *Ekologia Polska A* 19: 57–71.

Chmielewski, W. 1991. Stored product mites (Acaroidea) in Polish bee hives. In F. Dusbabek and V. Bukva, eds., *Modern Acarology,* vol.1, pp. 615–619. Academia Prague and SPB Academic Publishing, The Hague.

Clark, T. B. 1977. *Spiroplasma* as a new pathogen in the honeybees. *Journal of Invertebrate Pathology* 29: 112–113.

Clark, T. B. 1980. A second microsporidian in the honey bee. *Journal of Invertebrate Pathology* 35: 290–294.

Clark, T. B. 1982. Entomopoxvirus-like particles in three species of bumble-bees. *Journal of Invertebrate Pathology* 39: 119–122.

Clark, T. B., R. F. Whitcomb, J. G. Tully, G. Mouches, C. Saillard, J. M. Bové, H. Wroblewski, P. Carle, D. L. Rose, R. B. Henegar, and D. L. Williamson. 1985. *Spiroplasma melliferum,* a new species from the honeybee (*Apis mellifera*). *International Journal of Systematic Bacteriology* 35: 296–308.

Clausen, C. P. 1923. The biology of *Schizaspidia tenuicornis* Ashm., a eucharid parasite of *Camponotus. Annals of the Entomological Society of America* 16: 195–217.

Clausen, C. P. 1940a. *Entomophagous Insects.* McGraw-Hill, New York.

Clausen, C. P. 1940b. The oviposition habits of the Eucharidae (Hymenoptera). *Journal of the Washington Academy of Sciences* 30: 504–516.

Cole, B. J. 1983. Multiple mating and the evolution of social behavior in the Hymenoptera. *Behavioural Ecology and Sociobiology* 12: 191–291.

Cole, B. J. 1984. Colony efficiency and the reproductivity effect in *Leptothorax allardycei* (Mann). *Insectes Sociaux* 31: 403–407.

Cole, L. K. 1974. *Antifungal, Insecticidal, and Potential Chemotherapeutic Properties of Ant Venom Alkaloids and Ant Alarm Pheromones.* Ph.D. diss., University of Georgia, Athens.

Collins, H. L., and G. P. Markin. 1971. Inquilines and other arthropods collected from nests of the imported fire ant, *Solenopsis saevissima richteri. Annals of the Entomological Society of America* 64: 1376–1380.

Collins, N. M. 1981. Populations, age structure and survivorship of colonies of *Macrotermes bellicosus* (Isoptera: Termitidae). *Journal of Animal Ecology* 50: 293–311.

Contel, E. P. B., and W. E. Kerr. 1976. Origin of males in *Melipona subnitida* estimated from data of an isozymic polymorphic system. *Genetica* 46: 271–277.

Cook, J. M., and R. H. Crozier. 1995. Sex determination and population biology in the hymenoptera. *Trends in Ecology and Evolution* 10: 281–286.

Coombe, D. R., P. L. Ey, and C. R. Jenkin. 1984. Self/non-self recognition in invertebrates. *Quarterly Review of Biology* 59: 231.

Corso, C. R., and A. Serzedello. 1981. A study of multiple mating habits in *Atta laevigata* based on the DNA content. *Comparative Biochemistry and Physiology, B* 69: 901–902.

Côté, I. M., and R. Poulin. 1995. Parasitism and group size in social mammals: A meta-analysis. *Behavioral Ecology* 6: 159–165.

Cowan, D. P. 1991. The solitary and presocial wasps. In K. G. Ross and R. W. Matthews, eds., *The Social Biology of Wasps,* pp. 33–73. Comstock Publishing Associates, Ithaca, New York.

Craig, R. 1983. Subfertility and the evolution of eusociality by kin selection. *Journal of theoretical Biology* 100: 379–399.

Craig, R., and R. H. Crozier. 1979. Relatedness in the polygynous ant *Myrmecia pilosula. Evolution* 33: 335–341.

Crawley, W. C., and H. A. Baylis. 1921. *Mermis* parasitic on ants of the genus *Lasius. Journal of the Royal Microscopic Society London* (1921): 353–364.

Creighton, W. S. 1950. The ants of North America. *Bulletin of the Museum of Comparative Zoology at Harvard University* 104: 1–585.

Crespi, B. J. 1992. Eusociality in Australian gall thrips. *Nature* 359: 724–726.

Crespi, B. J., and D. Yanega. 1995. The definition of eusociality. *Behavioral Ecology* 6: 109–115.

Crosland, M. W. J. 1988. Effect of a gregarine parasite on the color of *Myrmecia pilosula* (Hymenoptera: Formicidae). *Annals of the Entomological Society of America* 81: 481–484.

Crosland, M. W. J. 1990. The influence of the queen, colony size and worker ovarian development on nestmate recognition in the ant *Rhytidoponera confusa. Animal Behaviour* 39: 413–425.

Crozier, R. H. 1968. Cytotaxonomic studies on some Australian Dolichoderinae ants (Hym. Formicidae). *Caryologia* 21: 241–259.

Crozier, R. H. 1970. Karyotypes of twenty-one ant species (Hymenoptera: Formicidae) with reviews of the known ant karyotypes. *Canadian Journal of Genetics and Cytology* 12: 109–128.

Crozier, R. H. 1977. Genetic differentiation between populations of the ant *Aphaenogaster 'rudis'* in the southeastern United States. *Genetica* 47: 17–36.

Crozier, R. H. 1979. Genetics of sociality. In H. R. Hermann, ed., *Social Insects,* vol. 1, pp. 223–286. Academic Press, New York.

Crozier, R. H. 1987. Genetic aspects of kin recognition: Concepts, models, and synthesis. In D. J. C. Fletcher and C. D. Michener, eds., *Kin Recognition in Animals,* pp. 55–74. John Wiley and Sons, Chichester, U.K.

Crozier, R. H., and D. Brückner. 1981. Sperm clumping and the population genetics of hymenoptera. *American Naturalist* 117: 561–563.

Crozier, R. H., and P. C. Consul. 1976. Conditions for genetic polymorphism in social hymenoptera under selection at the colony level. *Theoretical Population Biology* 10: 1–9.

Crozier, R. H., and M. W. Dix. 1979. Analysis of two genetic models for the innate components of colony odor in social Hymenoptera. *Behavioural Ecology and Sociobiology* 4: 217–224.

Crozier, R. H., and R. E. Page. 1985. On being the right size: Male contributions and multiple mating in social hymenoptera. *Behavioural Ecology and Sociobiology* 18: 105–116.

Crozier, R. H., and P. Pamilo. 1980. Asymmetry in relatedness: Who is related to whom? *Nature* 283: 604.

Crozier, R. H., and P. Pamilo. 1996. *Evolution of Social Insect Colonies.* Oxford University Press, Oxford.

Crozier, R. H., B. H. Smith, and Y. Crozier. 1987. Relatedness and population structure of the primitively eusocial bee *Lasioglossum zephyrum* (Hymenoptera: Halictidae) in Kansas. *Evolution* 41: 902–910.

Cumber, R. A. 1949. Humble-bee parasites and commensals found within a thirty mile radius of London. *Transactions of the Royal Entomological Society London, A* 24: 119–127.

Cumber, R. A. 1953. Some aspects of the biology and ecology of bumble-bees bearing upon the yields of red-clover seed in New Zealand. *New Zealand Journal of Science and Technology* 11: 227–240.

Davidson, D. W. 1978. Size in worker caste and competitive environment. *American Naturalist* 112: 523–532.

Davidson, E. W. 1973. Ultrastructure of American foulbrood: Disease pathogenesis in larvae of the worker honeybee, *Apis mellifera. Journal of Invertebrate Pathology* 21: 53–61.

Davies, C. R., J. M. Ayres, C. Dye, and L. M. Deane. 1991. Malaria infection rate of Amazonian primates increases with body weight and group size. *Functional Ecology* 5: 655–662.

Davis, R. E., T.-A. Chen, and J. F. Worley. 1981. Corn stunt spiroplasma. In D. T. Gordon, J. K. Knoke and G. E. Scott, eds., *Virus and Virus-like Diseases of Maize in the United States.* Southern Cooperation Series, Bulletin 247, June 1981, pp. 40–50. Ohio Agricultural Research Development Center, Wooster.

Davis, R. E., I.-M. Lee, and L. K. Basciano. 1979. Spiroplasmas: Serological grouping of strains associated with plants and insects. *Canadian Journal of Microbiology* 25: 861–866.

Dawkins, R. 1982. *The Extended Phenotype.* W. H. Freeman, Oxford.

Day, K. P., F. Karamalis, J. Thompson, D. A. Barnes, C. Peterson, H. Brown, G. V. Brown, and D. J. Kemp. 1993. Genes necessary for expression of a virulence determinant and for transmission of *Plasmodium falciparum* are located on a 0.3-megabase region of chromosome 9. *Proceedings of the National Academy of Sciences USA* 89: 6015–6019.

Day, K. P., J. C. Koella, S. Nee, S. Gupta, and A. F. Read. 1992. Population genetics and dynamics of *Plasmodium falciparum*—an ecological view. *Parasitology (suppl.)* 104: S35–S52.

De Andrade, C. F. S. 1980. Epizootia natural causada por *Cordyceps unilateralis* (Hypocreales, Euascomycetes) em adultos de *Camponotus* sp. (Hymenoptera, Formicidae) na regiao de Manaus, Amazonas, Brazil. *Acta Amazonia* 10: 617–677.

De Graaf, D. C., H. Raes, and F. J. Jacobs. 1994. Spore dimorphism in *Nosema apis* (Microsporida, Nosematidae) developmental cycle. *Journal of Invertebrate Pathology* 63: 92–94.

De Jong, D., and P. H. De Jong. 1983. Longevity of Africanized honeybees (Hymenoptera, Apidae) infested with *Varroa jacobsoni* (Parasitiformes, Varroidea). *Journal of Economic Entomology* 76: 766–768.

De Jong, D., L. S. Goncalves, and R. A. Morse. 1984. Dependence on climate of the virulence of *Varroa jacobsoni. Bee World* 65: 117–121.

De Jong, D., R. A. Morse, and G. C. Eickwort. 1982. Mite pests of honey bees. *Annual Review of Entomology* 27: 229–252.

Dew, H. E., and C. D. Michener. 1981. Division of labor among workers of *Polistes metricus* (Hymenoptera:Vespidae): Laboratory foraging activities. *Insectes Sociaux* 28: 87–101.

Dietert, R. R., R. L. Taylor, and M. E. Dietert. 1990. The chicken major histocompatibility complex: Structure and impact on immune function, disease resistance and productivity. *Monographs in Animal Immunology* 1: 7–26.

Disney, R. H. L., and D. H. Kistner. 1988. Phoridae collected from termite and ant colonies in Sulawesi (Diptera; Isoptera, Termitidae; Hymenoptera, Formicidae). *Sociobiology* 14: 361–369.

Dobrzanska, J. 1959. Studies on the division of labor in ants genus *Formica. Acta Biologiae Experimentalis* (Warsaw) 19: 57–81.

Dobson, A. P. 1988. The population dynamics of parasite-induced changes in host behavior. *Quarterly Review of Biology* 63: 139–165.

Donzé, G., and P. M. Guerin. 1994. Behavioral attributes and parental care of *Varroa* mites parasitizing honeybee brood. *Behavioural Ecology and Sociobiology* 34: 305–319.

Doull, K. M. 1961. A theory of the causes of development of epizootics of *Nosema* disease of the honey bee. *Journal of Insect Pathology* 3: 297–309.

Doull, K. M., and K. M. Cellier. 1961. A survey of the incidence of *Nosema* disease (*Nosema apis* Zander) in honey bees in South Australia. *Journal of Insect Pathology* 3: 280–288.

Dowell, R. V., H. G. Basham, and R. E. McCoy. 1981. Influence of five spiroplasma strains on growth rate and survival of *Galleria monella* (Lepidoptera, Pyralidae) larvae. *Journal of Invertebrate Pathology* 37: 231–235.

Drees, B. M., R. W. Miller, S. B. Vinson, and R. Georgis. 1992. Susceptibility and behavioral response of red imported fire ant (Hymenoptera: Formicidae) to selected entomogenous nematodes (Rhabditida: Steinernematidae and Heterorhabditidae). *Journal of Economic Entomology* 85: 365–370.

Drif, L., and M. Brehélin. 1993. Structure, classification and functions of insect haemocytes. In J. P. N. Pathak, ed., *Insect Immunity,* pp. 1–14. Kluwer Academic Publishers, Dordrecht.

Drum, N. H., and W. C. Rothenbuhler. 1985. Differences in non-stinging aggressive responses of worker honeybees to diseased and healthy bees in May and July. *Journal of Apicultural Research* 24: 184–187.

Duchateau, M. J., and H. H. W. Velthuis. 1988. Development and reproductive strategies in *Bombus terrestris* colonies. *Behaviour* 107: 186–207.

Duffy, J. E. 1996. Eusociality in a coral-reef shrimp. *Nature* 381: 512–514.

Dunkle, S. W. 1979. Sexual competition for space of the parasite *Xenos pallidus* Brues in male *Polistes annularis* (L.) (Strepsiptera, Stylopidae, and Hymenoptera, Vespidae). *Psyche* 86: 327–336.

Dunn, P. E. 1990. Humoral immunity in insects. *BioScience* 40: 738–744.

Durrer, S. 1996. Parasite load and assemblages of bumblebee species. Ph.D. diss., Dept. of Environmental Sciences, ETH, Zurich.

Durrer, S., and P. Schmid-Hempel. 1994. Shared use of flowers leads to horizontal pathogen transmission. *Proceedings of the Royal Society London, B* 258: 299–302.

Durrer, S., and P. Schmid-Hempel. 1995. Parasites and the regional distribution of bumble bee species. *Ecography* 18: 114–122.

Durrer, S., and P. Schmid-Hempel. 1996. On species diversity in *Bombus* spp. In prep.

Dye, C., C. R. Davies, and J. D. Lines. 1990. When are parasites clonal? *Nature* 348: 120.

Ebert, D. 1994a. Genetic differences in the interactions of a microsporidian parasite and four clones of its cyclically parthenogenetic host. *Parasitology* 108: 11–16.

Ebert, D. 1994b. Virulence and local adaptation of a horizontally transmitted parasite. *Science* 265: 1084–1086.

Ebert, D., and W. D. Hamilton. 1996. Sex against virulence: The evolution of parasitic diseases. *Trends in Ecology and Evolution* 11: 79–82.

Eckert, J. E. 1950. The development of resistance to American foulbrood by honey bees in Hawaii. *Journal of Economic Entomology* 43: 562–564.

Eickwort, G. C. 1990. Mites associated with social insects. *Annual Review of Entomology* 35: 469–488.

Eickwort, G. C. 1994. Evolution and life-history patterns of mites associated with bees. In M. A. Houck, ed., *Mites,* pp. 218–251. Chapman and Hall, New York.

Eischen, F. A., W. T. Wilson, J. S. Pettis, A. Suarez, D. Cardoso-Tamez, D. L. Maki, A. Dietz, J. Vargas, C. Garza de Estrada, and W. L. Rubink. 1990. The spread of *Acarapis woodi* (Acari: Tarsonemidae) in Northeastern Mexico. *Journal of the Kansas Entomological Society* 63: 375–384.

Eisner, T., and G. M. Happ. 1962. The infrabuccal pocket of a formicine ant: A social filtration device. *Psyche* 69: 107–116.

Elmes, G. W., and L. Keller. 1993. Distribution and ecology of queen number in ants of the genus *Myrmica*. In L. Keller, ed., *Queen Number and Sociality in Insects,* pp. 294–307. Oxford University Press, Oxford.

Elmes, G. W., and J. C. Wardlaw. 1982. A population study of the ants *Myrmica sabuleti* and *Myrmica scabrinodis,* living at two sites in the south of England. I. A comparison of colony populations. *Journal of Animal Ecology* 51: 651–664.

Ennesser, C. A., and A. J. Nappi. 1984. Ultrastructural study of the encapsulation response of the American cockroach, *Periplaneta americana. Journal of Ultrastructural Research* 87: 31–45.

Estoup, A., A. Scholl, A. Pouvreau, and M. Solignac. 1995. Monandry and polyandry in bumble bees (Hymenoptera, Bombinae) as evidenced by highly variable microsatellites. *Molecular Ecology* 4: 89–93.

Estoup, A., M. Solignac, and J.-M. Cornuet. 1994. Precise assessment of the number of patrilines and of genetic relatedness. *Proceedings of the Royal Society London, B* 258: 1–7.

Estoup, A., M. Solignac, J.-M. Cornuet, J. Goudet, and A. Scholl. 1996. Genetic differentiation of continental and island populations of *Bombus terrestris* (Hymenoptera: Apidae) in Europe. *Molecular Ecology* 5: 19–31.

Evans, G. O. 1992. *Principles of Acarology.* CAB International, Wallingford, U.K.

Evans, H. C. 1982. Entomogenous fungi in tropical forest ecosystems: An appraisal. *Ecological Entomology* 7: 47–60.

Evans, H. C. 1989. Mycopathogens of insects of epigeal and aerial habitats. In N. Wilding, N. M. Collins, P. M. Hammond, and J. F. Webber, eds., *Insect-Fungus Interactions,* pp. 205–238. Academic Press, London.

Evans, H. C., and R. A. Samson. 1982. *Cordyceps* spp. and their anamorph pathogenic on ants (Formicidae) in tropical forest ecosystems. I. The *Cephalotes* (Myrmicinae) complex. *Transactions of the British Mycological Society* 79: 431–453.

Evans, H. C., and R. A. Samson. 1984. *Cordyceps* species and their anamorphs pathogenic on ants (Formicidae) in tropical forest ecosystems. II. The *Camponotus* (Formicinae) complex. *Transactions of the British Mycological Society* 82: 127–150.

Evans, H. F., and P. F. Entwistle. 1987. Viral diseases. In J. R. Fuxa and Y. Tanada, eds., *Epizootiology of Insect Diseases,* pp. 257–322. John Wiley and Sons, New York.

Evers, C. A., and T. D. Seeley. 1986. Kin discrimination and aggression in honey bee colonies with laying workers. *Animal Behaviour* 34: 924–925.

Evesham, E. J. M., and M.-C. Cammaerts. 1984. Preliminary studies of worker behaviour towards alien queens in the ant *Myrmica rubra* L. *Biology of Behavior* 9: 131–143.

Ewald, P. W. 1987. Transmission modes and the evolution of the parasitism-mutualism continuum. *Annals of the New York Academy of Sciences* 503: 175–206.

Ewald, P. W. 1993. The evolution of virulence. *Scientific American* 268 (April): 56–62.

Ewald, P. W. 1994. *The Evolution of Infectious Disease.* Oxford University Press, Oxford.

Falconer, D. S. 1989. *Introduction to Quantitative Genetics,* 3d ed. John Wiley and Sons, New York.

Fantham, H. B., and A. Porter. 1913. The pathogenicity of *Nosema apis* to insects other than hive bees. *Annals of Tropical Medicine and Parasitology* 7: 569–579.

Fantham, H. B., A. Porter, and L. R. Richardson. 1941. Some microsporidia found in certain fish and insects in eastern Canada. *Parasitology* 38: 186–208.

Farish, D. J. 1972. The evolutionary implications of qualitative variation in the grooming behaviour of the Hymenoptera (Insecta). *Animal Behaviour* 20: 662–676.

Faye, I., and D. Hultmark. 1993. The insect immune proteins and the regulation of their genes. In N. E. Beckage, S. N. Thompson, and B. A. Federici, eds., *Parasites and Pathogens of Insects,* vol. 2: *Pathogens,* pp. 25–54. Academic Press, London.

Feener, D. H. 1981. Competition between ant species: Outcome controlled by parasitic flies. *Science* 214: 815–817.

Feener, D. H. 1987. Size-selective oviposition in *Pseudaceton crawfordi* (Diptera: Phoridae), a parasite of fire ants. *Annals of the Entomological Society of America* 80: 148–151.

Feener, D. H. 1988. Effects of parasites on foraging and defense behavior of termitophagous ants, *Pheidole titanis* Wheeler (Hymenoptera: Formicidae). *Behavioural Ecology and Sociobiology* 22: 421–427.

Fenner, F., M. F. Day, and G. M. Woodroofe. 1956. Epidemiological consequences of the mechanical transmission of myxomatosis by mosquitoes. *Journal of Hygiene* 54: 284–303.

Fenner, F., and F. N. Ratcliffe. 1965. *Myxomatosis.* Cambridge University Press, Cambridge, U.K.

Finlay, B. B., and S. Falkow. 1989. Common themes in microbial pathogenicity. *Microbiological Reviews* 53: 210–230.

Fischer, F. M. J. 1964. *Nosema* as a source of juvenile hormone in parasitized insects. *Biological Bulletin* 126: 235–252.

Fisher, R. A. 1958. *The Genetical Theory of Natural Selection.* 2d ed. Dover Publications, New York.

Fisher, R. M. 1988. Observations on the behaviours of three Euopean cuckoo bumble bee species (*Psithyrus*). *Insectes Sociaux* 35: 341–354.

Fisher, R. M. 1989. Incipient colony manipulation, *Nosema* incidence and colony productivity of the bumble bee *Bombus terrestris* (Hymenoptera, Apidae). *Journal of the Kansas Entomological Society* 62: 581–589.

Fletcher, D. J. C., and C. D. Michener, eds. 1987. *Kin Recognition in Animals.* John Wiley and Sons, New York.

Folstad, I., and A. J. Karter. 1992. Parasites, bright males, and the immunocompetence handicap. *American Naturalist* 139: 603–622.

Foster, R. L. 1992. Nestmate recognition as an inbreeding avoidance mechanism in bumble bees (Hymenoptera: Apidae). *Journal of the Kansas Entomological Society* 65: 238–243.

Fowler, H. G. 1985. Alloethism in the ant *Camponotus pennsylvanicus* (Hymenoptera: Formicidae). *Entomologia Generalis* 11: 69–76.

Fowler, H. G. 1986. Polymorphism and colony ontogeny in North American Carpenter ants (Hymenoptera: Formicidae). *Zoologisches Jahrbuch für Physiologie* 90: 297–316.

Fowler, J. L., and E. L. Reeves. 1974. Spore dimorphism in a microsporidian isolate. *Journal of Protozoology* 10: 644–673.

Fox-Wilson, G. 1946. Factors affecting populations of social wasps, *Vespula* spp., in England. *Proceedings of the Royal Entomological Society London* 21: 17–27.

Frank, S. A. 1996. Host-symbiont conflict over the mixing of symbiotic lineages. *Proceedings of the Royal Society London, B* 263: 339–344.

Franks, N. R. 1985. Reproduction, foraging efficiency and worker polymorphism in army ants. In B. Hölldobler and M. Lindauer, eds., *Experimental Behavioural Ecology and Sociobiology,* pp. 91–107. Sinauer Associates, Sunderland, Mass.

Franks, N. R., K. J. Healey, and L. Byrom. 1991. Studies on the relationship between the ant ectoparasite *Antennophorus grandis* (Acarina: Antennophoridae) and its host *Lasius flavus* (Hymenoptera: Formicidae). *Journal of Zoology* 225: 59–70.

Franks, N. R., B. Ireland, and A. F. G. Bourke. 1990. Conflicts, social economics and life history strategies in ants. *Behavioural Ecology and Sociobiology* 27: 175–181.

Free, J. B. 1958. The drifting of honey bees. *Journal of Agricultural Sciences* 51: 294–306.

Freeland, W. J. 1976. Pathogens and the evolution of primate sociality. *Biotropica* 8: 12–24.

Freeland, W. J. 1979. Primate social groups as biological islands. *Ecology* 60: 719–728.

Fries, I., G. Ekbohm, and E. Villumstad. 1984. *Nosema apis:* Sampling techniques and honey yield. *Journal of Apicultural Research* 23: 102–105.

Fries, I., R. R. Granados, and R. A. Morse. 1992. Intracellular germination of spores of *Nosema apis* Z. *Apidologie* 23: 61–70.

Frumhoff, P. C., and S. S. Schneider. 1987. The social consequences of honey bee polyandry: The effects of kinship on worker interactions within colonies. *Animal Behaviour* 35: 255–262.

Frumhoff, P. C., and P. S. Ward. 1992. Individual-level selection, colony-level selection, and the association between polygyny and worker monomorphism in ants. *American Naturalist* 139: 559–590.

Fuchs, S. 1992. Choice in *Varroa jacobsoni* Oud. between honey bee drone or worker brood cells for reproduction. *Behavioural Ecology and Sociobiology* 31: 429–435.

Furgala, B. 1962. The effect of the intensity of *Nosema* inoculum on queen supersedure. *Journal of Insect Pathology* 4: 429–432.

Fyg, W. 1964. Anomalies and diseases of the queen honey bee. *Annual Review of Entomology* 9: 207–224.

Gadagkar, R. 1991. *Belonogaster, Mischocyttarus, Parapolybia* and independent-founding *Ropalidia.* In K. G. Ross and R. W. Matthews, eds., *The Social Biology of Wasps,* pp. 149–190. Comstock Publishing Associates, Ithaca, New York.

Gadagkar, R., K. Chandrashekara, S. Chandran, and S. Bhagavan. 1993. Serial polygyny in the primitively eusocial wasp *Ropalidia marginata:* Implications for the study of sociality. In L. Keller, ed., *Queen Number and Sociality in Insects,* pp. 188–214. Oxford University Press, Oxford.

Gamboa, G. J., B. J. Heacock, and S. L. Wiltjer. 1978. Division of labor and subordinate longevity in foundress associations of the paper wasp *Polistes metricus* (Hymenoptera: Vespidae). *Journal of the Kansas Entomological Society* 51: 343–352.

Gamboa, G. J., H. K. Reeve, and D. W. Pfennig. 1986. The evolution and ontogeny of nestmate recognition in social wasps. *Annual Review of Entomology* 31: 431–454.

Gandon, S., Y. Capowiez, Y. Dubois, Y. Michalakis, and I. Olivieri. 1996. Local adaptation and gene for gene coevolution in a metapopulation model. *Proceedings of the Royal Society London, B* 263: 1003–1009.

Ganem, D., and H. E. Varmus. 1987. The molecular biology of hepatitis B viruses. *Annual Review of Biochemistry* 56: 651–693.

Gardner, W. A. 1985. Effects of temperature on the susceptibility of *Heliothis zea* larvae to *Nomureae rileyi*. *Journal of Invertebrate Pathology* 46: 348–349.

Garofalo, C. A. 1978. Bionomics of *Bombus* (*Fervidobombus*) *morio*. 2. Body size, length of life of workers. *Journal of Apicultural Research* 17: 130–136.

Gary, N. E., and R. E. Page. 1987. Phenotypic variation in susceptibility of honey bees, *Apis mellifera,* to infestation by tracheal mites, *Acarpis woodi*. *Experimental and Applied Acarology* 3: 291–305.

Gary, N. E., R. E. Page, and K. Lorenzen. 1989. Effect of age of worker honey bees (*Apis mellifera*) on tracheal mite (*Acarapis woodi*) infestation. *Experimental and Applied Acarology* 7: 153–160.

Gauld, I., and B. Bolton. 1988. *The Hymenoptera*. Oxford University Press, Oxford.

Gentry, J. B. 1974. Response to predation by colonies of the Florida harvester ant, *Pogonomyrmex badius*. *Ecology* 55: 1328–1338.

Gertsch, P., P. Pamilo, and S. L. Varvio. 1995. Microsatellites reveal high genetic diversity within colonies of *Camponotus* ants. *Molecular Ecology* 4: 257–260.

Getz, W. M., D. Brückner, and T. R. Parisian. 1982. Kin structure and the swarming behaviour of the honey bee *Apis mellifera*. *Behavioural Ecology and Sociobiology* 10: 265–270.

Getz, W. M., D. Brückner, and K. B. Smith. 1986. Conditioning honeybees to discriminate between heritable odors from full and half sisters. *Journal of Comparative Physiology A* 159: 251–256.

Getz, W. M., and K. B. Smith. 1983. Genetic kin recognition: Honey bees discriminate between full and half sisters. *Nature* 302: 147–148.

Gilliam, M. 1973. Are yeasts present in adult woker honey bees as a consequence of stress? *Annals of the Entomological Society of America* 66: 1176.

Gilliam, M. 1978. Fungi. In R. E. Morse, ed., *Honeybee Pests, Predators and Diseases,* pp. 78–101. Cornell University Press, Ithaca, New York.

Gilliam, M., D. W. Roubik, and B. J. Lorenz. 1990. Microorganisms associated with pollen, honey, and brood provisions in the nest of the stingless bee, *Melipona fasciata*. *Apidologie* 21: 89–97.

Gilliam, M., L. J. Wickerham, H. L. Morton, and R. D. Martin. 1974. Yeasts isolated from honey bees, *Apis mellifera,* fed 2, 4-D and antibiotics. *Journal of Invertebrate Pathology* 24: 349–356.

Glancey, B. M., R. K. Vander Meer, A. Glover, C. S. Lofgren, and S. B. Vinson. 1981. Filtration of microparticles from liquids ingested by the red imported fire ant, *Solenopsis invicta* Buren. *Insectes Sociaux* 28: 395–401.

Glare, T. R., M. O'Callaghan, and P. J. Wigley. 1993. Checklist of naturally occurring entomophilous microbes and nematodes in New Zealand. *New Zealand Journal of Zoology* 20: 95–120.

Glunn, F. J., D. F. Howard, and W. R. Tschinkel. 1981. Food preferences in colonies of the fire ant, *Solenopsis invicta*. *Insectes Sociaux* 28: 217–222.

Godfray, H. C. J. 1994. *Parasitoids—Behavioral and Evolutionary Ecology*. Princeton University Press, Princeton, New Jersey.

Godfray, H. C. J., and A. Grafen. 1988. Unmatedness and the evolution of sociality. *American Naturalist* 131: 303–305.

Goettel, M. S., J. D. Vandenberg, G. M. Duke, and G. B. Schaalje. 1993. Susceptibility to chalkbrood of alfalfa leafcutter bees, *Megachile rotundata,* reared on natural and artificial provisions. *Journal of Invertebrate Pathology* 61: 58–61.

Goldblatt, J. W. 1984. Parasites and parasitization rates in bumble bee queens *Bombus* spp. (Hymenoptera, apidae) in Southwestern Virginia. *Environmental Entomology* 13: 1661–1665.

Goldblatt, J. W., and R. D. Fell. 1987. Adult longevity of workers of the bumblebees *B. fervidus, B. pennsylvanicus. Canadian Journal of Zoology* 65: 2349–2353.

Goldman, I. F., J. Arnold, and B. Carlton. 1986. Selection for resistance to *Bacillus thuringensis* subspecies *israelensis* in field and laboratory populations of the mosquito *Aedes aegypti. Journal of Invertebrate Pathology* 47: 317–324.

Golley, F. B., and J. B. Gentry. 1964. Bioenergetics of the southern harvester ant, *Pogonomyrmex badius. Ecology* 45: 217–225.

Goncalves, L. S., D. De Jong, and R. H. Nogueria. 1982. Infestation of feral honey bee colonies in Brazil by *Varroa jacobsoni. American Bee Journal* 122: 249–251.

Goodwin, R. H., R. J. Milner, and C. D. Beaton. 1991. Entomopoxvirinae. In J. R. Adams and J. R. Bonami, eds., *Atlas of Invertebrate Viruses,* pp. 259–285. CRC Press, Baton Rouge, Louisiana.

Gordon, D. M. 1986. The dynamics of the daily round of the harvester ant colony (*Pogonomyrmex badius*). *Animal Behaviour* 34: 1402–1419.

Gordon, D. M. 1989. Dynamics of task switching in harvester ants. *Animal Behaviour* 38: 194–204.

Gordon, D. M. 1991. Behavioral flexibility and the foraging ecology of seed-eating ants. *American Naturalist* 138: 379–411.

Gordon, D. M. 1992. How colony growth affects forager intrusion in neighboring harvester ant colonies. *Behavioural Ecology and Sociobiology* 31: 417–427.

Gordon, D. M. 1995. The development of an ant colony's foraging range. *Animal Behaviour* 49: 649–659.

Gordon, D. M., and B. Hölldobler. 1987. Worker longevity in harvester ants (*Pogonomyrmex*). *Psyche* 94: 341–346.

Gösswald, K. 1930. Weitere Beiträge zur Verbreitung der Mermithiden bei Ameisen. *Zoologischer Anzeiger* 84: 202–204.

Gösswald, K. 1950. Pflege des Ameisenparasiten *Tamiclea globula* Meig. (Diptera) durch den Wirt mit Bemerkungen über den Stoffwechsel in der parasitierten Ameise. *Verhandlungen der Deutschen Zoologischen Gesellschaft* (Mainz) 1949: 256–264.

Gottwald, T. R., and W. L. Tedders. 1984. Colonization, transmission and longevity of *Beauveria bassiana* and *Metarhizium anisopliae* (Deuteromycotina: Hyphomycetes) on pecan weevil larvae (Coleoptera: Curculinoidae) in the soil. *Environmental Entomology* 13: 557–560.

Grafen, A. 1986. Split sex ratios and the evolutionary origins of eusociality. 122: 95–121.

Grafen, A. 1990. Do animals really recognize kin? *Animal Behaviour* 39: 42–54.

Graur, D. 1985. Eusociality and genetic variability: A re-evaluation. *Evolution* 39: 200–201.

Greenberg, L. 1979. Genetic component of bee odor in kin recognition. *Science* 206: 1095–1097.

Greene, A. 1991. *Dolichovespula* and *Vespula.* In K. G. Ross and R. W. Matthews, eds.,

The Social Biology of Wasps, pp. 263–308. Comstock Publishing Associates, Ithaca, New York.

Greenslade, P. J. M. 1975a. Dispersion and history of a population of the meat ant *Iridomyrmex purpureus* (Hymenoptera: Formicidae). *Australian Journal of Zoology* 23: 495–510.

Greenslade, P. J. M. 1975b. Short-term change in a population of the meat ant *Iridomyrmex purpureus* (Hymenoptera: Formicidae). *Australian Journal of Zoology* 23: 511–522.

Grenfell, B. T., and A. P. Dobson, eds. 1995. *Ecology and Infectious Diseases in Natural Populations.* Cambridge University Press, Cambridge, U.K.

Grosberg, R. K., and J. F. Quinn. 1988. The evolution of selective aggression based on allorecognition specificity. *Evolution* 43: 504–515.

Gu-Xiang, L., and D. Zi-Vong. 1990. Biology of *Coptotermes formosanus* Shiraki in China. In G. K. Veeresh, B. Mallik, and C. A. Viraktamath, eds., *Social Insects and the Environment,* pp. 47–48. Oxford University Press and IBH Publishing, New Delhi.

Gupta, A. P. 1985. Cellular elements in the hemolymph. In G. A. Kerkut and L. I. Gilbert, eds., *Comprehensive Insect Physiology, Biochemistry and Pharmacology,* vol. 3: *Integument, Respiration and Circulation,* pp. 401–451. Pergamon Press, New York.

Gupta, A. P. 1986. *Hemocytic and Humoral Immunity in Arthropods.* John Wiley and Sons, New York.

Haas, A. 1949. Arttypische Flugbahnen bei Hummelmännchen. *Zeitschrift für Vergleichende Physiologie* 31: 281–307.

Hadorn, E., and I. Walker. 1960. *Drosophila* and *Pseudeucoila.* I. Selektionsversuche zur Steigerung der Abwehrreaktion des Wirtes gegen den Parasiten. *Revue Suisse de Zoologie* 67: 216–225.

Haldane, J. B. S. 1949. Disease and evolution. *La Ricerca Scientifica* 19 (Suppl.): 68–76.

Halverson, D. D., J. Wheeler, and G. C. Wheeler. 1976. Natural history of the sandhill ant, *Formica bradleyi* (Hymenoptera: Formicidae). *Journal of the Kansas Entomological Society* 49: 280–303.

Halvorsen, O. 1986. On the relationship between social status of the host and risk of parasitic infection. *Oikos* 47: 71–74.

Hamilton, W. D. 1964. The genetical evolution of social behavior. *Journal of theoretical Biology* 7 (I): 1–16; (II): 17–32.

Hamilton, W. D. 1972. Altruism and related phenomena, mainly in social insects. *Annual Review of Ecology and Systematics* 3: 193–232.

Hamilton, W. D. 1980. Sex vs. non-sex vs. parasite. *Oikos* 35: 282–290.

Hamilton, W. D. 1987. Kinship, recognition, disease, and intelligence: Constraints of social evolution. In Y. Itô, J. L. Brown and J. Kikkawa, eds., *Animal Societies: Theory and Facts,* pp. 81–102. Japanese Scientific Society Press, Tokyo.

Hartl, D. L., and A. G. Clark. 1989. *Principles of Population Genetics.* Sinauer Associates, Sunderland, Mass.

Harvey, P. H., and M. D. Pagel. 1991. *The Comparative Method in Evolutionary Biology.* Oxford University Press, Oxford.

Harvey, T. L., and D. E. Howell. 1965. Resistance of the house fly to *Bacillus thuringensis* Berliner. *Journal of Invertebrate Pathology* 7: 92–100.

Haseman, L. 1961. How long can spores of American foulbrood live? *American Bee Journal* 101: 298–299.

Haskins, C. P., and E. F. Haskins. 1950. Notes on the biology and social behavior of the archaic ponerine ants of the genera *Myrmecia* and *Promyrmecia*. *Annals of the Entomological Society of America* 43: 461–491.

Haskins, C. P., and E. F. Haskins. 1979. Worker compatibilities within and between populations of *Rhytidoponera metallica*. *Psyche* 86: 299–312.

Haskins, C. P., and E. F. Haskins. 1980. Notes on female and worker survivorship in the archaic ant genus *Myrmecia*. *Insectes Sociaux* 27: 345–350.

Haskins, C. P., and E. F. Haskins. 1983. Situation and location-specific factors in the compatibility response in *Rhytidoponera metallica* (Hymenoptera: Formicidae: Ponerinae). *Psyche* 90: 163–174.

Haskins, C. P., and R. M. Whelden. 1954. Note on the exchange of ingluvial food in the genus *Myrmecia*. *Insectes Sociaux* 1: 33–37.

Hassanein, M. H. 1951. The influence of *Nosema apis* on the larval honey-bee. *Annals of Applied Biology* 38: 844–846.

Hassanein, M. H. 1952. Some studies on amoeba disease. *Bee World* 33: 109–112.

Hattingen, R. 1956. Beiträge zur Biologie von *Sphaerularia bombi* Léon Dufour (1837). *Zentralblatt für Bakteriologie, Parasitenkunde, Infektionskrankheiten und Hygiene* (Abt. II) 109: 236–249.

Hauschteck-Jungen, E., and H. Jungen. 1983. Ant chromosomes. II. Karyotypes of western palearctic species. *Insectes Sociaux* 30: 149–164.

Hausfater, G., and D. F. Watson. 1976. Social and reproductive correlates of parasite ova transmission by baboons. *Nature* 262: 688–689.

Haverty, M. I. 1979. Soldier production and maintenance of soldier proportions of laboratory experimental groups of *Coptotermes formosanus* (Shiraki). *Insectes Sociaux* 26: 69–84.

Haverty, M. I., and R. W. Howard. 1981. Production of soldiers and maintenance of soldier proportions by laboratory experimental groups of *Reticulitermes flavipes* (Kollar) and *Reticulitermes virginicus* (Banks) (Isoptera: Rhinotermitidae). *Insectes Sociaux* 28: 32–39.

Heath, L. A. F. 1982. Development of chalk brood in a honey bee colony: A review. *Bee World* 63: 119–130.

Hedrick, P. 1994. Evolutionary genetics of the major histocompatibility complex. *American Naturalist* 143: 945–964.

Hefetz, A. 1987. The role of Dufour's gland secretions in bees. *Physiological Entomology* 12: 243–253.

Heinrich, B. 1979. *Bumblebee Economics*. Harvard University Press, Cambridge, Mass.

Hellmich, R. L., J. M. Kulincevic, and W. C. Rothenbuhler. 1985. Selection for high- and low-pollen hoarding honey bees. *Journal of Heredity* 76: 155–158.

Heraty, J. M. 1985. A revision of the nearctic Eucharitinae (Hymenoptera: Chalcidoidea: Eucharitidae). *Proceedings of the Entomological Society of Ontario* 116: 61–103.

Herbers, J. M. 1979. Caste-based polyethism in a mound-building ant species. *American Midlands Naturalist* 101: 69–75.

Herbers, J. M. 1980. On caste ratios in ant colonies: Population responses to changing environments. *Evolution* 34: 575–585.

Herbers, J. M. 1985. Seasonal structuring of a north temperate ant community. *Insectes Sociaux* 32: 224–240.

Herbers, J. M. 1986. Nest site limitation and facultative polygyny in the ant *Leptothorax longispinosus. Oecologia* 19: 115–122.

Herbers, J. M. 1993. Ecological determinants of queen number in ants. In L. Keller, ed., *Queen Number and Sociality in Insects,* pp. 262–293. Oxford University Press, Oxford.

Herre, E. A. 1993. Population structure and the evolution of virulence in nematode parasites of fig wasps. *Science* 259: 1442–1445.

Hieber, C. S., and G. W. Uetz. 1990. Colony size and parasitoid load in two species of colonial *Metepeira* spiders from Mexico (Aranea: Araneidae). *Oecologia* 82: 145–150.

Hillesheim, E., N. Koeniger, and R. F. A. Moritz. 1989. Colony performance in honey bees (*Apis mellifera capensis* Esch) depends on the proportion of subordinate and dominant workers. *Behavioural Ecology and Sociobiology* 24: 291–296.

Hirschfelder, H. 1964. Untersuchungen über Pollenernährung, Lebenslänge und *Nosema*-Befall bei der Honigbiene. *Bulletin Apicole* 7: 7–17.

Hitchcock, J. D. 1948. A rare gregarine parasite of the adult honey bee. *Journal of Economic Entomology* 41: 854–858.

Hitchcock, J. D., A. Stoner, W. T. Wilson, and D. M. Menapace. 1979. Pathogenicity of *Bacillus pulvifaciens* to honey bee larvae of various ages (Hymenoptera: Apidae). *Journal of the Kansas Entomological Society* 52: 238–246.

Hoage, T. R., and W. C. Rothenbuhler. 1966. Larval honeybee response to various doses of *Bacillus larvae* spores. *Journal of Economic Entomology* 59: 42–45.

Hobbs, G. A. 1962. Further studies on the food-gathering behaviour of bumble bees (Hymenoptera: Apidae). *Canadian Entomologist.* 94: 538–541.

Hobbs, G. A. 1967. Ecology of species of *Bombus* (Hymenoptera, Apidae) in Southern Alberta. VI. Subgenus *Pyrobombus. Canadian Entomologist* 99: 1271–1292.

Hochberg, M. E., M. P. Hassell, and R. M. May. 1989. The dynamics of host-parasitoid-pathogen interactions. *American Naturalist* 135: 74–94.

Hochberg, M. E., Y. Michalakis, and T. De Meeus. 1992. Parasitism as a constraint on the rate of life-history evolution. *Journal of Evolutionary Biology* 5: 491–504.

Hoffmann, J. A., C. A. Janeway, and S. Natori, eds. 1994. *Phylogenetic Perspectives in Immunity: The Insect Host Defense.* R. G. Landes Company, Austin, Texas.

Hogendoorn, K., and H. H. W. Velthuis. 1988. Influence of multiple mating on kin recognition by worker honeybees. *Naturwissenschaften* 75: 412–413.

Hohorst, W., and G. Graefe. 1961. Ameisen—obligate Zwischenwirte des Lanzettegels (*Dicrocoelium dendriticum*). *Naturwissenschaften* 48: 229–230.

Hölldobler, B. 1976a. The behavioral ecology of mating in harvester ants (Hymenoptera: Formicidae: *Pogonomyrmex*). *Behavioural Ecology and Sociobiology* 1: 405–423.

Hölldobler, B. 1976b. Recruitment behaviour, home range orientation and territoriality in harvester ants, *Pogonomyrmex. Behavioural Ecology and Sociobiology* 1: 405–423.

Hölldobler, B., and S. H. Bartz. 1985. Sociobiology of reproduction in ants. In B. Hölldobler and M. Lindauer, eds., *Experimental Behavioral Ecology and Sociobiology,* pp. 237–258. Sinauer Associates, Sunderland, Mass.

Hölldobler, B., and N. F. Carlin. 1985. Colony founding, queen dominance and oligogyny in the Australian meat ant *Iridomyrmex purpureus*. *Behavioural Ecology and Sociobiology* 18: 45–58.

Hölldobler, B., and H. Engel-Siegel. 1984. The metapleural gland of ants. *Psyche* 91: 201–224.

Hölldobler, B., and E. O. Wilson. 1990. *The Ants*. Springer-Verlag, Berlin.

Holmes, W. G., and P. W. Sherman. 1982. The ontogeny of kin recognition in two species of ground squirrels. *American Zoologist* 22: 491–517.

Holst, E. C. 1945. An antibiotic from a bee pathogen. *Science* 102: 593–594.

Hoogland, J. L. 1979. Aggression, ectoparasitism and other possible costs of prairie dog coloniality. *Behaviour* 69: 1–35.

Hoogland, J. L., and P. W. Sherman. 1976. Advantages and disadvantages of bank swallow coloniality. *Ecological Monographs* 46: 33–58.

Horn, H., and J. Eberspächer. 1976. Die Waldtrachtkrankheit der Honigbiene. II. Nachweis von Bakterien in der Hämolymphe waldtrachtkranker Bienen und der zusätzliche Einfluss der Fütterung auf die Waldtrachtkrankheit. *Apidologie* 17: 307–324.

Hornitzky, M. A. Z. 1987. Prevalence of virus infections in Eastern Australia. *Journal of Apicultural Research* 26: 181–185.

Howick, L. D., and J. W. Creffield. 1980. Interspecific agonism in *Coptotermes acinaciformis*. *Bulletin of Entomological Research* 70: 17–23.

Huger, A. M., S. W. Skinner, and J. H. Werren. 1985. Bacterial infections associated with the son-killer trait in the parasitoid wasp *Nasonia* (=*Mormoniella*) *vitripennis* (Hymenoptera, Pteromalidae). *Journal of Invertebrate Pathology* 46: 277–280.

Hughes, C. R., D. C. Queller, J. E. Strassmann, C. R. Solis, J. A. Negron-Sotomayor, and K. R. Gastreich. 1993. The maintenance of high genetic relatedness in multi-queen colonies of social wasps. In L. Keller, ed., *Queen Number and Sociality in Insects*, pp. 151–170. Oxford University Press, Oxford.

Humber, R. A. 1981. *Erynia* (Zygomycetes: Entomophtorales): Validations and new species. *Mycotaxon* 13: 471–480.

Hunt, J. H. 1991. Nourishment and the evolution of the social vespidae. In K. G. Ross and R. W. Matthews, eds., *The Social Biology of Wasps*, pp. 426–450. Comstock Publishing Associates, Ithaca, New York.

Hunter, P. E., and R. W. Husband. 1973. *Pneumolaelaps* (Acarina: Laelapidae) mites from North America and Greenland. *Florida Entomologist* 56: 77–91.

Imai, H. T., C. Baroni Urbani, M. Kubota, G. P. Sharma, M. N. Narasimhanna, B. C. Das, A. K. Sharma, A. Sharma, G. B. Deodikar, V. G. Vaidya, and M. R. Rajasekarasetty. 1984. Karyological survey of Indian ants. *Japanese Journal of Genetics* 59: 1–32.

Imai, H. T., R. H. Crozier, and R. W. Taylor. 1977. Karyotype evolution in Australian ants. *Chromosoma* 37: 341–393.

Imai, H. T., and M. Kubota. 1972. Karyological studies of Japanese ants. III. Karyotypes of nine species in Ponerinae, Formicinae, and Myrmicinae. *Chromosoma* 37: 193–200.

Imai, H. T., R. W. Taylor, M. W. J. Crosland, and R. H. Crozier. 1988. Modes of spontaneous chromosomal mutation and karyotype evolution in ants with reference to the minimum interaction hypothesis. *Japanese Journal of Genetics* 63: 159–185.

Imai, H. T., R. W. Taylor, and R. H. Crozier. 1994. Experimental bases for the minimum interaction theory. I. Chromosome evolution in ants of the *Myrmecia pilosula* species complex (Hymenoptera: Formicidae: Myrmeciinae). *Japanese Journal of Genetics* 69: 137–182.

Imhoof, B., and P. Schmid-Hempel. 1998. Paterns of local adaptation of a protozoan parasite to its bumblebee host. *Oikos,* 82: 59–65.

Ishay, J., H. Bytinski-Salz, and A. Shulov. 1968. Contribution to the bionomics of the oriental hornet (*Vespa orientalis*). *Israel Journal of Entomology* 2: 45–106.

Ishikawa, Y., and S. Miyajima. 1968. Interaction between infectious flacherie virus and some bacteria in the silkworm, *Bombyx mori* L. *Journal of Sericultural Science of Japan* 37: 471–475.

Itô, Y. 1993a. *Behaviour and Social Evolution of Wasps.* Oxford University Press, Oxford.

Itô, Y. 1993b. The evolution of polygyny in primitively eusocial polistine wasps with special reference to the genus *Ropalidia.* In L. Keller, ed., *Queen Number and Sociality in Insects,* pp. 171–187. Oxford University Press, Oxford.

Jaenike, J. 1993. Rapid evolution of host specificity in a parasitic nematode. *Evolutionary Ecology* 7: 103–108.

Jaffe, K., and H. Puche. 1984. Colony-specific territorial marking with the metapleural gland secretion in the ant *Solenopsis geminata* (Fabr). *Journal of Insect Physiology* 30: 265–270.

Jay, S. C., and D. Warr. 1984. Sun position as a possibe factor in the disorientation of honeybees (*Apis mellifera*) in the southern hemisphere. *Journal of Apicultural Research* 23: 143–147.

Jean-Prost, P. 1956. Some vital statistics of bees in Provence. *Revue française d'Apiculture* 3: 1558–1561.

Jeanne, R. L. 1972. Social biology of the Neotropical wasp *Mischocyttarus drewseni. Bulletin of the Museum of Comparative Zoology, Harvard University* 144: 63–150.

Jeanne, R. L. 1986. The evolution of the organization of work in social insects. *Monitore Zoologico Italiano* (n.s.) 20: 119–133.

Jeanne, R. L. 1991. The swarm-founding Polistinae. In K. G. Ross and R. W. Matthews, eds., *The Social Biology of Wasps,* pp. 191–231. Comstock Publishing Associates, Ithaca, New York.

Jeanne, R. L., H. A. Downing, and D. C. Post. 1988. Age polyethism and individual variation in *Polybia occidentalis,* an advanced eusocial wasp. In R. L. Jeanne, ed., *Interindividual Behavioral Variability in Social Insects,* pp. 323–357. Westview Press, Boulder, Colorado.

Jeffree, E. P. 1955. Observations on the decline and growth of honey bee colonies. *Journal of Economic Entomology* 48: 723–726.

Jeffree, E. P., and M. D. Allen. 1956. The influence of colony size and *Nosema* disease on the rate of population loss in honeybee colonies in winter. *Journal of Economic Entomology* 49: 831–843.

Johnson, D. W. 1988. Eucharitidae (Hymenoptera: Chalcidoidea): Biology and potential for biological control. *Florida Entomologist* 71: 528–537.

Johnston, A. A., and E. O. Wilson. 1985. Correlates of variation in the major/minor ratio of the ant *Pheidole dentata* (Hymenoptera: Formicidae). *Annals of the Entomological Society of America* 78: 8–11.

Jones, J. C., and O. E. Tauber. 1952. Effects of hemorrhage, cauterization, ligation, desiccation and starvation on hemocytes of the mealworm larvae, *Tenebrio molitor. Iowa State College Journal Science* 26: 371–386.

Jouvenaz, D. P. 1983. Natural enemies of fire ants. *Florida Entomologist* 66: 111–121.

Jouvenaz, D. P. 1984. Some protozoa infecting fire ants, *Solenopsis* spp. In T. C. Cheng, ed., *Comparative Pathobiology,* vol. 7, pp. 195–203. Plenum Press, New York.

Jouvenaz, D. P. 1986. Diseases of fire ants: Problems and opportunities. In C. S. Lofgren and R. K. Vander Meer, eds., *Fire Ants and Leaf-Cutting Ants, Biology and Management,* pp. 327–338. Westview Press, Boulder, Colorado.

Jouvenaz, D. P. 1990. Studies on the endoparasitic yeasts of fire ants, *Solenopsis* spp. *Proceedings of the International Colloquium on Invertebrate Pathology and Microbial Control* 5: 346.

Jouvenaz, D. P., G. E. Allen, W. A. Banks, and D. P. Wojcik. 1977. A survey for pathogens of fire ants, *Solenopsis* spp., in the Southeastern United States. *Florida Entomologist* 60: 275–279.

Jouvenaz, D. P., and D. W. Anthony. 1979. *Mattesia geminata* sp. n. (Neogregarinida: Phrocystidae) a parasite of the tropical fire ant, *Solenopsis geminata* (Fabr.). *Journal of Protozoology* 26: 354–356.

Jouvenaz, D. P., W. A. Banks, and J. D. Atwood. 1980. Incidence of pathogens in fire ants, *Solenopsis* spp., in Brazil. *Florida Entomologist* 63: 345–346.

Jouvenaz, D. P., W. A. Banks, and C. S. Lofgren. 1974. Fire ants: Attraction of workers to queen secretions. *Annals of the Entomological Society of America* 67: 442–444.

Jouvenaz, D. P., M. S. Blum, and J. G. MacConnell. 1972. Anti-bacterial activity of venom alkaloids from the imported fire ant, *Solenopsis invicta* Buren. *Antimicrobial Agents and Chemotherapy* 2: 291–293.

Jouvenaz, D. P., and E. A. Ellis. 1986. *Vairimorpha invictae* n. sp. (Microspora: Microsporidia), a parasite of the red imported fire ant, *Solenopsis invicta* Buren (Hymenoptera: Formicidae). *Journal of Protozoology* 33: 457–461.

Jouvenaz, D. P., and E. I. Hazard. 1978. New family, genus and species of *Microsporidia* (Protozoa: Microsporidia) from the tropical fire ant *Solenopsis geminata* (Fabr.) (Insecta: Formicidae). *Journal of Protozoology* 25: 24–29.

Jouvenaz, D. P., and J. W. Kimbrough. 1991. *Myrmecomyces annellisae* gen. nov., sp. nov. (Deuteromycotina: Hypomycetes), an endoparasitic fungus of fire ants, *Solenopsis* spp. (Hymenoptera: Formicidae). *Mycological Research* 95: 1395–1401.

Jouvenaz, D. P., and C. S. Lofgren. 1984. Temperature-dependent spore dimorphism in *Burenella dimorpha* (Microspora: Microsporida). *Journal of Protozoology* 31: 175–177.

Jouvenaz, D. P., C. S. Lofgren, and G. E. Allen. 1981. Transmission and infectivity of spores of *Burenella dimorpha* (Microsporidae, Burenellidae). *Journal of Invertebrate Pathology* 37: 265–268.

Jouvenaz, D. P., and W. R. Martin. 1992. Evaluation of the nematode *Steinernema carpocapsae* to control fire ants in nursery stock. *Florida Entomologist* 75: 148–151.

Jouvenaz, D. P., and D. P. Wojcik. 1990. Parasitic nematode observed in the fire ant, *Solenopsis richteri,* in Argentina. *Florida Entomologist* 73: 674–675.

Junca, P., C. Saillard, J. Tully, O. Garcia-Jurado, J.-R. Degorce-Dumas, C. Mouches, J.-C. Vignault, R. Vogel, R. McCoy, R. Whitcomb, D. Williamson, J. Latrille, and J.

Bove. 1980. Charactérisation de spiroplasmes isolés d'insectes et des fleures de France continentale, de Corse et du Maroc. Proposition pour une classification des spiroplasmes. *Comptes Rendus de l'Académie des Sciences, Paris, Série D,* 290: 1209–1212.

Jutsum, A. R., T. S. Saunders, and J. M. Cherrett. 1979. Intraspecific aggression in the leaf-cutting ant *Acromyrmex octospinosus. Animal Behaviour* 27: 839–844.

Kaiser, H. 1986. Ueber Wechselbeziehungen zwischen Nematoden (Mermithidae) und Ameisen. *Zoologischer Anzeiger* 217: 156–177.

Kalavati, C. 1976. Four new species of microsporidia from termites. *Acta Protozoologica* 15: 411–422.

Kalmus, H., and C. R. Ribbands. 1952. The origin of the odours by which honeybees distinguish their companions. *Proceedings of the Royal Society London, B* 140: 50–59.

Karmo, E., and O. Morgenthaler. 1939. The development of *Nosema apis* at various temperatures. *Bee World* 20: 57–58.

Karp, R. D., and L. E. Duwel-Eby. 1991. Adaptive immune responses in insects. In G. W. Warr and N. Cohen, eds., *Phylogenesis of Immune Functions,* pp. 1–18. CRC Press, Boca Raton, Florida.

Kaschef, A. H., and T. Abou-Zeid. 1987. The pathogenicity of the entomogenous fungus *Metarhizium anisopliae* to *Cryptotermes brevis* and its intestinal protozoa. In J. Eder and H. Rembold, eds., *Chemistry and Biology of Social Insects,* p. 638. Verlag J. Pepperny, Munich.

Kaspari, M., and E. L. Vargo. 1995. Colony size as a buffer against seasonality: Bergmann's rule in social insects. *American Naturalist* 145: 610–632.

Kathirithamby, J. 1991. Strepsiptera. In Melbourne University Press, eds., *Insects of Australia,* vol. 2, pp. 684–695. Melbourne University Press, Melbourne.

Kathirithamby, J., and J. S. Johnston. 1992. Stylopization of *Solenopsis invicta* (Hymenoptera: Formicidae) by *Caenochloax fenyesi* (Strepsiptera: Myrmecolacidae) in Texas. *Annals of the Entomological Society of America* 85: 293–297.

Kaya, H. K., and R. Gaugler. 1993. Entomopathogenic nematodes. *Annual Review of Entomology* 38: 181–206.

Kaya, H. K., J. M. Marston, J. E. Lindegren, and Y. S. Peng. 1982. Low susceptibility of the honey bee, *Apis mellifera* L. (Hymenoptera: Apidae), to the entomogenous nematode, *Neoaplectana carpocapsae* Weiser. *Environmental Entomology* 11: 920–924.

Kaya, H. K., R. D. Moon, and P. L. Witt. 1979. Influence of the nematode *Heterotylenchus autumnalis,* on the behaviour of face fly, *Musca autumnalis. Environmental Entomology* 8: 537–540.

Keating, S. T., and W. G. Yendol. 1987. Influence of selected host plants on gypsy moth (Lepidoptera: Lymantriidae) larval mortality caused by baculovirus. *Environmental Entomology* 16: 459–462.

Keck, C. B. 1949. The disappearance of American foulbrood in Hawaii. *American Bee Journal* 89: 514–515, 542.

Keller, L. 1995. Parasites, worker polymorphism, and queen number in social insects. *American Naturalist* 145: 842–847.

Keller, L., and L. Passera. 1993. Incest avoidance, fluctuating asymmetry, and the consequences of inbreeding in *Iridomyrmex humilis,* an ant with multiple queen colonies. *Behavioural Ecology and Sociobiology* 33: 191–199.

Keller, L., and H. K. Reeve. 1994a. Genetic variability, queen number, and polyandry in social hymenoptera. *Evolution* 48: 694–704.

Keller, L., and H. K. Reeve. 1994b. Partitioning of reproduction in animal societies. *Trends in Ecology and Evolution* 9: 98–102.

Keller, L., and H. K. Reeve. 1995. Why do females mate with multiple males? The sexually selected sperm hypothesis. *Advances in the Study of Behavior* 24: 291–315.

Kent, D. S., and J. A. Simpson. 1992. Eusociality in the beetle *Austroplatypus incompertus* (Coleoptera: Curculionidae). *Naturwissenschaften* 79: 86–87.

Kermack, W. O., and A. G. McKendrick. 1927. A contribution to the mathematical theory of epidemics. *Proceedings of the Royal Society, A* 115: 700–721.

Kermarrec, A., H. Mauleon, and D. Marival. 1990. Comparison of susceptibility of *Acromyrmex octospinosus* Reich (Attini, Formicidae) to two insect parasitic nematodes of the Genera *Heterorhabditis* and *Neoplectana* (Rhabditina, Nematoda). In R. K. Vander Meer, K. Jaffe, and C. Cedeno, eds., *Applied Myrmecology, a World Perspective,* pp. 638–644. Westview Press, Boulder, Colordao.

Kerr, W. E. 1950. Genetic determination of castes in the genus *Melipona. Genetics* 35: 143–152.

Kerr, W. E. 1969. Some aspects of the evolution of social bees (Apidae). *Evolutionary Biology* 3: 119–175.

Kerr, W. E. 1974. Genetik des Polymorphismus bei Bienen. In G. H. Schmidt, ed., *Sozialpolymorphismus bei Insekten,* pp. 94–109. Wissenschaftliche Verlagsgesellschaft, Stuttgart.

Kerr, W. E., and W. E. Nielsen. 1966. Evidence that genetically determined *Melipona* queens can become workers. *Genetics* 54: 859–866.

Kerr, W. E., R. Zucchi, J. T. Nauadaira, and J. E. Butolo. 1962. Reproduction in the social bees (Hymenoptera, Apidae). *Journal of the New York Entomologcial Society* 70: 265–276.

Keymer, A. E., R. M. May, and P. H. Harvey. 1990. Parasite clones in the wild. *Nature* 346: 109–110.

Keymer, A. E., and A. F. Read. 1991. Behavioural ecology: The impact of parasitism. In C. A. Toft, A. Aeschlimann, and L. Bolis, eds., *Parasite-Host Associations: Coexistence or Conflict?,* pp. 37–61. Oxford University Press, Oxford.

Kilbourne, E. D. 1994. Host determination of viral evolution: A tautology. In S. S. Morse, ed., *The Evolutionary Biology of Viruses,* pp. 253–271. Raven Press, New York.

Kinn, D. N. 1984. Life cycle of *Dendrolaelaps neodisetus* (Mesostigmata, Digamasellidae), a nematodophagous mite associated with the pine bark beetle (Coleoptera: Scolytidae). *Environmental Entomology* 13: 1141–1144.

Kistner, D. H. 1982. The social insects' bestiary. In H. R. Hermann, ed., *Social Insects,* vol. 3, pp. 1–244. Academic Press, London.

Klahn, J. E. 1979. Philopatric and non-philopatric foundress associations in the social wasp *Polistes fuscatus. Behavioural Ecology and Sociobiology* 5: 417–424.

Klein, M. G. 1990. Efficacy against soil-inhabiting insect pests. In R. Gaugler and H. K.

Kaya, eds., *Entomopathogenic Nematodes in Biological Control,* pp. 195–214. CRC Press, Boca Raton, Florida.

Kloft, W. 1951. Pathologische Untersuchungen an einem Wespenweibchen, infiziert durch einen Gordioiden (Nematomorpha). *Zeitschrift für Parasitenkunde* 15: 134–147.

Knell, J. D., G. E. Allen, and E. I. Hazard. 1977. Light and electron microscope study of *Thelohania solenopsae* n.sp. (Microsporidia: Protozoa) in the red imported fire ant. *Journal of Invertebrate Pathology* 29: 192–200.

Koella, J. C. 1993. Ecological correlates of chiasma frequency and recombination index of plants. *Biological Journal of the Linnean Society* 48: 227–238.

Koenig, J. P., G. M. Boush, and E. H. Erickson. 1987. Effects of spore introduction and ratio of adult bees to brood on chalkbrood disease in honeybee colonies. *Journal of Apicultural Research* 25: 58–62.

Koeniger, G. 1991. Diversity in *Apis* mating systems. In D. R. Smith, ed., *Diversity in the Genus Apis,* pp. 265. Westview Press, Boulder Colorado.

Koeniger, N., G. Koeniger, and N. H. P. Wijayagunasekara. 1981. Beobachtungen über die Anpassung von *Varroa jacobsoni* an ihren natürlichen Wirt *Apis cerana* in Sri Lanka. *Apidologie* 12: 37–40.

Kolmes, S. A. 1986. Have hymenopteran societies evolved to be ergonomically efficient? *Journal of the New York Entomological Society* 94: 447–457.

Kolmes, S. A., and M. L. Winston. 1988. Division of labor among worker honey bees in demographically manipulated colonies. *Insectes Sociaux* 35: 262–270.

Komeili, A. B., and J. T. Ambrose. 1990. Biology, ecology and damage of tracheal mites on honey bees (*Apis mellifera*). *American Bee Journal* 130: 193–199.

König, C., and P. Schmid-Hempel. 1995. Foraging activity and immunocompetence in workers of the bumble bee, *Bombus terrestris* L. *Proceedings of the Royal Society London, B* 260: 225–227.

Kovac, H., and K. Vrailsheim. 1988. Lifespan of *Apis mellifera carnica* Pollm. infested by *Varroa jacobsoni* Oud. in relation to season and extent of infestation. *Journal of Apicultural Research* 27: 230–238.

Kramm, K. R., and D. F. West. 1982. Termite pathogens: Effects of ingested *Metarhizium, Beauveria,* and *Gliocladium* conidia on worker termites (*Reticulitermes* sp.). *Journal of Invertebrate Pathology* 40: 7–11.

Kramm, K. R., D. F. West, and P. G. Rockenbach. 1982. Termite pathogens: Transfer of the entomophagous *Metarhizium anisoplia* between *Reticulitermes* sp. termites. *Journal of Invertebrate Pathology* 40: 1–6.

Kraus, B., N. Koeniger, and S. Fuchs. 1986. Unterscheidung zwischen Bienen verschiedenen Alters durch *Varroa jacobsoni* Oud. und Bevorzugung von Ammenbienen im Sommervolk. *Apidologie* 17: 257–266.

Kraus, B., and R. E. Page. 1995. The impact of *Varroa jacobsoni* (Mesostigmata, Varroidae) on feral *Apis mellifera* (Hymenoptera, Apidae) in California. *Environmental Entomology* 24: 1473–1480.

Krieg, A. 1987. Diseases caused by bacteria and other prokaryotes. In J. R. Fuxa and Y. Tanada, eds., *Epizootiology of Insect Diseases,* pp. 323–355. John Wiley and Sons, New York.

Krieg, A., A. Gröner, J. Huber, and G. Zimmermann. 1981. Inaktivierung von verschiedenen Insektenpathogenen durch ultraviolette Strahlen. *Zeitschrift für Pflanzenkrankheiten und Pflanzenschutz* 88: 38–48.

Krishna, K., and F. M. Weesner, eds. 1969. *Biology of Termites.* Vol. 1. Academic Press, New York.

Krishna, K., and F. M. Weesner, eds.. 1970. *Biology of Termites.* Vol. 2. Academic Press, New York.

Krombein, K. V. 1967. *Trap-nesting Wasps and Bees: Life Histories, Nests and Associates.* Smithsonian Institution Press, Washington, D.C.

Kukuk, P. F., M. D. Breed, A. Sobti, and W. J. Bell. 1977. The contributions of kinship and conditioning to nest recognition and colony member recognition in a primitevly eusocial bee, *Lasioglossum zephyrum. Behavioural Ecology and Sociobiology* 2: 319–327.

Kulincevic, J. M. 1986. Breeding accomplishments with honey bees. In T. E. Rinderer, ed., *Bee Genetics and Breeding,* pp. 391–414. Academic Press, Orlando, Florida.

Kulincevic, J. M., and W. C. Rothenbuhler. 1975. Selection for resistance and susceptibility to hairless-black syndrome in the honey bee. *Journal of Invertebrate Pathology* 25: 289–295.

Kumbkarni, C. G. 1965. Cytological studies in hymenoptera. III. Cytology of parthenogenesis in the formicid ant, *Camponotus compressus. Caryologica* 18: 305–312.

Kutter, H. 1969. Die sozialparasitischen Ameisen der Schweiz. *Neujahrsblatt der Naturforschenden Gesellschaft Zürich* 171: 1–62.

Lachaud, J. P., and L. Passera. 1982. Données sur la biologie de trois Diapriidae myrmécophiles: *Plagiopria passerai* Masner, *Solenopsia imitatrix* Wasmann et *Lepidopria pedestris* Kiefer. *Insectes Sociaux* 29: 561–567.

Lackie, A. M. 1988. Immune mechanisms in insects. *Parasitology Today* 4: 98–109.

Laidlaw, H. H. J., and R. E. Page. 1984. Polyandry in honeybees (*Apis mellifera*): Sperm utilization and intracolony genetic relationships. *Genetics* 108: 985–997.

Landerkin, G. B., and H. Katznelson. 1959. Organisms associated with septicemia in the honey bee, *Apis mellifera. Canadian Journal of Microbiology* 5: 169–172.

L'Arrivée, J. C. M. 1965a. Sources of *Nosema* infection. *American Bee Journal* 105: 246–248.

L'Arrivée, J. C. M. 1965b. Tolerance of honey bees to *Nosema* disease. *Journal of Invertebrate Pathology* 7: 408–413.

Laverty, T. M., and R. C. Plowright. 1985. Comparative bionomics of temperate and tropical bumblebees with special reference to *Bombus ephippiatus* (Hymenoptera: Apidae). *Canadian Entomologist* 117: 467–474.

Leatherdale, D. 1970. The arthropod hosts of entomogenous fungi in Britain. *Entomophaga* 15: 419–435.

Lenoir, A. 1984. Brood-colony recognition in *Cataglyphis cursor* worker ants. *Animal Behaviour* 32: 942–944.

Lenski, R. E., and R. M. May. 1994. The evolution of virulence in parasites and pathogens: Reconciliation between two competing hypotheses. *Journal of theoretical Biology* 169: 253–265.

Leston, D. 1973. The ant mosaic-tropical tree crops and the limiting of pests and diseases. *Proceedings of the National Academy of Sciences USA* 19: 311–341.

Levin, B. R., and C. Svanborg Eden. 1990. Selection and evolution of virulence in bacteria: An ecumenical excursion and a modest suggestion. *Parasitology* (Suppl.) 100: S103–S115.

Levin, S. A., and D. Pimentel. 1981. Selection of intermediate rates of increase in parasite-host systems. *American Naturalist* 117: 308–315.

Liersch, S., and P. Schmid-Hempel. 1998. Genetic variation within social insect colonies reduces parasite load. *Proceedings of the Royal Society London B* 265:221–225.

Lin, N. 1964. Increased parasite pressure as a major factor in the evolution of social behavior in halictine bees. *Insectes Sociaux* 11: 187–192.

Lin, N., and C. D. Michener. 1974. Evolution of sociality in insects. *Quarterly Review of Biology* 47: 131–159.

Linsenmair, K. E. 1985. Individual and family recognition in subsocial arthropods, in particular the desert isopod, *Hemilepistus reaumuri*. In B. Hölldobler and M. Lindauer, eds., *Experimental Behavioral Ecology and Sociobiology*,, pp. 411–436. Sinauer associates, Sunderland, Mass.

Lipa, J. J., and O. Triggiani. 1996. *Apicystis* n.gen., *Apicystis bombi* (Liu, Macfarlane & Pengelly 1974) n.comb., a cosmopolitan pathogen of *Apis* and *Bombus* (Hymenoptera, Apidae). *Apidologie* 27: 29–34.

Lipsitch, M., E. A. Herre, and M. A. Nowak. 1995a. Host population structure and the evolution of virulence: A law of diminishing returns. *Evolution* 49: 743–748.

Lipsitch, M., and M. A. Nowak. 1995. The evolution of virulence in sexually transmitted HIV/AIDS. *Journal of theoretical Biology* 174: 427–440.

Lipsitch, M., M. A. Nowak, D. Ebert, and R. M. May. 1995b. The population dynamics of vertically and horizontally transmitted parasites. *Proceedings of the Royal Society London, B* 260: 321–327.

Lipsitch, M., S. Siller, and M. A. Nowak. 1996. The evolution of virulence in pathogens with vertical and horizontal transmission. *Evolution* 50: 1729–1741.

Litte, M. 1977. Aspects of the social biology of the bee *Halictus ligatus* in New York state (Hymenoptera, Halictidae). *Insectes Sociaux* 24: 9–36.

Litte, M. 1981. Social biology of the polistine wasp *Mischocyttarus labiatus:* Survival in a Colombian rainforest. *Smithsonian Contributions to Zoology* 327: 1–27.

Lively, C. M. 1987. Evidence from a New Zealand snail for the maintenance of sex by parasitism. *Nature* 328: 519–521.

Lockhart, A. B., P. H. Thrall, and J. Antonovics. 1996. Sexually transmitted diseases in animals: Ecological and evolutionary implications. *Biological Reviews* 71: 415–471.

Loehle, C. 1995. Social barriers to pathogen transmission in wild animal populations. *Ecology* 76: 326–335.

Loos-Frank, B., and G. Zimmermann. 1976. Über eine dem *Dicrocoelium*-Befall analoge Verhaltensänderung bei Ameisen der Gattung *Formica* durch einen Pilz der Gattung *Entomophthora. Zeitschrift für Parasitenkunde* 49: 281–289.

Lotmar, R. 1943. Über den Einfluss der Temperatur auf den Parasiten *Nosema apis. Beihefte zur Schweizerischen Bienenzeitung* 1: 261–284.

Low, B. S. 1988. Pathogen stress and polygyny in humans. In L. Betzig, M. Bergerhoff-Mulder, and P. Turke, eds., *Human Reproductive Behaviour: A Darwinian Perspective,* pp. 115–127. Cambridge University Press, Cambridge, U.K.

Lubbock, R. 1980. Clone-specific cellular recognition in a sea anemone. *Proceedings of the National Academy of Sciences USA* 77: 6667–6669.

Lundberg, H., and B. G. Svensson. 1975. Studies on the behaviour of *Bombus* Latr. species (Hymenoptera: Apidae) parasitized by *Sphaerularia bombi* Dufour (Nematoda) in an alpine area. *Norwegian Journal of Entomology* 22: 129–134.

Lüscher, M. 1955. Der Sauerstoffverbrauch bei Termiten und die Ventilation des Nestes bei *Macrotermes natalensis. Acta Tropica* 12: 289–307.

Luykx, P. 1985. Genetic relations among castes in lower termites. In J. A. L. Watson, B. M. Okot-Kotber, and C. Noirot, eds., *Caste Differentiation in Social Insects,* pp. 17–15. Pergamon Press, New York.

Mabelis, A. A. 1979. Wood ant wars: The relationship between aggression and predation in the red wood ant (*Formica polyctena* Forst.). *Netherlands Journal of Zoology* 29: 451–620.

MacFarlane, R. P., and R. P. Griffin. 1990. New Zealand distribution and seasonal incidence of the nematode *Sphaerularia bombi* Dufour, a parasite of bumble bees. *New Zealand Journal of Zoology* 17: 191–199.

MacFarlane, R. P., J. J. Lipa, and H. J. Liu. 1995. Bumble bee pathogens and internal enemies. *Bee World* 76: 130–148.

MacFarlane, R. P., and R. L. Palma. 1987. The first record of *Melittobia australica* Girault in New Zealand and new host records for *Melittobia* (Eulophidae). *New Zealand Journal of Zoology* 14: 423–425.

MacFarlane, R. P., and D. H. Pengelly. 1974. Conopidae and Sarcophagidae (Diptera) as parasites of adult Bombinae (Hymenoptera) in Ontario. *Proceedings of the Entomological Society of Ontario* 105: 55–59.

Madel, G. 1966. Beiträge zur Biologie von *Sphaerularia bombi* Leon Dufour 1837. *Zeitschrift für Parasitenkunde* 28: 99–107.

Madel, G. 1973. Zur Biologie des Hummelparasiten *Sphaerularia bombi* Leon Dufour 1886 (Nematoda, Tylenchida). *Bonner Zoologische Beiträge* 24: 134–151.

Marikovsky, P. I. 1962. On some features of behavior of the ants *Formica rufa* L. infected with fungus disease. *Insectes Sociaux* 9: 173–179.

Martin, M. M. 1970. The biochemical basis of the fungus-attine ant symbiosis. *Science* 169: 16–20.

Maschwitz, U. 1974. Vergleichende Untersuchungen zur Funktion der Ameisenmetathorakaldrüse. *Oecologia* 16: 303–310.

Maschwitz, U., K. Koob, and H. Schildknecht. 1970. Ein Beitrag zur Funktion der Metathorakaldrüse der Ameisen. *Journal of Insect Physiology* 16: 387–404.

Maschwitz, U., S. Lenz, and A. Buschinger. 1986. Individual specific trails in the ant *Leptothorax affinis* (Formicidae: Myrmicinae). *Experientia* 42: 1173–74.

Masson, C., and G. Arnold. 1984. Ontogeny, maturation and plasticity of the olfactory system in the worker bee. *Journal of Insect Physiology* 30: 7–14.

Matsuura, M. 1984. Comparative biology of the five Japanese species of the genus *Vespa* (Hymenoptera, Vespidae). *Bulletin of the Faculty of Agriculture, Mie University* 69: 1–131.

Matsuura, M. 1991. *Vespa* and *Provespa.* In K. G. Ross and R. W. Matthews, eds., *The Social Biology of Wasps,* pp. 232–262. Cornell University Press, Ithaca, New York.

Matthews, R. W. 1991. Evolution of social behavior in sphecid wasps. In K. G. Ross and R. W. Matthews, eds., *The Social Biology of Wasps,* pp. 570–602. Comstock Publishing, Ithaca, New York.

May, R. M. 1985. Regulation of populations with nonoverlapping generations by microparasites: A purely chaotic system. *American Naturalist* 125: 573–584.

May, R. M., and R. M. Anderson. 1983. Parasite-host coevolution. In D. J. Futuyma and M. Slatkin, eds., *Coevolution,* pp. 186–206. Sinauer Associates, Sunderland, Mass.

May, R. M., and R. M. Anderson. 1984. Spatial heterogeneity and the design of immunisation programmes. *Mathematical Biosciences* 72: 83–111.

McAllister, M. K., B. D. Roitberg, and K. L. Weldon. 1990. Adaptive suicide in pea aphids: Decisions are cost-sensitive. *Animal Behaviour* 40: 167–175.

McCleskey, C. S., and R. M. Melampy. 1939. Bactericidal properties of royal jelly of the honeybee. *Journal of Economic Entomology* 32: 581–587.

McCorquodale, D. B. 1989. Nest defense in single- and multifemale nests of *Cerceris antipodes* (Hymenoptera: Sphecidae). *Journal of Insect Behavior* 2: 267–276.

McCoy, C. W., R. A. Samson, and D. G. Boucias. 1988. Entomogenous fungi. In C. M. Ignoffo, ed., *CRC Handbook of Natural Pesticides,* vol. 5: *Microbial Insecticides,* part A *Entomogenous Protozoa and Fungi,* pp. 237–236. CRC Press, Boca Raton, Florida.

McIvor, C. A., and L. A. Melone. 1995. *Nosema bombi,* a microsporidian pathogen of the bumble bee *Bombus terrestris* L. *New Zealand Journal of Zoology* 22: 25–31.

Metcalf, R. A. 1980. Sex ratios, parent-offspring conflict, and local mate competition for mates in the social wasps *Polistes metricus* and *P. variatus. American Naturalist* 116: 642–654.

Michener, C. D. 1964. Reproductive efficiency in relation to colony size in hymenopterous societies. *Insectes Sociaux* 11: 317–341.

Michener, C. D. 1965. The life-cycle and social organisation of bees of the genus *Exoneura* and their parasite, *Inquilina. University of Kansas Science Bulletin* 46: 317–358.

Michener, C. D. 1969. Comparative social behavior of bees. *Annual Review of Entomology* 14: 299–342.

Michener, C. D. 1970. Social parasites among African allodapine bees. *Zoological Journal of the Linnean Society* 49: 199–215.

Michener, C. D. 1974. *The Social Behavior of the Bees.* Harvard University Press, Cambridge, Mass.

Michener, C. D. 1985. From solitary to eusocial: Need there be a series of intervening species? In B. Hölldobler and M. Lindauer, eds., *Experimental Behavioural Ecology,* pp. 293–305. Fischer Verlag, Stuttgart.

Michener, C. D., and D. J. Brothers. 1974. Were workers of eusocial Hymenoptera initially altruistic or oppressed? *Proceedings of the National Academy of Sciences USA* 71: 671–674.

Michener, C. D., R. J. McGinley, and B. N. Danforth. 1994. *The Bee Genera of North and Central America (Hymenoptera: Apoidea).* Smithsonian Institution Press, Washington, D.C.

Michener, C. D., and B. H. Smith. 1987. Kin recognition in primitively eusocial insects. In D. J. C. Fletcher and C. D. Michener, eds., *Kin Recognition in Animals,* pp. 209–242. Wiley and Sons, New York.

Michod, R. E. 1993. Inbreeding and the evolution of social behavior. In N. W. Thornhill, ed., *The Natural History of Inbreeding and Outbreeding,* pp. 74–96. University of Chicago Press, Chicago.

Mikkola, K. 1978. Spring migrations of wasps and bumble bees in the southern coasts of Finland (Hymneoptera, Vespidae and Apidae). *Annales Entomologici Fennica* 44: 10–26.

Mikkola, K. 1984. Migration of wasp and bumble bee queens across the Gulf of Finland (Hymenoptera: Vespidae and Apidae). *Notulae Entomologicae* 64: 125–128.

Miller, S. R., and L. R. Brown. 1983. Studies on microbial fire ant pathogens. *Developments in Industrial Microbiology* 24: 443–450.

Milne, C. P. J. 1983. Honey bee (Hymenoptera, Apidae) hygienic behavior and resistance to chalkbrood. *Annals of the Entomological Society of America* 76: 384–387.

Milne, C. P. J. 1985. An estimate of the heritability of worker longevity or length of life in the honeybee. *Journal of Apicultural Research* 24: 140–143.

Milner, R. J. 1973. *Nosema whitei,* a microsporidian pathogen of some species of *Tribolium.* IV. The effect of temperature, humidity and larval age on pathogenicity for *T. castaneum. Entomophaga* 18: 305–315.

Minchella, D. J. 1985. Host life-history variation in response to parasitation. *Parasitology* 90: 205–216.

Mintzer, A. 1982. Nestmate recognition and incompatibility between colonies of the acacia-ant *Pseudomyrmex ferruginea. Behavioural Ecology and Sociobiology* 10: 165–168.

Mintzer, A. 1989. Incompatibility between colonies of acacia ants: Genetic models and experimental results. In M. Breed and R. E. Page, eds., *The Genetics of Social Evolution,* pp. 163–182. Westview Press, Boulder, Colorado.

Mintzer, A., and B. S. Vinson. 1985. Kinship and incompatibility between colonies of the acacia ant *Pseudomyrmex ferruginea. Behavioural Ecology and Sociobiology* 17: 75–78.

Mitchell, G. B., and D. P. Jouvenaz. 1985. Parasitic nematode observed in the tropical fire ant, *Solenopsis geminata* (F.) (Hymenoptera: Formicidae). *Florida Entomologist* 68: 492–493.

Miyano, S. 1980. Life tables of colonies and workers in a paper wasp, *Polistes chinensis antennalis* in central Japan (Hymenoptera, Vespidae). *Research in Population Ecology* 22: 69–88.

Mohamed, A. K. A., P. P. Sikorowski, and J. V. Bell. 1977. Susceptibility of *Heliothis zea* larvae to *Nomuraea riley* at various temperaturs. *Journal of Invertebrate Pathology* 30: 414–417.

Møller, A. P. 1987. Advantages and disadvantages of coloniality in the swallow *Hirundo rustica. Animal Behaviour* 35: 819–832.

Molyneux, D. H., G. Takle, E. A. Ibrahim, and G. A. Ingram. 1986. Insect immunity to trypanosomatidae. In A. M. Lackie, ed., *Immune Mechanisms in Invertebrate Hosts,* pp. 117–144. Clarendon Press, Oxford.

Moore, J. 1984. Altered behavioral responses in intermediate hosts—an acantocephalan parasite strategy. *American Naturalist* 123: 572–577.

Moore, J., D. Simberloff, and M. Freehling. 1988. Relationships between bobwhite quail social-group size and intestinal helminth parasitism. *American Naturalist* 131: 22–32.

Morand, S., S. D. Manning, and M. E. J. Woolhouse. 1996. Parasite host coevolution and geographic patterns of parasite infectivity and host susceptibility. *Proceedings of the Royal Society of London, B* 263: 119–128.

Moretto, G., L. S. Gonçalves, D. De Jong, and M. Z. Bichuette. 1991. The effects of climate and bee race on *Varroa jacobsoni* Oud infestations in Brazil. *Apidologie* 22: 197–203.

Morgenthaler, O. 1926. Bienenkrankheiten im Jahre 1925. *Schweizerische Bienenzeitung* 49: 176–224.

Morgenthaler, O. 1941. Einwinterung und *Nosema. Schweizerische Bienenzeitung* 64: 401–404.

Moritz, R. F. A. 1985. The effects of multiple mating on the worker-queen conflict in *Apis mellifera* L. *Behavioural Ecology and Sociobiology* 16: 375–377.

Moritz, R. F. A., and E. E. Southwick. 1992. *Bees as Superorganisms—An Evolutionary Reality.* Springer-Verlag, Berlin.

Morris, R. F. 1976. Influence of genetic changes and other variables on the encapsulation of parasites by *Hyphantria cunea. Canadian Entomologist* 108: 673–684.

Mossadegh, M. S. 1991. Geographical distribution, levels of infestation and population density of the mite *Euvarroa sinhai* Delfinado and Baker (Acarina: Mesostigmata) in *Apis florea* F colonies in Iran. *Apdiologie* 22: 127–134.

Mouches, C., J. M. Bové, and J. Albisetti. 1984. Pathogenicity of *Spiroplasma apis* and other spiroplasmas for honey-bees in southwestern France. *Annales de microbiologie* 135: 151–155.

Mueller, J. F. 1974. The biology of Spirometra. *Journal of Parasitology* 60: 3–14.

Muldrew, J. A. 1953. The natural immunity of the larch sawfly (*Pristiphora erichsonii* Htg.) to the introduced parasite *Mesoleius tenthredinis* Morley, in Manitoba and Saskatchewan. *Canadian Journal of Zoology* 31: 313–332.

Müller, C. B. 1993. The Impact of Conopid Parasitoids on Life History Variation and Behavioural Ecology of Bumblebees. Ph.D. diss., Phil.Naturw. Fakultät der Universität Basel, Basel, Switzerland.

Müller, C. B. 1994. Parasite-induced digging behaviour in bumble bee workers. *Animal Behaviour* 48: 961–966.

Müller, C. B., T. M. Blackburn, and P. Schmid-Hempel. 1996. Field evidence that host selection by conopid parasitoids is related to host body size. *Insectes Sociaux* 43: 227–233.

Müller, C. B., and P. Schmid-Hempel. 1993a. Correlates of reproductive success among field colonies of *Bombus lucorum* L.: The importance of growth and parasites. *Ecological Entomology* 17: 343–353.

Müller, C. B., and P. Schmid-Hempel. 1993b. Exploitation of cold temperature as defense against parasitoids in bumblebees. *Nature* 363: 65–67.

Müller, C. B., and R. Schmid-Hempel. 1992a. To die for host or parasite? *Animal Behaviour* 44: 177–179.

Müller, C. B., and P. Schmid-Hempel. 1992b. Variation in worker mortality and reproductive performance in the bumble bee, *B. lucorum. Functional Ecology* 6: 48–56.

Mundt, J. O. 1961. Occurrence of enterococci: Bud, blossom and soil studies. *Applied Microbiology* 9: 541–544.

Nappi, A. J., and Y. Carton, eds. 1986. *Cellular Immune Responses and Their Genetic Aspects in Drosophila.* Springer-Verlag, Berlin and Heidelberg.

Nappi, A. J., and M. Silvers. 1984. Cell surface changes associated with cellular immune reactions in *Drosophila. Science* 225: 1166–1168.

Neigel, J. E., and J. C. Avise. 1983. Histocompatibility bioassays of population structure in marine sponges. Clonal structure in *Verongia longissima* and *Iotrochota birotulata*. *Journal of Heredity* 74: 134–140.

Nelson, J. M. 1968. Parasites and symbionts of nests of *Polistes* wasps. *Annals of the Entomological Society of America* 61: 1528–1539.

Nickle, W. R., and G. L. Ayre. 1966. *Caenorhabditis dolichura* (A. Schneider, 1866) Dougherty (Rhabditidae, Nematoda) in the head glands of the ants *Camponotus herculeanus* (L.) and *Acanthomyops claviger* (Roger) in Ontario. *Proceedings of the Entomological Society of Ontario* 96: 96–98.

Nixon, H. L., and C. R. Ribbands. 1952. Food transmission within the honeybee community. *Proceedings of the Royal Society London, B* 140: 43–50.

Nixon, M. 1983. World maps of *Varroa jacobsoni* and *Tropilaelaps clarae* with additional records for honeybee diseases and parasites previously mapped. *Bee World* 64: 124–131.

Nnakumusane, E. S. 1987. Histological studies of *Cordyceps myrmecophila* infection in the ant *Paltothyreus tarsata* Fab. (Formicidae, Ponerinae). *Applied Entomology and Zoology* 22: 1–6.

Nonacs, P. 1986. Ant reproductive strategies and sex allocation theory. *Quarterly Review of Biology* 61: 1–21.

Nonacs, P. 1988. Queen number in colonies of social hymenoptera as a kin-selected adaptation. *Evolution* 42: 566–580.

Noonan, K. C. 1986. Recognition of queen larvae by worker honey bees (*Apis mellifera*). *Ethology* 73: 295–306.

Norris, K., M. Anwar, and A. F. Read. 1994. Reproductive effort influences the prevalence of haematozoan parasites in great tits. *Journal of Animal Ecology* 63: 601–610.

Nowak, M. A. 1991. The evolution of viruses: Competition between horizontal and vertical transmission of mobile genes. *Journal of theoretical Biology* 150: 339–347.

Nowak, M. A., R. M. Anderson, A. R. McLean, T. F. W. Wolfs, J. Goudsmit, and R. M. May. 1991. Antigenic diversity thresholds and the development of AIDS. *Science* 245: 963–969.

Nowak, M. A., and R. M. May. 1994. Superinfection and the evolution of parasite virulence. *Proceedings of the Royal Society London, B* 255: 81–89.

Nutting, W. L. 1969. Flight and colony foundation. In K. Krishna and F. M. Weesner, eds., *Biology of Termites,* Vol. 1, pp. 233–282.

Obin, M. S. 1986. Nestmate recognition cues in laboratory and field colonies of *Solenopsis invicta* Buren (Hymenoptera: Formicidae). Effect of environment and the role of cuticular hydrocarbons. *Journal of Chemical Ecology* 12: 1965–1975.

Obin, M. S., and R. K. Vander Meer. 1985. Gaster flagging by fire ants (*Solenopsis* spp.): Functional significance of venom dispersal behavior. *Journal of Chemical Ecology* 11: 1757–1768.

Obin, M. S., and R. K. Vander Meer. 1988. Sources of nestmate recognition cues in the imported fire ant, *Solenopsis invicta* Buren (Hymenoptera: Formicidae). *Animal Behaviour* 36: 1361–1370.

Obin, M. S., and R. K. Vander Meer. 1989. Nestmate recognition in fire ants (*Solenopsis invicta* Buren). Do queens label workers? *Ethology* 80: 255–264.

O'Connor, B. M. 1994. Life-history modifications in astigmatid mites. In M. A. Houck, ed., *Mites—Ecology and Evolutionary Analyses of Life-History Patterns,* pp. 136–159. Chapman and Hall, New York.

Oertel, E. 1965. Gregarines found in several honeybee colonies. *American Bee Journal* 105: 10–11.

Ogloblin, A. A. 1939. The Strepsiptera parasites of ants. *Verhandlungen des VII Internationalen Kongress für Entomologie* 2: 1277–1285.

Oi, D. H., and R. M. Pereira. 1993. Ant behavior and microbial pathogens (Hymenoptera: Formicidae). *Florida Entomologist* 76: 63–73.

Oldroyd, B. P., T. E. Rinderer, J. R. Harbo, and S. M. Buco. 1992. Effects of intracolonial genetic diversity on honey bee (Hymenoptera: Apidae) colony performance. *Annals of the Entomological Society of America* 85: 335–343.

Olmsted, R. A., R. S. Baric, B. A. Sawyer, and R. E. Johnston. 1984. Sindbis virus mutants selected for rapid growth in cell culture display attenuated virulence in animals. *Science* 225: 424–426.

Oster, G. F., and E. O. Wilson. 1978. *Caste and Ecology in the Social Insects.* Princeton University Press, Princeton, New Jersey.

Otis, G. W., and C. D. Scott-Dupree. 1992. Effects of *Acarapis woodi* on overwintered colonies of honey bees (Hymenoptera: Apidae) in New York. *Journal of Economic Entomology* 85: 40–46.

Owen, R. E. 1985. Difficulties with the interpretation of patterns of genetic variation in the eusocial hymenoptera. *Evolution* 39: 201–205.

Owen, R. E., and R. C. Plowright. 1982. Worker-queen conflict and male parentage in bumble bees. *Behavioural Ecology and Sociobiology* 11: 91–99.

Pacala, S. W., D. M. Gordon, and J. H. C. Godfray. 1996. Effects of social group size on information transfer and task allocation. *Evolutionary Ecology* 10: 127–165.

Packer, L. 1986. The social organization of *Halictus ligatus* (Hymenoptera, Halictidae) in Southern Ontario. *Canadian Journal of Zoology* 64: 2317–2324.

Packer, L. 1993. Multiple-foundress associations in sweat bees. In L. Keller, ed., *Queen Number and Sociality in Insects,* pp. 215–233. Oxford University Press, Oxford.

Packer, L., and G. Knerer. 1985. Social evolution and its correlates in bees of the subgenus *Evylaeus* (Hymenoptera: Halictidae). *Behavioural Ecology and Sociobiology* 17: 143–149.

Page, R. E. 1986. Sperm utilization in social insects. *Annual Review of Entomology* 31: 297–320.

Page, R. E., and E. H. Erickson. 1984. Selective rearing of queens by worker honey bees: Kin or nestmate recognition? *Annals of the Entomological Society of America* 77: 578–580.

Page, R. E., and N. E. Gary. 1990. Genotypic variation in susceptibility of honey bees (*Apis mellifera*) to infestation by tracheal mites (*Acarapis woodi*). *Experimental and Applied Acarology* 8: 275–283.

Page, R. E., and R. A. Metcalf. 1982. Multiple mating, sperm utilization, and social evolution. *American Naturalist* 119: 263–281.

Page, R. E., G. E. Robinson, and M. K. Fondrk. 1989. Genetic specialists, kin recognition, and nepotism in honey-bee colonies. *Nature* 338: 576–579.

Page, R. E., G. E. Robinson, M. K. Fondrk, and M. E. Nasr. 1995. Effects of worker genotypic diversity on honey bee colony development and behavior (*Apis mellifera* L). *Behavioural Ecology and Sociobiology* 36: 387–396.

Palm, N. B. 1948. Normal and pathological histology of the ovaries of *Bombus* Latr. (Hymenoptera). *Opusc. Entomol.* (suppl. 7): 1–101.

Palmieri, J. R. 1982. Be fair to parasites. *Nature* 298: 220.

Palomeque, T., E. Chica, and R. D. Delaguardia. 1993. Karyotype evolution and chromosomal relationships between several species of the genus *Aphaenogaster* (Hymenoptera: Formicidae). *Caryologia* 46: 25–40.

Pamilo, P. 1981. Genetic organization of *Formica sanguinea* populations. *Behavioural Ecology and Sociobiology* 9: 45–50.

Pamilo, P. 1982a. Genetic population structure in polygynous *Formica* ants. *Heredity* 48: 95–106.

Pamilo, P. 1982b. Multiple mating in *Formica* ants. *Hereditas* 97: 37–45.

Pamilo, P. 1984. Genetic correlation and regression in social groups: Multiple alleles, multiple loci, and subdivided populations. *Genetics* 107: 307–320.

Pamilo, P. 1991a. Evolution of colony characteristics in social insects. 2. Number of reproductive individuals. *American Naturalist* 137: 83–107.

Pamilo, P. 1991b. Life span of queens in the ant *Formica exsecta. Insectes Sociaux* 38: 111–119.

Pamilo, P. 1993. Polyandry and allele frequency differences between the sexes in the ant *Formica aquilonia. Heredity* 70: 472–480.

Pamilo, P., and R. H. Crozier. 1982. Measuring genetic relatedness in natural populations: Methodology. *Theoretical Population Biology* 21: 171–193.

Pamilo, P., L. Sundström, W. Fortelius, and R. Rosengren. 1994. Diploid males and colony-level selection in *Formica* ants. *Ethology, Ecology and Evolution* 6: 221–235.

Pamilo, P., and S.-L. Varvio-Aho. 1979. Genetic structure of nests in the ant *Formica sanguinea. Behavioural Ecology and Sociobiology* 6: 91–98.

Pape, T. 1986. Tardigrades, mites, and insects from a bumble bee nest in Greenland. *Entomologiske Meddelelser* 53: 75–81.

Passera, L. 1976. Origine des intercastes dans les sociétés de *Pheidole pallidula* (Nyl.) (Hymenoptera, Formicidae) parasitées par *Mermis* sp. (Nematoda, Mermithidae). *Insectes Sociaux* 23: 559–575.

Passera, L., E. Roncin, B. Kaufmann, and L. Keller. 1996. Increased soldier production in ant colonies exposed to intraspecific competition. *Nature* 379: 630–632.

Pathak, J. M. N. 1993. *Insect Immunity.* Kluwer Academic Publishers, Dordrecht.

Patterson, R. S., and J. Briano. 1990. *Thelohania solenopsae,* a microsporidian of fire ants: its effect on indigenous populations in Argentina. In G. K. Veeresh, B. Mallik, and C. A. Viraktamath, eds., *Social Insects and the Environment,* pp. 625–626. Oxford and IBH Publishing, New Delhi.

Paulian, R. 1949. *Une Naturaliste en Côte d'Ivoire.* Editions Stock, Paris.

Pearson, B. 1983. Intra-colonial relatedness amongst workers in a population of nests of the polygynous ant *Myrmica rubra* Latreille. *Behavioural Ecology and Sociobiology* 12: 1–14.

Peeters, C. 1993. Monogyny and polygyny in ponerine ants with or without queens. In L. Keller, ed., *Queen Number and Sociality in Insects,* pp. 234–261. Oxford University Press, Oxford.

Peng, Y. S., Y. Fang, S. Xu, and L. Ge. 1987a. The resistance mechanism of the Asian honeybee, *Apis cerana* Fabr., to an ectoparasitic mite, *Varroa jacobsoni* Oudemans. *Journal of Invertebrate Pathology* 49: 54–60.

Peng, Y. S. C., Y. Fang, S. Xu, L. Ge, and M. E. Nasr. 1987b. Response of foster Asian honeybee (*Apis cerana* Fabr.) colonies to the brood of European honeybee (*Apis mellifera* L.) infested with parasitic mite, *Varroa jacobsoni* Oudemans. *Journal of Invertebrate Pathology* 49: 259–264.

Pesquero, M. A., S. Campiolo, and H. G. Fowler. 1993. Phorids (Diptera: Phoridae) associated with mating swarms of *Solenopsis saevissima* (Hymenoptera, Formicidae). *Florida Entomologist* 76: 179–181.

Pettis, J. S., A. Dietz, and F. A. Eischen. 1989. Incidence rates of *Acarapis woodi* (Rennie) in queen honey bees of various ages. *Apidologie* 20: 69–75.

Pettis, J. S., W. T. Wilson, and F. A. Eischen. 1992. Nocturnal dispersal by female *Acarapis woodi* in honey bee (*Apis mellifera*) colonies. *Experimental and Applied Acarology* 15: 99–108.

Pfennig, D. W., and H. K. Reeve. 1989. Neighbor recognition and context-dependent aggression in a solitary wasp, *Sphecius speciosus* (Hymenoptera, Sphecidae). *Ethology* 80: 1–18.

Pfennig, D. W., H. K. Reeve, and J. S. Shellman. 1983. Learned component of nestmate discrimination in workers of a social wasp, *Polistes fuscatus* (Hymenoptera: Vespidae). *Animal Behaviour* 31: 412–416.

Pickles, W. 1940. Fluctuations in the populations, weights and biomasses of ants in Thornhill, Yorkshire, from 1935 to 1939. *Transactions of the Royal Entomological Society London* 90: 467–485.

Pimentel, D., and M. Shapiro. 1962. The influence of environment on a virus-host relationship. *Journal of Insect Pathology* 4: 77–87.

Plateau-Quénu, C. 1962. Biology of *Halictus marginatus* Brullé. *Journal of Apicultural Research* 1: 41–51.

Plowright, W. 1982. The effects of Rinderpest and Rinderpest control on wildlife in Africa. In M. A. Edwards and U. McDonnell, eds., *Animal Disease in Relation to Animal Conservation.* Symposium of the Zoological Society London, vol. 50, pp. 1–28. Academic Press, London.

Plurad, S. B., and P. A. Hartmann. 1965. The fate of bacterial spores ingested by adult honey bees. *Journal of Invertebrate Pathology* 7: 449–454.

Poinar, G. O. 1975. *Entomogenous Nematodes—A Manual and Host List of Insect-Nematode Associations.* E. J. Brill, Leiden.

Poinar, G. O., and P. A. van der Laan. 1972. Morphology and life history of *Sphaerularia bombi*. *Nematologica* 18: 239–252.

Ponten, A., and W. Ritter. 1992. Influence of Acute Paralysis Virus attacks on brood care in honeybees. *Apidologie* 23: 363–365.

Porter, S. D., and C. D. Jorgensen. 1981. Foragers of the harvester ant *Pogonomyrmex owyheei*—a disposable caste? *Behavioural Ecology and Sociobiology* 9: 247–256.

Porter, S. D., and C. D. Jorgensen. 1988. Longevity of harvester ant colonies in southern Idaho. *Journal of Range Management* 41: 104–107.

Porter, S. D., and W. R. Tschinkel. 1985. Fire ant polymorphism: The ergonomics of brood production. *Behavioural Ecology and Sociobiology* 16: 323–336.

Post, D. C., and R. L. Jeanne. 1982. Recognition of former nestmates during colony founding in the social wasp *Polistes fuscatus*. *Behavioural Ecology and Sociobiology* 11: 283–285.

Potts, W. K., C. J. Manning, and E. K. Wakeland. 1991. Mating patterns in seminatural populations of mice influenced by MHC genotype. *Nature* 352: 619–621.

Pouvreau, A. 1962. Contribution à l'étude de *Sphaerularia bombi* (Nematoda, Tylenchida), parasite des reines de bourdons. *Année Abeille* 5: 181–199.

Pouvreau, A. 1964. Observations d'une infestation précoce des reines de bourdons (Hymenoptera, Apoidea, *Bombus*) par *Sphaerularia bombi* (Nematoda, Tylenchida, Allantonematidae). *Bulletin de la Société Zoologique de France* 89: 717–719.

Pouvreau, A. 1974. Les enemies des bourdons. II. Organismes affectant les adultes. *Apidologie* 5: 39–62.

Prell, H. 1926. Beiträge zur Kenntnis der Amöbenseuche der erwachsenen Honigbiene. *Archiv für Bienenkunde* 7: 113–121.

Prest, D. B., M. Gilliam, S. Taber, and J. P. Mills. 1974. Fungi associated with discolored honey bee, *Apis mellifera,* larvae and pupae. *Journal of Invertebrate Pathology* 24: 253–255.

Provost, E. 1985. Etude de la fermeture de la société chez les fourmis. I. Analyse des interactions entre ouvrières de sociétés differentes, lors de rencontres experimentales, chez des fourmis du genre *Leptothorax* et chez *Camponotus lateralis* O. L. *Insectes Sociaux* 32: 445–462.

Provost, E. 1987. Role of the queen in the intra-colonial aggressivity and nestmate recognition in *Leptothorax lichtensteini* ants. In J. Eder and H. Rembold, eds., *Chemistry and Biology of Social Insects,* pp. 479. Peperny Verlag, Munich.

Purvis, A., and A. Rambaut. 1995. Comparative analysis by independent contrasts (CAIC): An Apple Macintosh application for analysing comparative data. *Computer Applications in Biosciences* 11: 247–251.

Pusey, A., and M. Wolf. 1996. Inbreeding avoidance in animals. *Trends in Ecology and Evolution* 11: 201–206.

Queller, D. C. 1992. Does population viscosity promote kin selection? *Trends in Ecology and Evolution* 7: 322–324.

Queller, D. C. 1993a. Genetic relatedness and its components in polygynous colonies of social insects. In L. Keller, ed., *Queen Number and Sociality in Insects,* pp. 132–152. Oxford University Press, Oxford.

Queller, D. C. 1993b. Worker control of sex ratios and selection for extreme multiple mating by queens. *American Naturalist* 142: 346–351.

Queller, D. C., and K. F. Goodnight. 1989. Estimating relatedness using genetic markers. *Evolution* 43: 258–275.

Queller, D. C., J. A. Negron-Sotomayor, J. E. Strassmann, and C. R. Hughes. 1993. Queen number and genetic relatedness in a neotropical wasp, *Polybia occidentalis. Behavioral Ecology* 4: 7–13.

Queller, D. C., and J. E. Strassmann. 1988. Reproductive success and group nesting in the paper wasp, *Polistes annularis.* In T. H. Clutton-Brock, ed., *Reproductive Success,* pp. 76–98. University of Chicago Press, Chicago.

Queller, D. C., J. E. Strassmann, and C. R. Hughes. 1988. Genetic relatedness in colonies of tropical wasps with multiple queens. *Science* 242: 1155–1157.

Queller, D. C., J. E. Strassmann, and C. R. Hughes. 1992. Genetic relatedness and population structure in primitively eusocial wasps in the genus *Mischocyttarus. Journal of Hymenopteran Research* 1: 115–145.

Raju, B. C., G. Nyland, T. Meikle, and A. H. Purcell. 1981. Helical, motile mycoplasmas associated with flowers and honeybees in California. *Canadian Journal of Microbiology* 27: 249–253.

Ramirez, B. W., and G. W. Otis. 1986. Developmental phases in the life cycle of *Varroa jacobsoni,* an ectoparasitic mite on honeybees. *Bee World* 67: 92–97.

Ransome, H. M. 1986. *The Sacred Bee.* Butler and Tanner, London.

Ratcliffe, N. A., S. J. Gagen, A. F. Rowley, and A. R. Schmit. 1976. Studies on insect cellular defence mechanisms and aspects of the recognition of foreignress. In T. A. Agnus, P. Faulkner, and A. Rosenfield, eds., *Proceedings of the First International Colloquium on Invertebrate Pathology,* pp. 210–214. Queens University, Kingston, Ontario.

Rath, W., and W. Drescher. 1990. Response of *Apis cerana* Fabr. towards brood infested with *Varroa jacobsoni* Oud and infestation rate of colonies in Thailand. 21: 311–321.

Ratner, S., and S. B. Vinson. 1983. Encapsulation reactions *in vitro* by haemocytes of *Heliothis virescens. Journal of Insect Physiology* 29: 855–863.

Ratnieks, F.L.W., and J. J. Boomsma. 1995. Facultative sex allocation by workers and the evolution of polyandry by queens in social Hymenoptera. *American Naturalist* 145: 969–993.

Ratnieks, F.L.W. 1988. Reproductive harmony via mutual policing by workers in eusocial hymenoptera. *American Naturalist* 132: 217–236.

Ratnieks, F.L.W. 1990. The evolution of polyandry by queens in social Hymenoptera: The significance of the timing of removal of diploid males. *Behavioural Ecology and Sociobiology* 26: 343–348.

Ratnieks, F. L. W., and J. Nowakowski. 1989. Honeybee swarms accept bait hives contaminated with American foulbrood disease. *Ecological Entomology* 14: 475–478.

Read, A. F., S. D. Albon, J. Antonovics, V. Apanius, G. Dwyer, R. D. Holt, O. Judson, C. M. Lively, A. Martin-Löf, A. R. McLean, J. A. J. Metz, P. Schmid-Hempel, P. H. Thrall, S. Via, and K. Wilson. 1995. Genetics and evolution of infectious diseases in natural populations. In B. T. Grenfell and A. P. Dobson, eds., *Ecology of Infectious Diseases in Natural Populations,* pp. 450–478. Cambridge University Press, Cambridge, U.K.

Reeve, H. K. 1991. *Polistes.* In K. G. Ross and R. W. Matthews, eds., *The Social Biology of Wasps,* pp. 99–148. Comstock Publishing, Ithaca, New York.

Régnière, J. 1984. Vertical transmission of diseases and population dynamics of insects with discrete generations: A model. *Journal of theoretical Biology* 107: 287–301.

Rehm, S. M., and W. Ritter. 1989. Sequence of the sexes in the offspring of *Varroa jacobsoni* and the resulting consequences for the calculation of the developmental period. *Apidologie* 20: 339–343.

Reilly, L. M. 1987. Measurements of inbreeding and average relatedness in a termite population. *American Naturalist* 130: 339–349.

Rennie, J. 1992. Trends in parasitology—living together. *Scientific American,* January, pp. 105–113.

Rettenmeyer, C. W., and R. D. Akre. 1968. Ectosymbiosis between phorid flies and army ants. *Annals of the Entomological Society of America* 61: 1317–1326.

Richards, K. W. 1973. Biology of *Bombus polaris* Curtis and *B. hyperboreus* Schönherr at Lake Hazen, Northwest Territories (Hymenoptera: Bombini). *Quaestiones Entomologicae* 9: 115–157.

Richards, L. A., and K. W. Richards. 1976. Parasitid mites associated with bumble bees in Alberta, Canada (Acarina: Parasitidae; Hymenoptera: Apidae). II. Biology. *University of Kansas Science Bulletin* 51: 1–18.

Richards, M. H., L. Packer, and J. Seger. 1995. Unexpected patterns of parentage and relatedness in a primitively eusocial bee. *Nature* 373: 239–241.

Riek, E. F. 1970. Strepsiptera. In CSIRO, ed., *The Insects of Australia,* pp. 622–35. Melbourne University Press, Carlton, Victoria.

Rinderer, T. E., and K. D. Elliott. 1977. Worker honey bee responses to infection with *Nosema apis:* Influence of diet. *Journal of Economic Entomology* 70: 431–433.

Rinderer, T. E., and K. D. Elltiott. 1977. Influence of nosematosis on the hoarding behaviour of the honeybees. *Journal of Invertebrate Pathology* 30: 110–111.

Rinderer, T. E., and T. J. Green. 1976. Serological relationship between chronic bee paralysis virus and the virus causing hairless-black syndrome in the honeybee. *Journal of Invertebrate Pathology* 27: 403–405.

Rinderer, T. E., W. C. Rothenbuhler, and T. A. Gochnauer. 1974. The influence of pollen on the susceptibility of honeybee larvae to *Bacillus* larvae. *Journal of Invertebrate Pathology* 23: 347–350.

Rissing, S. W., R. A. Johnson, and G. B. Pollock. 1986. Natal nest distribution and pleometrosis in the desert leaf-cutter ant *Acromyrmex versicolor* (Pergande) (Hymenoptera: Formicidae). *Psyche* 93: 177–186.

Rissing, S. W., and G. B. Pollock. 1987. Queen aggression, pleometrotic advantage and brood raiding in the ant *Veromessor pergandei* (Hymenoptera: Formicidae). *Animal Behaviour* 35: 975–981.

Ritchie, J. 1915. Some observations and deductions regarding the habits and biology of the common wasp. *Scottish Naturalist* 47: 318–331.

Ritter, W. 1988. *Varroa jacobsoni* in Europe, the tropics and subtropics. In G. R. Needham, ed., *Africanized Honey Bees and Bee Mites,* pp. 349–359. Ellis Horwood, Chichester, U.K.

Ritter, W., and D. De Jong. 1984. Reproduction of *Varroa jacobsoni* Oud in Europe, the Middle East and tropical South America. *Zeitschrift für Angewandte Entomologie* 98: 55–57.

Ritter, W., and W. Schneider-Ritter. 1988. Differences in biology and means of controlling *Varroa jacobsoni* and *Tropilaelaps clareae,* two novel parasitic mites of *Apis mellifera.* In G. R. Needham, ed., *Africanized Honey Bees and Bee Mites,* pp. 387–395. Ellis Horwood, Chichester, U.K.

Robinson, G. E. 1992. Regulation of division of labor in insect societies. *Annual Review of Entomology* 37: 637–665.

Robinson, G. E., and R. E. Page. 1988. Genetic determination of guarding and undertaking in honey-bee colonies. *Nature* 333: 356–358.

Robinson, G. E., and R. E. Page. 1989a. Genetic basis for division of labor in an insect society. In M. D. Breed and R. E. Page, eds., *The Genetics of Social Evolution,* pp. 61–80. Westview Press, Boulder, Colorado.

Robinson, G. E., and R. E. Page. 1989b. Genetic determination of nectar foraging, pollen foraging, and nest-site scouting in honey bee colonies. *Behavioural Ecology and Sociobiology* 24: 317–323.

Roisin, Y. 1987. Polygyny in *Nasutitermes* species: Field data and theoretical approaches. *Experientia (Supplement)* 54: 379–404.

Roisin, Y. 1993. Selective pressures on pleometrosis and secondary polygyny: A comparison of termites and ants. In L. Keller, eds., *Queen Number and Sociality in Insects,* pp. 402–421. Oxford University Press, Oxford.

Roisin, Y., and J. M. Pasteels. 1986. Reproductive mechanisms in termites: Polycalism and polygyny in *Nasutitermes polygynus* and *N. costalis. Insectes Sociaux* 33: 149–167.

Rose, R. I., and J. D. Briggs. 1969. Resistance to American foulbrood. IX. Effects of honey bee larval food on the growth and viability of *Bacillus larvae. Journal of Invertebrate Pathology* 13: 74–80.

Röseler, P.-F. 1973. Die Anzahl Spermien im Receptaculum seminis von Hummelköniginnen (Hymenoptera, Apidae, Bombinae). *Apidologie* 4: 267–274.

Rosengaus, R. B., and J. F. A. Traniello. 1993. Disease risk as a cost of outbreeding in the termite *Zootermopsis angusticollis. Proceedings of the National Academy of Sciences USA* 90: 6641–6645.

Rosengren, R. 1977. Foraging strategy of wood ants (*Formica rufa* group). I. Age polyethism and topographic traditions. *Acta Zoologica Fennici* 149: 1–30.

Rosengren, R., and D. Chérix. 1981. The pupa-carrying test as a taxonomic tool in the *Formica rufa* group. In P. E. Howse and J.-L. Clément, eds., *Biosystematics of Social Insects,* pp. 263–281. Academic Press, New York.

Rosengren, R., L. Sundström, and W. Fortelius. 1993. Monogyny and polygyny in *Formica* ants: The result of alternative dispersal tactics. In L. Keller, ed., *Queen Number and Sociality in Insects,* pp. 308–333. Oxford University Press, Oxford.

Ross, K. G. 1986. Kin selection and the problem of sperm utilization in social insects. *Nature* 323: 798–800.

Ross, K. G. 1988. Differential reproduction in multiple-queen colonies of the fire ant *Solenopsis invicta* (Hymenoptera: Formicidae). *Behavioural Ecology and Sociobiology* 23: 341–355.

Ross, K. G. 1993. The breeding system of the fire ant *Solenopsis invicta:* Effects on colony genetic structure. *American Naturalist* 141: 554–576.

Ross, K. G., and J. M. Carpenter. 1991. Population genetic structure, relatedness, and breeding systems. In K. G. Ross and R. W. Matthews, eds., *The Social Biology of Wasps,* pp. 451–479. Comstock Publishing, Ithaca, New York.

Ross, K. G., and D. J. C. Fletcher. 1985. Comparative study of genetic and social structure in two forms of the fire ant *Solenopsis invicta* (Hymenoptera, Formicidae). *Behavioural Ecology and Sociobiology* 17: 349–356.

Ross, K. G., and R. W. Matthews, eds. 1990. *The Social Biology of Wasps.* Cornell University Press, Ithaca, New York.

Rothenbuhler, W. C. 1958. Genetics and breeding of the honey bee. *Annual Review of Entomology* 3: 161–180.

Rothenbuhler, W. C. 1964a. Behavior genetics of nest cleaning honeybees. IV. Responses of F1 and backcross generations to disease killed brood. *American Zoologist* 4: 111–123.

Rothenbuhler, W. C. 1964b. Behavior genetics of nest cleaning in honey bees. I. Response of four inbred lines to disease-killed brood. *Animal Behaviour* 112: 578–583.

Rothenbuhler, W. C., and V. C. Thompson. 1956. Resistance to American foulbrood in honeybees. I. Differential survival of larvae of different genetic lines. *Journal of Economic Entomology* 49: 470–475.

Roubik, D., L. A. Weigt, and M. A. Bonilla. 1996. Population genetics, diploid males, and limits to social evolution of euglossine bees. *Evolution* 50: 931–935.

Roubik, D. W. 1982. Obligate necrophagy in a social bee. *Science* 217: 1059–1060.

Roubik, D. W. 1989. *Ecology and Natural History of Tropical Bees.* Cambridge University Press, Cambridge, U.K.

Royce, L. A., and P. A. Rossignol. 1990a. Epidemiology of honey bee parasites. *Parasitology Today* 6: 348–353.

Royce, L. A., and P. A. Rossignol. 1990b. Honey bee (*Apis mellifera*) mortality due to tracheal mite (*Acarapis woodi*) infestation. *Parasitology* 100: 147–151.

Royce, L. A., P. A. Rossignol, D. M. Burgett, and B. A. Stringer. 1991. Reduction of tracheal mite parasitism of honey bees by swarming. *Philosophical Transactions of the Royal Society London, B* 331: 123–129.

Rupp, L. 1987. The genus *Volucella* (Diptera: Syrphidae) as commensals and parasites in bumblebee and wasp nests. In J. Eder and D. H. Rembold, eds., *Chemistry and Biology of Social Insects,* pp. 642–643. Peperny Verlag, Munich.

Ryan, R. E., T. C. Cornell, and G. J. Gamboa. 1985. Nestmate recognition in the bald-faced hornet, *Dolichovespula maculata* (Hymenoptera: Vespidae). *Zeitschrift für Tierpsychologie* 69: 19–26.

Ryan, R. E., and G. J. Gamboa. 1986. Nestmate recognition between males and gynes of the social wasp *Polistes fuscatus* (Hymenoptera: Vespidae). *Annals of the Entomological Society of America* 79: 572–575.

Sakagami, S. F. 1976. Species differences in the bionomic characters of bumblebees. A comparative review. *Journal of the Faculty of Science, Hokkaido University, ser. 6, Zoology* 20: 390–447.

Sakagami, S. F., and K. Hayashida. 1961. Biology of the primitive social bee *Halictus duplex* Dalla Torre. III. Activities in the spring solitary phase. *Journal of the Faculty of Science, Hokkaido University, ser. 6, Zoology* 14: 639–682.

Sakofski, F. 1990. Quantitative investigations on transfer of *Varroa jacobsoni.* In W. Ritter, ed., *Proceedings of the International Symposium on Recent Research on Bee Pathology,* pp. 70–72. International Federation of Beekepeers Association, Gent, Belgium.

Sakofski, F., and N. Koeniger. 1988. Natural transfer of *Varroa jacobsoni* between honeybee colonies in autumn. In R. Cavalloro, ed., *European Research on Varroatosis Control,* pp. 81–83. Balkema, Rotterdam.

Sakofski, F., N. Koeniger, and S. Fuchs. 1990. Seasonality of honey bee colony invasion by *Varroa jacobsoni* Oud. *Apidologie* 21: 547–550.

Salt, G. 1970. *The Cellular Defence Reactions of Insects.* Cambridge University Press, Cambridge, U.K.

Salt, G., and R. van den Bosch. 1967. The defence reactions of three species of *Hypera* (Coleoptera: Curculionidae) to an ichneumon wasp. *Journal of Invertebrate Pathology* 9: 164–177.

Salzemann, A., and L. Plateaux. 1987. Reduced egg laying by workers of the ant *Leptothorax nylanderi* in the presence of workers parasitized by a cestoda. In J. Eder and H. Rembold, eds., *Chemistry and Biology of Social Insects,* p. 45. Peperny Verlag, Munich.

Samson, R. A., H. C. Evans, and E. S. Hoekstra. 1982. Notes on entomogenous fungi from Ghana. VI. The genus *Cordyceps. Proceedings of the Koninklijke Nederlandse Akadmie van Wetenschappen, ser. C,* 85: 589–605.

Samson, R. A., H. C. Evans, and J. P. Latgé. 1988. *Atlas of Entomopathogenic Fungi.* Springer-Verlag, Berlin.

Samson, R. A., H. C. Evans, and G. Van de Klashorts. 1981. Notes on entomopathogenic fungi from Ghana. V. The genera *Stilbella* and *Polycephalomyces. Proceedings of the Koninklijke Nederlandse Akademie van Wetenschappen, ser. C,* 84: 289–301.

Sanchez-Pena, S. R., A. Buschinger, and R. A. Humber. 1993. *Myrmicinosporidium durum,* an enigmatic fungal parasite of ants. *Journal of Invertebrate Pathology* 61: 90–96.

Sands, W. A. 1965. Mound population movements and fluctuations in *Trinervitermes ebenerianus* Sjöstedt (Isoptera, Termitidae, Nasutitermitinae). *Insectes Sociaux* 12: 49–58.

Sasaki, A., and Y. Iwasa. 1991. Optimal growth schedule of pathogens within a host: Switching between lytic and latent cycles. *Theoretical Population Biology* 39: 201–239.

Schad, G. A., and R. M. Anderson. 1985. Predisposition to hookworm in man. *Science* 228: 1537–1540.

Schaub, G. A. 1994. Pathogenicity of trypanosomatids on insects. *Parasitology Today* 10: 463–468.

Scherba, G. 1958. Reproduction, nest orientation and population structure of an aggregation of mound nests of *Formica ulkei* Emery (Formicidae). *Insectes Sociaux* 5: 201–213.

Scherba, G. 1961. Nest structure and reproduction in the mound-building ant *Formica opaciventris* Emery in Wyoming. *Journal of the New York Entomological Society* 69: 71–87.

Scherba, G. 1963. Population characteristics among colonies of the ant *Formica opaciventris* Emery (Hymenoptera: Formicidae). *Journal of the New York Entomological Society* 71: 219–232.

Schildknecht, H., P. B. Reed, F. D. Reed, and K. Koob. 1973. Auxin activity in the symbiosis of leaf-cutting ants and their fungus. *Insect Biochemistry* 3: 439–442.

Schmid-Hempel, P. 1983. *Foraging Ecology and Colony Structure of Two Sympatric Species of Desert Ants,* Cataglyphis bicolor *and* Cataglyphis albicans. Verlag Studentenschaft, Zürich.

Schmid-Hempel, P. 1990. Reproductive competition and the evolution of work load in social insects. *American Naturalist* 135: 501–526.

Schmid-Hempel, P. 1991a. The ergonomics of worker behavior in social insects. *Advances in the Study of Behaviour* 20: 87–134.

Schmid-Hempel, P. 1991b. Worker caste and adaptive demography. *Journal of Evolutionary Biology* 5: 1–12.

Schmid-Hempel, P. 1994a. Infection and colony variability in social insects. *Philosophical Transactions of the Royal Society London, B* 346: 313–321.

Schmid-Hempel, P. 1995. Parasites and social insects. *Apidologie* 26: 255–271.

Schmid-Hempel, P., and S. Durrer. 1991. Parasites, floral resources and reproduction in natural populations of bumblebees. *Oikos* 62: 342–350.

Schmid-Hempel, P., and D. Heeb. 1991. Worker mortality and colony development in bumblebees, *B. lucorum* L. *Mitteilungen der Schweizerischen Entomologischen Gesellschaft* 64: 93–108.

Schmid-Hempel, P., C. Müller, R. Schmid-Hempel, and J. A. Shykoff. 1990. Frequency and ecological correlates of parasitism by conopid flies (Conopidae, Diptera) in populations of bumblebees. *Insectes Sociaux* 37: 14–30.

Schmid-Hempel, P., K. Pur, N. Krüger, C. Rebel and R. Schmid-Hempel. 1998. Fitness of *an intestinal* parasite, *C. bombi,* in relation to transmission and host genotype. In prep.

Schmid-Hempel, P., and R. Schmid-Hempel. 1984. Life duration and turnover of foragers in the ant *Cataglyphis bicolor* (Hymenoptera, Formicidae). *Insectes Sociaux* 31: 345–360.

Schmid-Hempel, P., and R. Schmid-Hempel. 1993. Transmission of a pathogen in *Bombus terrestris,* with a note on division of labour in social insects. *Behavioural Ecology and Sociobiology* 33: 319–327.

Schmid-Hempel, P., M. L. Winston, and R. C. Ydenberg. 1993. Invitation paper (The C. P. Alexander Fund): The foraging behavior of individual workers in relation to colony state in the social hymenoptera. *Canadian Entomologist* 126: 129–160.

Schmid-Hempel, R. 1994b. Evolutionary Ecology of a Host-Parasite Interaction. Ph.D. diss., Universität Basel, Basel, Switzerland.

Schmid-Hempel, R., and C. B. Müller. 1991. Do parasitized bumblebees forage for their colony? *Animal Behaviour* 41: 910–912.

Schmid-Hempel, R., and P. Schmid-Hempel. 1989. Superparasitism and larval competition in conopid flies (Dipt., Conopidae) parasitizing bumblebees (Hym., Apidae). *Mitteilungen der Schweizerischen Entomologischen Gesellschaft* 62: 279–289.

Schmid-Hempel, R., and P. Schmid-Hempel. 1996. Larval development of two parasitic flies (Conopidae) in the common host *Bombus pascuorum. Ecological Entomology* 21: 63–70.

Schmid-Hempel, R., and P. Schmid-Hempel. 1998. Colony performance and immunocompetence of a social insect, *Bombus terrestris,* in poor and variable environments. *Functional Ecology* 12: 22–30.

Schmit, A. R., and N. A. Ratcliffe. 1978. The encapsulation of araldite implants and recognition of foreignness in *Clitumnus extradentatus. Journal of Insect Physiology* 24: 511–521.

Schneider, G., and W. Hohorst. 1971. Wanderungen der Metacercarien des Lanzett-Egels in Ameisen. *Naturwissenschaften* 58: 327–328.

Schneider, P., and W. Drescher. 1988. Die Folgen eines unterschiedlich hohen *Varroa*-Befalls während der Puppenentwicklung auf die erwachsene Biene; parts 1–3. *Allgemeine Deutsche Imkerzeitung* 22: 16–18, 54–56, 87–91.

Schousboe, C. 1986. On the biology of *Scutacarus acarorum* Goeze (Acarina: Trombidiformes). *Acarologia* 27: 151–158.

Schousboe, C. 1987. Deutonymphs of *Parasitellus* phoretic on Danish bumblebees (Parasitidae, Mesostigmata; Apidae, Hymenoptera). *Acarology* 28: 37–41.

Schulz-Langner, E. 1958. Untersuchungen über die Anwesenheit und Verbreitung des Ämobenparasiten (*Malphigamoeba mellificae*) in einem Bienenvölkchen von April bis Juli 1957. *Zeitschrift für Bienenforschung* 9: 381–389.

Schwarz, H. H., and K. Huck. 1997. Phoretic mites use flowers to transfer between foraging bumblebees. *Insectes Sociaux* 44: 303–310.

Schwarz, H. H., K. Huck, and P. Schmid-Hempel. 1996. Prevalence and host preferences of mesostigmatic mites (Acari: Anactinochaeta) phoretic on Swiss bumble bees (Hymenoptera: Apidae). *Journal of the Kansas Entomological Society* 69: 35–42.

Schwarz, M. P. 1987. Intra-colony relatedness and sociality in the allodapine bee, *Exoneura bicolor. Behavioural Ecology and Sociobiology* 21: 387–392.

Scofield, V. L., J. M. Schlumpberger, L. A. West, and I. L. Weisman. 1982. Protochordate allorecognition is controlled by a MHC-like gene system. *Nature* 295: 499–502.

Seeley, T. D. 1978. Life history strategy of the honey bee, *Apis mellifera. Oecologia* 32: 109–118.

Seeley, T. D. 1982. Adaptive significance of the age polyethism schedule in honeybee colonies. *Behavioural Ecology and Sociobiology* 11: 287–293.

Seeley, T. D. 1985. *Honeybee Ecology.* Princeton University Press, Princeton, New Jersey.

Seeley, T. D., and P. K. Visscher. 1985. Survival of honeybees in cold climates: The critical timing of colony growth and reproduction. *Ecological Entomology* 10: 81–88.

Seelinger, G., and U. Seelinger. 1983. On the social organisation, alarm and fighting in the primitive cockroach *Cryptocercus puntculatus* Scudder. *Zeitschrift für Tierpsychologie* 61: 315–333.

Seger, J. 1983. Partial bivoltinism may cause alternating sex-ratio biases that favour eusociality. *Nature* 301: 59–62.

Seger, J. 1993. Opportunities and pitfalls in co-operative reproduction. In L. Keller, ed., *Queen Number and Sociality in Insects,* pp. 1–15. Oxford University Press, Oxford.

Seger, J., and N. A. Moran. 1996. Snapping social swimmers. *Nature* 381: 473–474.

Shapiro, M. 1967. Pathological changes in the blood of the greater wax moth, *Galleria mellonella,* during the course of nucleopolyhedrosis and starvation. 1. Total hemocyte count. *Journal of Invertebrate Pathology* 9: 111–113.

Sherman, P. W. 1979. Insect chromosome number and eusociality. *American Naturalist* 113: 925–935.

Sherman, P. W., J. U. M. Jarvis, and R. D. Alexander, eds. 1991. *The Biology of the Naked Mole-Rat.* Princeton University Press, Princeton, New Jersey.

Sherman, P. W., E. A. Lacey, H. K. Reeve, and L. Keller. 1995. The eusociality continuum. *Behavioral Ecology* 6: 102–108.

Sherman, P. W., T. D. Seeley, and H. K. Reeve. 1988. Parasites, pathogens, and polyandry in social Hymenoptera. *American Naturalist* 131: 602–610.

Sherman, P. W., and J. S. Shellman-Reeve. 1994. Ant sex ratios. *Nature* 370: 257.

Shoemaker, D. D., J. T. Costa, and K. G. Ross. 1992. Estimates of heterozygosity in two social insects using a large number of electrophoretic markers. *Heredity* 69: 573–582.

Showers, R. E., A. Jones, and F. E. Moeller. 1967. Cross-incoculation of the bumble bee *Bombus fervidus* with the microsporidian *Nosema apis* from the honey bee. *Journal of Economic Entomology* 60: 774–777.

Shykoff, J. A., and P. Schmid-Hempel. 1991a. Genetic relatedness and eusociality: Parasite-mediated selection on the genetic composition of groups. *Behavioural Ecology and Sociobiology* 28: 371–376.

Shykoff, J. A., and P. Schmid-Hempel. 1991b. Parasites and the advantage of genetic variability within social insect colonies. *Proceedings of the Royal Society London, B* 243: 55–58.

Shykoff, J. A., and P. Schmid-Hempel. 1992. Parasites delay worker reproduction in the bumblebee: Consequences for eusociality. *Behavioral Ecology* 2: 242–248.

Sidorov, N. G., V. S. Kuptev, and I. A. Mugalimov. 1975. Resistance of honey bees to *Nosema* disease and a genetic method of control. *Veterinariya (Moscow)* 7: 63–65.

Siva-Jothy, M. 1995. Immunocompetence: Conspicuous by its absence. *Trends in Ecology and Evolution* 10: 205–206.

Skaife, S. H. 1953. Subsocial bees of the genus *Allodape* Lep. & Serb. *Journal of the Entomological Society of Southern Africa* 16: 3–16.

Skou, J. P., and S. N. Holm. 1980. Occurrence of melanosis and other diseases in the queen honeybee, and risk of their transmission during instrumental insemination. *Journal of Apicultural Research* 19: 133–143.

Skou, J. P., S. N. Holm, and H. Haas. 1963. Preliminary investigations on diseases in bumble bees (*Bombus* Latr.). *Kongelige Veterinaerskole og Landbohoejskole Denmark* (1963): 27–41.

Skou, J. P., and J. King. 1984. *Ascosphaera osmophila* sp. nov., an Australian spore cyst fungus. *Journal of Botany* 32: 225–231.

Smith, B. H. 1983. Recognition of female kin by male bees through olfactory signals. *Proceedings of the National Academny of Sciences USA* 80: 4551–4553.

Smith, B. H., and M. Ayasse. 1987. Kin-based male mating preferences in two species of halictine bees. *Behavioural Ecology and Sociobiology* 20: 313–318.

Smith, G., and A. H. Hamm. 1914. Studies in the experimental analysis of sex. II. On stylops and stylopization. *Quarterly Journal of Microscopic Sciences* 60: 435–461.

Smith, M. R. 1946. Ant hosts of the fungus, *Laboulbenia formicarum* Thaxter. *Proceedings of the Entomological Society of Washington* 48: 29–31.

Smith Trail, D. R. 1980. Behavioral interactions between parasites and hosts: Host suicide and the evolution of complex life cycles. *American Naturalist* 116: 77–91.

Snelling, R. R. 1981. Systematics of social hymenoptera. In H. R. Hermann, ed., *Social Insects,* vol. 2, pp. 369–453. Academic Press, New York.

Soper, R. S., A. J. Delyzer, and L. F. R. Smith. 1976. The genus *Massospora* entomopathogenic for cicadas. II: Biology of *Massospora levispora* and its host *Okanagana rimosa. Annals of the Entomological Society of America* 69: 89–95.

Spiess, E. B. 1987. Discrimination among prospective mates in *Drosophila.* In D. J. C. Fletcher and C. D. Michener, eds., *Kin Recognition in Animals,* pp. 75–119. John Wiley and Sons, Chichester, U.K.

Spradbery, J. P. 1973. *Wasps: An Account of the Biology and Natural History of Solitary and Social Wasps.* University of Washington Press, Seattle.

Spradbery, J. P. 1991. Evolution of queen number and queen control. In K. G. Ross and R. W. Matthews, eds., *The Social Biology of Wasps,* pp. 336–388. Comstock Publishing, Ithaca, New York.

Sprague, V. 1977. The zoological distribution of the microsporidia. In L. A. J. Bulla and T. C. Cheng, eds., *Comparative Pathobiology: Systematics of the Microsporidia,* vol. 2, pp. 335–446. Plenum Press, New York.

Stairs, G. R. 1965. Quantitative differences in susceptibility to nuclear-polyhedrosis virus among larval instars of the forest tent caterpillar, *Malacosoma disstria* (Hübner). *Journal of Invertebrate Pathology* 7: 427–429.

Starr, C. K. 1984. Sperm competition, kinship, and sociality in the aculeate hymenoptera. In R. L. Smith, ed., *Sperm Competition and the Evolution of Animal Mating Systems,* pp. 428–464. Academic Press, New York.

Stearns, S. C. 1992. *Life History Evolution.* Oxford University Press, Oxford.

Steche, W. 1965. Observations sur l'ontogenèse de *Nosema apis* Zander dans l'intestin moyen de l'abeille ouvrière. *Bulletin d'Apiculture* 8: 181–209.

Steiger, V., H. E. Lamparter, C. Sandri, and K. Akert. 1969. Virus-ähnliche Partikel im Zytoplasma von Nerven- und Gliazellen der Waldameise. *Archiv für die gesamte Virusforschung* 26: 271–282.

Stein, G., and E. Lohmar. 1972. Über die Infektion verschiedener Hummelarten mit *Sphaerularia bombi* Leon Dufour 1837 im Raum Bonn, Frühjahr 1970. *Decheniana* 124: 135–140.

Steinhaus, E. A. 1949. *Principles of Insect Pathology.* McGraw-Hill, New York.

Steinhaus, E. A. 1958. Stress as a factor in insect disease. *Proceedings of the 9th International Congress of Entomology* 4: 725–730.

Steinhaus, E. A., eds. 1963. *Insect Pathology: An Advanced Treatise.* Vol. 2. Academic Press, New York.

Stejskal, M. 1955. Gregarines found in the honey-bee (*Apis mellifera* Linnaeus) in Venezuela. *Journal of Protozoology* 2: 185–188.

Stephen, W. P., and B. L. Fichter. 1990a. Chalkbrood (*Ascosphaera aggregata*) resistance in the leaf-cutter bee (*Megachile rotunda*). II. Random matings of resistant lines to wild type. *Apidologie* 21: 221–231.

Stephen, W. P., and B. L. Fichter. 1990b. Chalkbrood (*Asocsphaera aggregata*) resistance on the leaf-cutting bee (*Megachile rotunda*). I. Challenge of selected lines. *Apidologie* 21: 209–219.

Stille, M., B. Stille, and P. Douwes. 1991. Polygyny, relatedness and nest founding in the polygynous myrmicine ant *Leptothorax acervorum* (Hymenoptera; Formicidae). *Behavioural Ecology and Sociobiology* 28: 91–96.

Stoffolano, J. G. J. 1986. Nematode-induced host responses. In A. P. Gupta, ed., *Hemocytic and Humoral Immunity in Arthropods,* pp. 117–155. John Wiley and Sons, New York.

Strand, M. R. 1986. The physiological interactions of parasitoids with their hosts and their influence on reproductive strategies. In J. Waage and D. Greathead, eds., *Insect Parasitoids,* pp. 97–136. Academic Press, London.

Strassmann, J. E. 1981a. Evolutionary implications of early male and satellite nest production in *Polistes exclamans* colony cycles. *Behavioural Ecology and Sociobiology* 8: 55–64.

Strassmann, J. E. 1981b. Parasitoids, predators and group size in the paper wasp *Polistes exclamans. Ecology* 62: 1225–1233.

Strassmann, J. E. 1985. Worker mortality and the evolution of castes in the social wasp *Polistes exclamans. Insectes Sociaux* 32: 275–287.

Strassmann, J. E., C. R. Hughes, D. C. Queller, S. Turillazzi, R. Cervo, S. K. Davis, and K. J. Goodnight. 1989. Genetic relatedness in primitively eusocial wasps. *Nature* 342: 268–270.

Strassmann, J. E., and D. C. Meyer. 1983. Gerontocracy in the social wasp *Polistes exclamans. Animal Behaviour* 31: 431–438.

Strassmann, J. E., D. C. Queller, and C. R. Hughes. 1988. Predation and the evolution of sociality in the paper wasp *Polistes bellicosus. Ecology* 69: 1497–1505.

Strick, H., and G. Madel. 1988. Transmission of the pathogenic bacterium *Hafnia alvei* to honeybees by the ectoparasitic mite *Varroa jacobsoni.* In G. R. Needham, ed., *Africanized Honey Bees and Bee Mites,* pp. 462–466. Ellis Horwood, Chichester, U.K.

Stuart, R. J. 1987a. Individual workers produce colony-specific nestmate recognition cues in the ant, *Leptothorax curvispinosus. Behavioural Ecology and Sociobiology* 21: 229–235.

Stuart, R. J. 1987b. Individually-produced nestmate recognition cues and a colony odour 'gestalt' in Leptothoracine ants. In J. Eder and H. Rembold, eds., *Chemistry and Biology of Social Insects,* pp. 480. Peperny, Verlag Peperny Munich.

Stuart, R. J. 1987c. Transient nestmate recognition cues contribute to a multicolonial population structure in the ant, *Leptothorax curvispinosus. Behavioural Ecology and Sociobiology* 21: 229–235.

Stuart, R. J., and T. M. Alloway. 1988. Aberrant yellow ants: North American *Leptothorax* species as intermediate hosts of Cestodes. In J. C. Trager, ed., *Advances in Myrmecology,* pp. 537–545. E. J.Brill, Leiden.

Stubblefield, J. W., and J. Seger. 1994. Sexual dimorphism in the hymenoptera. In R. V. Short and E. Balaban, eds., *The Difference between the Sexes,* pp. 71–103. Cambridge University Press, Cambridge, U.K.

Stumper, R., and H. Kutter. 1950. Sur le stade ultime du parasitisme social chez les fourmis atteint par *Teleutomyrmex schneideri. Comptes Rendus de l'Académie des Sciences Paris* 231: 876–878.

Sturtevant, A. P. 1932. Relation of commercial honey to the spread of American foulbrood. *Journal of Agricultural Research* 45: 257–285.

Sturtevant, A. P., and I. L. Revell. 1953. Reduction of *Bacillus larvae* spores in liquid food of honey bees by the action of the honey stopper, and its relation to the development of American foulbrood. *Journal of Economic Entomology* 46: 855–860.

Summerlin, J. W., W. A. Banks, and K. H. Schroeder. 1975. Food exchange between mounds of the red imported fire ant. *Annals of the Entomological Society of America* 68: 863–866.

Sundström, L. 1993. Genetic population structure and sociogenetic organisation in *Formica truncorum* (Hymenoptera: Formicidae). *Behavioural Ecology and Sociobiology* 33: 345–354.

Sundström, L. 1994. Sex ratio bias, relatedness asymmetry and queen mating frequency in ants. *Nature* 367: 266–268.

Sundström, L., M. Chapuisat, and L. Keller. 1996. Conditional manipulation of sex ratios by ant workers: A test of kin selection theory. *Science* 274: 993–995.

Sutcliffe, G. H. 1987. Resource Use and Population Dynamics in Captive Colonies of the Bumblebee *Bombus terricola* Kirby (Hymenoptera, Apidae). M.Sc. thesis, University of Toronto, Toronto, Canada.

Sutcliffe, G. H., and R. C. Plowright. 1988. The effects of food supply on adult size in the bumblebee *Bombus terricola* Kirby (Hymenoptera, Formicidae). *Canadian Entomologist* 120: 1051–1058.

Sutter, G. R., W. C. Rothenbuhler, and E. S. Raun. 1968. Resistance to American foulbrood in honey bees. VII. Growth of resistant and susceptible larvae. *Journal of Invertebrate Pathology* 12: 25–28.

Svadzhyan, P. K., and L. V. Frolkova. 1966. Ants as first and second intermediate hosts of some trematodes and cestodes (in Russian). *Zoologiceskij Zurnal* 45: 213–219.

Svanborg-Eden, C., and B. R. Levin. 1991. Infectious disease and natural selection in human populations: A critical re-examination. In A. C. Swedlund and G. J. Armelagos, eds., *Disease in Populations in Transition: Anthropological and Epidemiological Perspectives,* pp. 531–546. Bergin and Garvey, South Hadley, Mass.

Sweeney, A. W., E. I. Hazard, and M. F. Graham. 1985. Intermediate host for an *Amblyospora* sp. (Microsporidia) infecting the mosquito, *Culex annulirostris. Journal of Invertebrate Pathology* 46: 98–102.

Sylvester, H. A., and T. E. Rinderer. 1978. Assessing longevity, hoarding behaviour, and response to *Nosema* in honey bees. *American Bee Journal* December: 806–807.

Taber, S. 1955. Sperm distribution in the spermathecae of multiple-mated queens. *Journal of Economic Entomology* 48: 522–525.

Taber, S. 1982. Breeding for disease resistance. *American Bee Journal* 122: 177–179.

Taber, S. 1984. Acarine disease. *American Bee Journal* 124: 794–795.

Taber, S. W., J. C. Cokendolpher, and O. F. Francke. 1988. Karyological study of North American *Pogonomyrmex* (Hymenoptera: Formicidae). *Insectes Sociaux* 35: 47–60.

Talbot, M. 1957. Population studies of the slave-making ant *Leptothorax duloticus* and its slave *Leptothorax curvispinosus. Ecology* 38: 449–456.

Talbot, M. 1961. Mounds of the ant *Formica ulkei* at the Edwin S. George Reserve, Livingston County, Michigan. *Ecology* 42: 202–205.

Tamashiro, M. 1960. The susceptibility of *Bracon*-paralyzed *Corcyra cephalonica* (Stainton) to *Bacillus thuringensis* var. *thuringensis* Berliner. *Journal of Insect Pathology* 2: 209–219.

Tanada, Y. 1959. Synergism between two viruses of the armyworm, *Pseudaletia unipuncta* Haworth (Lepidoptera, Noctuidae). *Journal of Insect Pathology* 1: 215–231.

Tanada, Y. 1967. Effect of high temperatures on the resistance of insects to infectious diseases. *Journal of Sericultural Science of Japan* 15: 333–339.

Tanada, Y., and H. K. Kaya. 1993. *Insect Pathology.* Academic Press, San Diego.

Taylor, P. D. 1988. Inclusive fitness models with two sexes. *Theoretical Population Biology* 34: 145–168.

Templeton, A. R. 1979. Chromosome number, quantitative genetics and eusociality. *American Naturalist* 113: 937–941.

Teson, A., and A.M.M. de Remes Lenicov. 1979. Estrepsipteros parasitoides de hymenopteros (Insecta—Strepsiptera). *Revista de la Sociedad Entomologica Argentina* 38: 115–122.

Tewarson, N. C., A. Singh, and W. Engels. 1992. Reproduction of *Varroa jacobsoni* in colonies of *Apis cerana indica* under natural and experimental conditions. *Apidologie* 23: 161–171.

Thomas, G. M., and A. Luce. 1972. An epizootic of chalkbrood, *Ascosphaera apis* (Maassen ex Calussen) Olive & Splitoir in the honey bee, *Apis mellifera* L. in California. *American Bee Journal* 112: 88–90.

Thompson, J. N., and J. J. Burdon. 1992. Gene-for-gene coevolution between plants and parasites. 360: 121–125.

Thorne, B. L. 1985. Termite polygyny: The ecological dynamics of queen mutualism. In B. Hölldobler and M. Lindauer, eds., *Experimental Behavioral Ecology and Sociobiology*, pp. 325–341. Sinauer Associates, Sunderland, Mass.

Thornhill, R., and J. Alcock. 1983. *The Evolution of Insect Mating Systems.* Harvard University Press, Cambridge, Mass.

Tibayrenc, M., F. Kjellberg, and F. Ayala. 1990. A clonal theory of parasitic protozoa: The population structures of *Entamoeba, Giardia, Leishmania, Naegleria, Plasmodium, Trichomonas,* and *Trypanosoma* and their medical and taxonomic consequences. *Proceedings of the National Academy of Sciences USA* 87: 2412–2418.

Tooby, J. 1982. Pathogens, polymorphism, and the evolution of sex. *Journal of theoretical Biology* 97: 557–576.

Treat, A. E. 1975. *Mites of Moths and Butterflies.* Cornell University Press, Ithaca, New York.

Trivers, R. L., and H. Hare. 1976. Haplodiploidy and the evolution of social insects. *Science* 191: 249–263.

Tschinkel, W. R. 1993. Sociometry and sociogenesis of colonies of the fire ant *Solenopsis invicta* during one annual cycle. *Ecologcial Monographs* 64: 425–457.

Turian, G., and J. Wuest. 1969. Mycoses à Entomophthoracées frappant des populations de fourmis et de Drosophiles. *Mitteilungen der Schweizerischen Entomologischen Gesellschaft* 42: 197–201.

Turillazzi, S. 1991. The Stenogastrinae. In K. G. Ross and R. W. Matthews, eds., *The Social Biology of Wasps,* pp. 74–98. Cornell University Press, Ithaca, New York.

Vandel, A. 1927. Modifications déterminées par un nématode du genre *Mermis* chez les ouvrières et les soldats de la fourmi *Pheidole pallidula* Nyl. *Bulletin Biologique de France et de Belge* 41: 38–48.

Vandel, A. 1934. Le cycle évolutif d' *Hexamermis* sp. parasite de la fourmi *Pheidole pallidula. Annales des Sciences Naturelles* 17: 47–58.

Vandenberg, J. D., B. L. Fitchter, and W. P. Stephen. 1980. Spore load of *Ascosphaera* species on emerging adults of the alfalfa leafcutting bee, *Megachile rotunda. Applied and Environmental Microbiology* 39: 650–655.

Vandenberg, J. D., and W. P. Stephen. 1982. Etiology and symptomatology of chalkbrood in the alfalfa leaf cutting bee, *Megachile rotundata. Journal of Invertebrate Pathology* 39: 133–137.

Van den Eijnde, J., and N. Vette. 1993. *Nosema* infection in honey bees (*Apis mellifera* L.) and bumble bees (*Bombus terrestris* L.). *Proceedings of the Section of Experimental and Applied Entomology of the Netherlands Entomological Society (N. E. V.)* 4: 205–208.

Van der Have, T. M., J. J. Boomsma, and S. B. J. Menken. 1988. Sex-investment ratios and relatedness in the monogynous ant *Lasius niger* L. *Evolution* 42: 160–172.

Vander Meer, R. K., D. P. Jouvenaz, and D. P. Wojcik. 1989. Chemical mimicry in a parasitoid (Hymenoptera: Eucharitidae) of fire ants (Hymenoptera: Formicidae). *Journal of Chemical Ecology* 15: 2247–2261.

Vander Meer, R. K., and D. P. Wojcik. 1982. Chemical mimicry in the myrmecophilous beetle, *Myrmecaphodius exavaticollis. Science* 218: 806–808.

VanDoorn, A. 1987. Investigations into the regulation of dominance behaviour and of the division of labour in bumblebee colonies (*Bombus terrestris*). *Netherlands Journal of Zoology* 37: 255–276.

VanDoorn, A., and J. Heringa. 1986. The ontogeny of a dominance hierarchy in colonies of the bumblebee *Bombus terrestris* (Hymenoptera, Apidae). *Insectes Sociaux* 33: 3–25.

VanHonk, C. G. J., and P. Hogeweg. 1981. The ontogeny of social structure in a captive *Bombus terrestris* colony. *Behavioural Ecology and Sociobiology* 9: 111–119.

Varis, A. L., B. V. Ball, and M. Allen. 1992. The incidence of pathogens in honey bee (*Apis mellifera* L.) colonies in Finland and Great Britain. *Apidologie*23: 133–137.

Veal, D. A., J. E. Trimble, and A. J. Beattie. 1992. Antimicrobial properties of secretions from the metapleural glands *Myrmecia gulosa* (the Australian bull ant). *Journal of Applied Bacteriology* 72: 188–194.

Verma, L. R., B. S. Rana, and S. Verma. 1990. Observations on *Apis cerana* colonies surviving from Thai sacbrood virus infestation. *Apidologie*21: 169–174.

Visscher, P. K. 1986. Kinship discrimination in queen rearing by honey bees (*Apis mellifera*). *Behavioural Ecology and Sociobiology* 18: 453–460.

Visscher, P. K., R. A. Morse, and T. D. Seeley. 1985. Honey bees choosing a home prefer previously occupied cavities. *Insectes Sociaux* 32: 217–220.

Vollrath, F. 1986. Eusociality and extraordinary sex ratios in the spider, *Anelosimus eximius* (Araneae: Theridiidae). *Behavioural Ecology and Sociobiology* 18: 283–287.

Waddington, K. D., and W. C. Rothenbuhler. 1976. Behaviour associated with hairless-black syndrome of adult honeybees. *Journal of Apicultural Research* 15: 35–41.

Wahab, A. 1962. Untersuchungen über Nematoden in den Drüsen des Kopfes der Ameisen (Formicidae). *Zeitschrift für Morphologie und Ökologie der Tiere* 52: 33–92.

Wakelin, D. 1984. *Immunity to Parasites.* Edward Arnold, London.

Waldman, B. 1981. Sibling recognition in toad tadpoles: Field evidence and implications. *Zeitschrift für Tierpsychologie* 56: 341–358.

Waldman, B. 1987. Mechanisms of kin recognition. *Journal of theoretical Biology* 128: 159–185.

Walker, I. 1962. *Drosophila* and *Pseudeucoila.* III. Selektionsversuche zur Steigerung der Resistenz des Parasiten gegen die Abwehrreaktion des Wirtes. *Revue Suisse de Zoologie* 69: 209–227.

Walker, J., and J. Stamps. 1986. A test of optimal caste ratio theory using the ant *Camponotus (Colobopsis) impressus. Ecology* 67: 1052–1062.

Wallis, D. I. 1962. Aggressive behavior in the ant *Formica fusca. Animal Behaviour* 10: 267–274.

Waloff, N., and R. E. Blackith. 1962. The growth and distribution of the mounds of *Lasius flavus* (Fabricius) (Hym: Formicidae) in Silwood Park, Berkshire. *Journal of Animal Ecology* 31: 421–437.

Waloff, N., and M. A. Jervis. 1987. Communities of parasitoids associated with leafhoppers and planthoppers in Europe. *Advances in Ecological Research* 17: 281–376.

Walther, B. A., P. Cotgreave, R. D. Price, R. D. Gregory, and D. H. Clayton. 1995. Sampling effort and parasite species richness. *Trends in Ecology and Evolution* 11: 306–310.

Wang, D. I., and F. E. Moeller. 1969. Histological comparisons of the development of hypopharyngeal glands in healthy and *Nosema*-infected worker honey bees. *Journal of Invertebrate Pathology* 14: 135–142.

Wang, D. I., and F. E. Moeller. 1970. The division of labor and queen attendance behavior of *Nosema*-infected worker honey bees. *Journal of Economic Entomology* 63: 1539–1541.

Watanabe, H. 1967. Development of resistance in the silkworm, *Bombyx mori,* to peroral infection of a cytoplasmicpolyhedrosis virus. *Journal of Invertebrate Pathology* 9: 474–479.

Watanabe, H., and S. Maeda. 1981. Genetically determined nonsusceptibility of the silkworm, *Bombyx mori,* to infection with densonucleosis virus (Densovirus). *Journal of Invertebrate Pathology* 38: 370–373.

Watanabe, H., and Y. Tanada. 1972. Infection of a nuclear-polyhedrosis virus in armyworm, *Pseudaletia unipuncta* (Lepidoptera: Noctuidae), reared at high temperature. *Applied Entomology and Zoology* 7: 43–51.

Watkins, J. F., and T. W. Cole. 1966. The attraction of army ant workers to secretions of their queens. *Texas Journal of Science* 18: 254–265.

Watson, J. A. L., B. M. Okot-Kotber, and C. Noirot, eds. 1984. *Caste Differentiation in Social Insects*. Pergamon Press, Oxford.

Wcislo, W. T. 1987. The roles of seasonality, host synchrony, and behaviour in the evolutions and distributions of nest parasites in Hymenoptera (Insecta), with special reference to bees (Apoidea). *Biological Review* 62: 515–543.

Weiser, J. 1961. Die Mikrosporidien als Parasiten der Insekten. *Monographien zur Angewandten Entomologie* 17: 1–149.

Weiser, J. 1969. Immunity of insects to protozoa. In G. J. Jackson, R. Herman and I. Singer, eds., *Immunity to Parasitic Animals*, Vol. 1, pp. 129–147. Appleton-Century Crofts, New York.

Weiss, K. 1984. *Bienen-Pathologie.* Ehrenwirth, Munich.

Weissflog, A., U. Maschwitz, R. H. L. Disney, and K. Rosciszewski. 1995. A fly's ultimate con. *Nature* 378: 137.

Welch, H. E. 1965. Entomophilic nematodes. *Annual Review of Entomology* 10: 275–302.

Werren, J. H., W. Zhang, and L. R. Guo. 1995. Evolution and phylogeny of *Wolbachia:* Reproductive parasites of arthropods. *Proceedings of the Royal Society London, B* 261: 55–71.

West-Eberhard, M. J. 1975. The evolution of social behavior by kin selection. *Quarterly Review of Biology* 50: 1–33.

Wheeler, D. E. 1986. Developmental and physiological determinants of caste in social hymenoptera: Evolutionary implications. *American Naturalist* 128: 13–34.

Wheeler, W. M. 1907. The polymorphism of ants, with an account of some singular abnormalities due to parasitism. *Bulletin of the American Museum of Natural History* 23: 1–93.

Wheeler, W. M. 1910. The effects of parasitic and other kinds of castration in insects. *Journal of Experimental Zoology* 8: 377–438.

Wheeler, W. M. 1928. *Mermis* parasitism and intercastes among ants. *Journal of Experimental Zoology* 50: 165–237.

Wheeler, W. M., and I. W. Bailey. 1920. The feeding habits of pseudomyrmine and other ants. *Transactions of the American Philosophical Society,* n.s., 22: 235–279.

Whitcomb, R. F., J. M. Bové, T. A. Chen, J. G. Tully, and D. L. Williamson. 1987. Proposed criteria for an interim serogroup classification for members of the genus *Spiroplasms* (Class Mollicutes). *International Journal for Systematic Bacteriology* 37: 82–84.

Whitcomb, R. F., J. G. Tully, J. M. Bové, and P. Saligo. 1973. Spiroplasmas and acholeplasmas: Multiplication in insects. *Science* 181: 1251–1253.

White, J. W. J., M. H. Subers, and A. Schepartz. 1963. The identification of inhibine, the antibacterial factor in honey, as hydrogen peroxide and its origin in a honey glucose-oxidase system. *Biochimica et Biophysica Acta* 73: 57–70.

Whitfield, J. B., and S. A. Cameron. 1993. Comparative notes on hymenopteran parasitoids in bumble bee and honey bee colonies (Hymenoptera: Apidae) reared adjacently. *Entomological News* 104: 240–248.

Widmer, A. 1996. Aspects of the Molecular Ecology of *Bombus (Hymenoptera: Apidae)* in Europe. Ph.D. diss., Department of Environmental Sciences, ETH Zurich, Zurich.

Wigley, P. J., and S. Dhana. 1988. Prospects for microbial control of the social wasps, *Vespula germanica* and *V. vulgaris. Australasian Invertebrate Pathology Working Group Newsletter* 9: 17.

Wilkes, A. 1966. Sperm utilization following multiple insemination in the wasp *Dahlbominus fuscipennis. Canadian Journal of Genetics and Cytology* 8: 451–461.

Wille, H., and L. Pinter. 1961. Untersuchungen über bakterielle Septikämie der erwachsenen Honigbiene in der Schweiz. *Bulletin d'Apiculture* 4: 141–179.

Williams, G. C., and R. M. Nesse. 1991. The dawn of Darwinian medicine. *Quarterly Review of Biology* 66: 1–22.

Williams, P. H. 1994. Phyolgenetic relationships among bumble bees (*Bombus* Latr.): A re-appraisal of morphological evidence. *Systematic Entomology* 19: 327–244.

Williams, R. M. C. 1959. Colony development in *Cubitermes ugandensis* Fuller (Isoptera: Termitidae). *Insectes Sociaux* 6: 291–304.

Williamson, D. L., and D. F. Poulson. 1979. Sex ratio organisms (spiroplasmas) of *Drosophila.* In R. F. Whitcom and J. G. Tully, eds., *The Mycoplasmas: Plant and Insect Mycoplasmas,* vol. 3, pp. 175–208. Academic Press, New York.

Wilson, D. S., and W. G. Knollenberg. 1987. Adaptive indirect effects: The fitness of burying beetles with and without phoretic mites. *Evolutionary Ecology* 1: 134–159.

Wilson, E. O. 1968. The ergonomics of caste in the social insects. *American Naturalist* 102: 41–66.

Wilson, E. O. 1971. *The Insect Societies.* Harvard University Press, Cambridge, Mass.

Wilson, E. O. 1975. *Sociobiology—The New Synthesis.* Harvard University Press, Cambridge, Mass.

Wilson, E. O. 1976. Behavioral discretization and the number of castes in an ant species. *Behavioural Ecology and Sociobiology* 1: 141–154.

Wilson, E. O. 1980a. Caste and division of labor in leaf-cutter ants (Hymenoptera: Formicidae: *Atta*). I. The overall pattern in *A. sexdens. Behavioural Ecology and Sociobiology* 7: 143–156.

Wilson, E. O. 1980b. Caste and division of labor in leaf-cutter ants (Hymenoptera: Formicidae: *Atta*). II. The ergonomic optimization of leaf-cutting. *Behavioural Ecology and Sociobiology* 7: 143–156.

Wilson, E. O. 1983. Caste and division of labor in leaf-cutter ants (Hymenoptera: Formicidae: *Atta*). III. Ergonomic resiliency in foraging by *A. cephalotes. Behavioural Ecology and Sociobiology* 14: 47–54.

Wilson, E. O., and B. Hölldobler. 1988. Dense heterarchies and mass communication as the basis of organisation in ant colonies. *Trends in Ecology and Evolution* 3: 65–68.

Wilson, G. G. 1979. Reduced spore production of *Nosema fumiferanae* (Microsporidia) in spruce budworm (*Choristoneura fumiferana*) at elevated temperature. *Canadian Journal of Zoology* 57: 1167–1168.

Wilson, W. T. 1972. Resistance to American foulbrood in honey bees. XII. Persistence of viable *Bacillus larvae* spores in faeces of adults permitted flight. *Journal of Invertebrate Pathology* 20: 165–169.

Winston, M. L. 1980. Swarming and afterswarm and reproductive rate of unmanaged honey bee colonies (*Apis mellifera*). *Insectes Sociaux* 27: 391–398.

Winston, M. L. 1987. *The Biology of the Honey Bee.* Harvard University Press, Cambridge, Mass.

Winter, U., and A. Buschinger. 1986. Genetically mediated queen polymorphism and caste determination in the slave-making ant, *Harpagoxenus sublaevis* (Hymenoptera: Formicidae). *Entomologia Generalis* 11: 1–15.

Wojcik, D. P. 1989. Behavioral interactions between ants and their parasites. *Florida Entomologist* 72: 43–51.

Wojcik, D. P. 1990. Behavioral interactions of fire ants and their parasites, predators and inquilines. In R. K. Vander Meer, K. Jaffe, and A. Cedeno, eds., *Applied Myrmecology: A World Perspective,* pp. 329–344. Westview Press, Boulder, Colorado.

Wojcik, D. P., D. P. Jouvenaz, W. A. Banks, and A. C. Pereira. 1987. Biological control agents of fire ants in Brazil. In J. Eder and D. H. Rembold, eds., *Chemistry and Biology of Social Insects,* pp. 627–628. Peperny Verlag, Munich.

Wolf, T. J., and P. Schmid-Hempel. 1990. On the integration of individual foraging strategies with colony ergonomics in social insects: Nectar-collection in honeybees. *Behavioural Ecology and Sociobiology* 27: 103–111.

Wood, S. N., and M. B. Thomas. 1996. Space, time and persistence of virulent pathogens. *Proceedings of the Royal Society London, B* 263: 673–680.

Woyciechowski, M. 1990. Mating behaviour in the ant *Myrmica rubra* (Hymenoptera: Formicidae). *Acta Zoologica Cracoviensia* 33: 565–574.

Woyciechowski, M., E. Krol, E. Figurny, M. Stachowicz, and M. Tracz. 1994. Genetic diversity of workers and infection by the parasite *Nosema apis* in honey bee colonies (*Apis mellifera*). In A. Lenoir, G. Arnold, and M. Lepage, eds., *Proceedings of the 12th Congress of the International Union for the Study of Social Insects,* p. 347. Université Paris-Nord, Paris.

Woyke, J. 1983. Dynamics of entry of spermatozoa into the spermatheca of instrumentally inseminated queen honeybees. *Journal of Apicultural Research* 22: 150–154.

Wright, S. 1922. Coefficients of inbreeding and relationship. *American Naturalist* 56: 330–338.

Wright, S. 1943. Isolation by distance. *Genetics* 28: 114–138.

Wu, W. 1994. Microevolutionary Studies on a Host-Parasite Interaction. Ph.D. diss., Phil.-Naturwiss. Fakultät, University of Basel, Basel, Switzerland.

Wülker, W. 1975. Parasite-induced castration and intersexuality in insects. In R. Reinboth, ed., *Intersexuality in the Animal Kingdom,* pp. 121–134. Springer-Verlag, New York.

Yamauchi, K., T. Furukawa, K. Kinomura, H. Takamine, and K. Tsuji. 1991. Secondary polygyny by inbred wingless sexuals in the dolichoderine ant *Technomyrmex albipes. Behavioural Ecology and Sociobiology* 29: 313–319.

Yamazaki, K., E. A. Boyse, V. Mike, H. T. Thaler, B. J. Mathieson, J. Abbot, J. Boyse, Z. A. Zayas, and L. Thomas. 1976. Control of mating preferences in mice by genes in the major histocompatibility types. *Journal of Experimental Medicine* 144: 1324–1335.

Yoshikawa, K. 1954. Ecological studies of *Polistes* wasps. I. On the nest evacuation. *Journal of the Institute of Polytechnics, Osaka,* 5: 9–17.

Yoshikawa, K. 1963. Introductory studies on the life economy of polistine wasps. III. Social stage. *Journal of Biology, Osaka City University,* 14: 63–66.

Youssef, N. N., and D. M. Hammond. 1971. The fine structure of the developmental stages of the microsporidian *Nosema apis. Tissue and Cell* 3: 283–294.

Zeh, J. A., and D. W. Zeh. 1996. The evolution of polyandry. I. Intragenomic conflict and genetic incompatibility. *Proceedings of the Royal Society London, B* 263: 1711–1717.

Subject Index

Host Taxonomic Index

Parasite Taxonomic Index

(see also alphabetical listing of parasites in Appendix 2)

Author Index

Paul Schmid-Hempel is Professor of Experimental Ecology at the ETH in Zurich, Switzerland, and the head of a group conducting research on the evolutionary ecology of host-parasite interactions.